MICROGRIDS AND METHODS OF ANALYSIS

MICROGRIDS AND METHODS OF ANALYSIS

GEVORK B. GHAREHPETIAN
Amirkabir University of Technology, Tehran, Iran

HAMID REZA BAGHAEE
Amirkabir University of Technology, Tehran, Iran

MASOUD M. SHABESTARY
University of Alberta, Edmonton, Canada

Academic Press is an imprint of Elsevier
125 London Wall, London EC2Y 5AS, United Kingdom
525 B Street, Suite 1650, San Diego, CA 92101, United States
50 Hampshire Street, 5th Floor, Cambridge, MA 02139, United States
The Boulevard, Langford Lane, Kidlington, Oxford OX5 1GB, United Kingdom

Library of Congress Cataloging-in-Publication Data
A catalog record for this book is available from the Library of Congress

British Library Cataloguing-in-Publication Data
A catalogue record for this book is available from the British Library

ISBN: 978-0-12-816172-2

For information on all Academic Press publications visit our website at https://www.elsevier.com/books-and-journals

Publisher: Joe Hayton
Acquisitions Editor: Lisa Reading
Editorial Project Manager: Mariana C. Henriques
Production Project Manager: Nirmala Arumugam
Cover Designer: Miles Hitchen

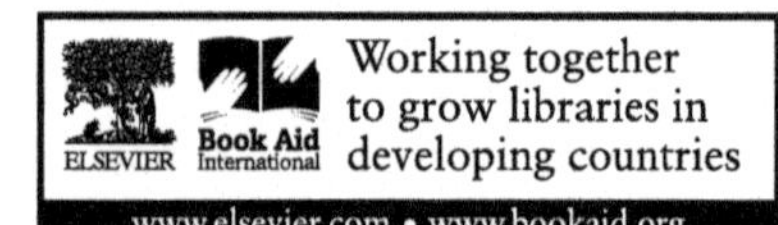

Typeset by TNQ Technologies

Contents

CHAPTER 1

Introduction

1. Microgrids

A microgrid (MG) is a geographically limited low-voltage (LV) distribution network, including localized energy resources, energy storage systems (ESSs), and loads that can operate synchronously with the main grid (macrogrid) or disconnected as an isolated grid considering its physical and/or economic operational conditions [1–4]. The following definition has been presented by the US Department of Energy Microgrid Exchange Group [4]:

> *A microgrid is a group of interconnected loads and distributed energy resources within clearly defined electrical boundaries that acts as a single controllable entity with respect to the grid. A microgrid can connect and disconnect from the grid to enable it to operate in both grid-connected or islanded operation mode.*

Therefore, an MG presents a framework for the application of different resources of distributed generation (DG) units, especially renewable energy resources (RER), and can have isolated and grid-connected operation modes. The ability to change its mode results in challenges in MG control and protection [2]. It can also satisfy the needs of the demand side in the form of electricity, heating, and cooling. This in turn leads to the possibility of substitution of energy carriers and improved energy efficiency due to utilization of waste heat in combined heat and power systems (or also, combined cooling heating and power systems).

It can be said that the US DOE MG Exchange Group [4] considers MG as a set of distributed energy resources (DERs), including DG units and ESSs, and loads that from the main grid side point of view, can be seen as a single controllable entity. This entity can be connected to and disconnected from the main grid considering the system conditions.

The EU research project [5] presents the MG as an LV distribution network with DERs, including fuel cells, microturbines, wind generation systems, photovoltaic panels, batteries, flywheels, etc., and flexible loads. This system should be able to operate either in on-grid (grid-connected) or

Microgrids and Methods of Analysis
ISBN 978-0-12-816172-2
https://doi.org/10.1016/B978-0-12-816172-2.00001-8

in off-grid (disconnected) operation modes. The operation of DERs, also known as microsources, can result in advantages for the performance of the overall system, in case of their efficient management and coordination.

Higher energy efficiency, minimized energy consumption, decreased environmental issues, and enhanced supply reliability are advantages, which can be offered by MGs to customers and utilities. Also, reduction of losses, congestion management, regulation of voltage, and system security improvement can be achieved using MGs. The realization of the mentioned privileges may prevent new investments in the main grid for the integration of RERs and reduce the generation and transmission system expansion costs.

It is well known that the smart grid concept has been presented to

- improve the operation of power systems, e.g., using phasor measurement units;
- increase grid–customer interactions, e.g., using demand response programs and smart meters; and
- integrate new distributed system players and entities, e.g., MGs.

Therefore, the MG can be considered as a new architecture in the distribution system, which can realize the concept of smart grid for customers and utilities, and using this framework, it is possible to use the advantages of integration of a huge amount of small-scale DERs into LV networks. Using the proper combination of DERs, it is possible to have a balance between carbon emissions reduction, cost reduction, and reliable energy supply for customers.

In on-grid operation mode, MG can offer ancillary services to the main grid in the retail energy market. Also, other possible revenue streams exist [6]. In off-grid operation mode, the real and reactive power balance in MG must be established between DERs and the loads. However, an MG may change its mode of operation between these two modes, considering energy market conditions, intentional islanding need for scheduled maintenance, degraded power quality of the main grid, shortage of generated energy, or unintentional islanding due to faults [7,8].

MGs can improve the system reliability, and also, the resilience of the network versus extreme weather conditions and natural disasters, which can result in huge damages to power systems [7,8]. Also, using multiobjective optimization algorithms, MG energy management system can enhance efficiency and resiliency and reduce the costs [9–11].

2. Challenges

To preserve the mentioned advantages and use the presented opportunities of MGs, their challenges must be studied and addressed in their protection and control systems design. To study conventional distribution systems, there are some typical assumptions, which are no longer valid for MG investigations, and these new conditions can result in new challenges. Also, some transmission system challenges can be observed in MGs, which must be considered in detail [7]. The most important challenges of MGs, which will be discussed in this book, are as follows:

- **Bidirectional power flow**: The conventional power flow in power systems is from the main grid to passive distribution networks. The integration of DG units in distribution systems, which can now be considered as an active distribution network, has changed this condition. Now, it is possible to have a reverse power flow. In this case, the contribution of DGs in fault current will affect the fault current pattern, and as a result, the coordination of protection relays and the performance of voltage control systems must be revised [7].
- **Stability**: The DERs control system and control systems of other devices installed in distribution systems such as reactive power controllers can have interactions and result in local oscillations, which can endanger the small-signal stability of the system. Also, MGs should operate in both operation modes, but a transition from isolated mode to on-grid mode can lead to transient stability issues for MG components [7,12]. Some researches on DC MG have shown that these MGs can improve the system performance better than AC MGs [13,14].
- **Modeling**: To overcome MG problems, they must be modeled and studied before practical implementation and application. Many solutions have been investigated and presented for power systems in the literature, which is based on verified and well-known assumptions. But these assumptions are not valid for MG modelings such as three-phase balanced conditions, inductive behavior of lines, and constant—power loads. As a result, new models should be suggested and used [7].
- **Low inertia**: The conventional synchronous generators can obtain enough inertia versus power system oscillations and transients. In MGs, an increase in penetration level of inverter-interfaced DERs (IIDERs) in distributions systems and MGs can result in the reduction of the system inertia. Especially in the isolated mode of operation of MGs, any disturbance can cause significant voltage and frequency

variations, if a proper control system is not designed [7], or ESSs such as battery or flywheel are not used [15]. To overcome this issue, the researchers have suggested the application of virtual synchronous generators, which are inverters mimicking synchronous generators' characteristics to provide virtual inertia to MGs.

- **Uncertainty**: Highly correlated variations of power system resources lead to less uncertainty in bulk power systems. But, in MGs, stochastic load behavior during different periods and intermittent weather conditions are two main uncertainties, which affect the reliable and economic operation of an MG. In islanded mode, uncertainties are more challenging. The power balance must be satisfied between demand and supply over an extended time horizon.

3. Purpose and target audiences

Considering the ever-increasing penetration level of IIDERs and MGs in power systems, researchers need a set of software and hardware tools similar to the ones used for bulk power systems for

- steady-state and fault studies,
- design of control and protection systems, and
- determination of optimal conditions.

This book aims to offer solutions for these studies in MGs. Therefore, it can be used by researchers and professional engineers, working in the electrical power industry, professors, and electrical engineering postgraduate students. Also, this book is useful for the BS students, working on their final thesis.

This book will be useful for utility engineers not only in the countries that are now planning to restructure their power system and implement MGs but also in the countries, whose MGs have been implemented before and now their systems are in operation. Moreover, all around the world, the researchers who are working in the field of the MGs can use the methods presented in this book for their new researches.

4. Benefits to the audiences

This book has two major benefits to the audiences, as follows:

- Using new analytical solutions for solving MGs operational problems
- Gaining new insight into MGs studies

This book aims to provide useful insight for the researchers in aspects of MG planning, operation, control, and protection. To form the backbone of this book, the information mentioned in the recently published papers and researches has been used in all the chapters. Also, discussion of related international standards has been provided in some sections of the book, to make it more useful for industry experts.

The focuses of previously published books in this field are mainly on architecture and control of MGs. However, the authors of this book have tried to provide analytical solutions to facilitate the analysis of MGs and cover some research gaps identified in the literature. This book addresses the technical challenges in MGs planning, operation, protection, and control.

5. Outline of the book

To achieve the aforementioned purposes, the rest of the book is organized as follows: In Chapter 2, the basic control strategies of MGs are reviewed. Also, we model an improved hierarchical control structure and discuss the basic challenges in the control of MGs. In Chapter 3, we present a three-phase AC/DC power flow algorithm for balanced and unbalanced MGs and extend it for harmonic and probabilistic power flow algorithm. Chapter 4 describes the MG fault analysis method along with current limiting strategies and fault ride-through (FRT) requirements. The operation of MGs in unbalanced situations is discussed in Chapter 5. In Chapter 6, we discuss the protection strategies for MGs. Chapter 7 describes the optimal sizing of hybrid MGs based on numerical algorithms, and finally, the power management strategies of MGs will be discussed in Chapter 8.

References

[1] N. Hatziargyriou, H. Asano, R. Iravani, C. Marnay, Microgrids, IEEE Power Energy Mag. 5 (4) (July 2007) 78–94.

[2] R. Venkatraman, S.K. Khaitan, A survey of techniques for designing and managing microgrids, in: Proc. IEEE Power & Energy Society General Meeting, Denver, CO, USA, July 2015, pp. 1–6.

[3] F. Katiraei, R. Iravani, N. Hatziargyriou, A. Dimeas, "Microgrids management," IEEE Power Energy Mag., vol. 6, no. 3, pp. 54 – 65, (M).

[4] reportDOE Microgrid Workshop Report, [online]: https://www.energy.gov/sites/prod/files/Microgrid%20Workshop%20Report%20August%202011.pdf.

[5] N. Hatziargyriou, Microgrids Architectures and Control, John Wiley and Sons Ltd, 2014, ISBN 978-1-118-72068-4, p. 4.

[6] M. Stadler, G. Cardoso, S. Mashayekh, T. Forget, N. DeForest, A. Agarwal, A. Schönbein, Value streams in microgrids: a literature review, Appl. Energy 162 (January 2016) 980–989.
[7] M. Saleh, Y. Esa, A. Mohamed, Communication-based control for DC microgrids, IEEE Trans. Smart Grid 10 (2) (March 2019) 2180–2195.
[8] D.E. Olivares, et al., Trends in microgrid control, IEEE Trans. Smart Grid 5 (4) (July 2014) 1905–1919.
[9] M. Jin, W. Feng, P. Liu, C. Marnay, C. Spanos, MOD-DR: microgrid optimal dispatch with demand response, Appl. Energy 187 (February 2017) 758–776.
[10] P. Tenti, T. Caldognetto, On microgrid evolution to local area energy network (E-LAN), IEEE Trans. Smart Grid 10 (2) (March 2019) 1567–1576.
[11] S. Mashayekh, M. Stadler, G. Cardoso, M. Heleno, A mixed integer linear programming approach for optimal DER portfolio, sizing, and placement in multi-energy microgrids, Appl. Energy 187 (February 2017) 154–168.
[12] M.S. Saleh, A. Althaibani, Y. Esa, Y. Mhandi, A.A. Mohamed, Impact of clustering microgrids on their stability and resilience during blackouts, in: Proc. International Conference on Smart Grid and Clean Energy Technologies (ICSGCE), October 2015, pp. 195–200. Offenburg, Germany.
[13] T. Dragicevic, X. Lu, J. Vasquez, J.M. Guerrero, DC microgrids–Part I: a review of control strategies and stabilization techniques, IEEE Trans. Power Electron. 31 (7) (July 2016) 4876–4891.
[14] T. Dragicevic, X. Lu, J. Vasquez, J.M. Guerrero, DC microgrids—Part II: a review of power architectures, applications, and standardization issues, IEEE Trans. Power Electron. 31 (5) (May 2016) 3528–3549.
[15] Y.S. Kim, E.S. Kim, S.I. Moon, Frequency and voltage control strategy of standalone microgrids with high penetration of intermittent renewable generation systems, IEEE Trans. Power Syst. 31 (1) (January 2016) 718–728.

CHAPTER 2

Microgrid control strategies

1. Introduction

The ever-increasing penetration of distributed energy resources (DERs), including renewable energy resources (RERs) and energy storage systems (ESSs), and also technoeconomic application of power electronics converters in distribution networks (DNs), result in a need for new frameworks such as active distribution networks (ADNs) [1], microgrids (MGs) [2–4], and virtual power plants (VPPs) [5]. These frameworks can realize the application of DERs, such as wind generation (WG) systems, photovoltaic (PV) generation units, fuel cells (FCs), microturbines [6,7], and also electric vehicles (EVs). Integration of RERs and ESSs into low-voltage DNs leads to many advantages such as improved power quality (PQ), enhanced voltage stability, and elevated reliability [8–10]. Also, they can reduce greenhouse gas emissions and other negative environmental impacts. However, the RERs also result in new challenges such as coordination of protection relays for both directions of current flow, islanding protection of MG, energy quality-based distribution, and energy management of MGs considering their intermittent generation. In general, the performance of distributed generation (DG) units in MGs depends on their operating point and the protection system settings [11–13].

In this chapter, we introduce basic architecture and discuss for challenges and trends in the operation, management, and control of future ADNs toward modern smart MGs.

2. Basic infrastructures for control and management of microgrids

Fig. 2.1 illustrates a typical multilayer block diagram representation for general structure of MG control system [15–18]. In the lowest layer, we have all the physical components, including interfacing power converters (PCs), DERs, and MG elements, such as loads, transformers, transmission lines, switchgears, etc.

Microgrids and Methods of Analysis
ISBN 978-0-12-816172-2
https://doi.org/10.1016/B978-0-12-816172-2.00002-X

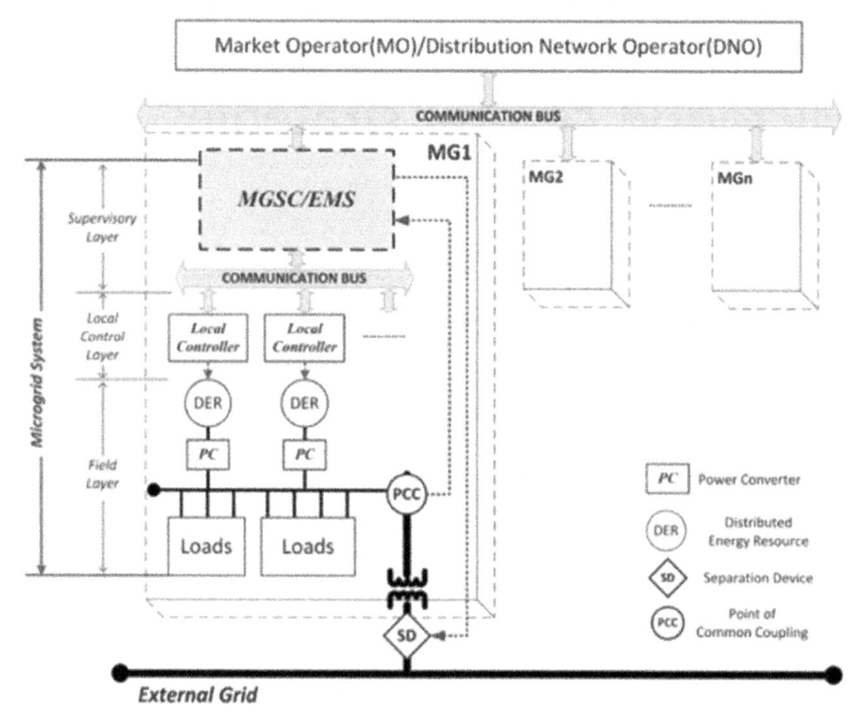

Figure 2.1 General structure of microgrid control system [14].

The basic functions such as local generation and consumption control and management of ESSs are realized by the local control layer, which usually follows the commands sent by upper-level controllers. The highest control level of MG fulfills the function of supervisory control and data acquisition (SCADA). This control level is also known as MG supervisory controller (MGSC), energy management system (EMS), or microgrid central controller (MCC). This control level should provide important tasks such as power quality control, support of ancillary services, involvement in the energy market, and optimization of system performance by enhancing its intelligence level [15]. As shown in Fig. 2.1, the higher-level operators, i.e., distribution network operator (DNO) and market operator (MO) affect the MGSC/EMS. This interaction needs the availability of information and communication technology (ICT). The requirements and characteristics of MGs bring more challenges to MGSC/EMS [15,16]. For example, energy management system must be able to provide intentional or unintentional smooth transition between off-grid (islanded) and on-grid (grid-connected) modes, handle demand side management programs, and manage RERs using advanced scheduling and dispatching strategies [19].

3. Microgrid control schemes

DG units in MGs can be controlled in grid-forming (voltage-controlled), grid-feeding (voltage-controlled), or grid-supporting (voltage/current-controlled) modes [20]. The control strategies used in grid-feeding inverters have been presented in Ref. [21]. This section focuses on the control strategies of grid-forming and grid-supporting DGs, which play the main role in power sharing and balancing of MGs. After discussion of these DGs converters, grid-feeding and grid-supporting converters will be presented as well.

The grid-forming control strategies can be classified considering their requirements for a communication system. They include master/slave control [4,10,22,23] centralized control [24,25], decentralized control [26,27], distributed control [26,27], and hierarchical control [6,9,17,28].

Essentially, there are four basic communication-assisted schemes for control of large-scale systems (LSSs). These systems almost include different coupled subsystems, which have some interactions together. Based on the theory of LSSs [29], they can be controlled based on centralized, decentralized, distributed, and hierarchical control structures. These control schemes are illustrated in Figs. 2.2A–2.6D, and their main salient features are summarized as follows.

3.1 Centralized control

Design of the centralized controllers for LSSs is complicated because of their intrinsic computational complexities, reliability constraints, and limitations of the communication bandwidth. As shown in Fig. 2.2A, all computations are performed in centralized controller (which have all the information of the system) and necessary commands are transmitted to the actuators. When the system (in this chapter, MG) become larger and larger, the amount of computations in the centralized controller will be extensively increased. So, the required computation time is increased and controller show slow and inefficient performance which is not acceptable in real-time applications. Moreover, communication between subsystems and centralized controllers demand fast two-way high-bandwidth communication (HBC) which results in increasing investment costs. More importantly, the centralized control strategy has low reliability because failure in the controller, will stop the operation of whole plant. Besides the mentioned disadvantages, if a controllable agent (such as DG/ESS unit or PHEV charging station) is added to the microgrid, the centralized controller should be redesigned or at least essential changes are required. To sum up, the benefits and disadvantages of centralized control structure are summarized in Table 2.1.

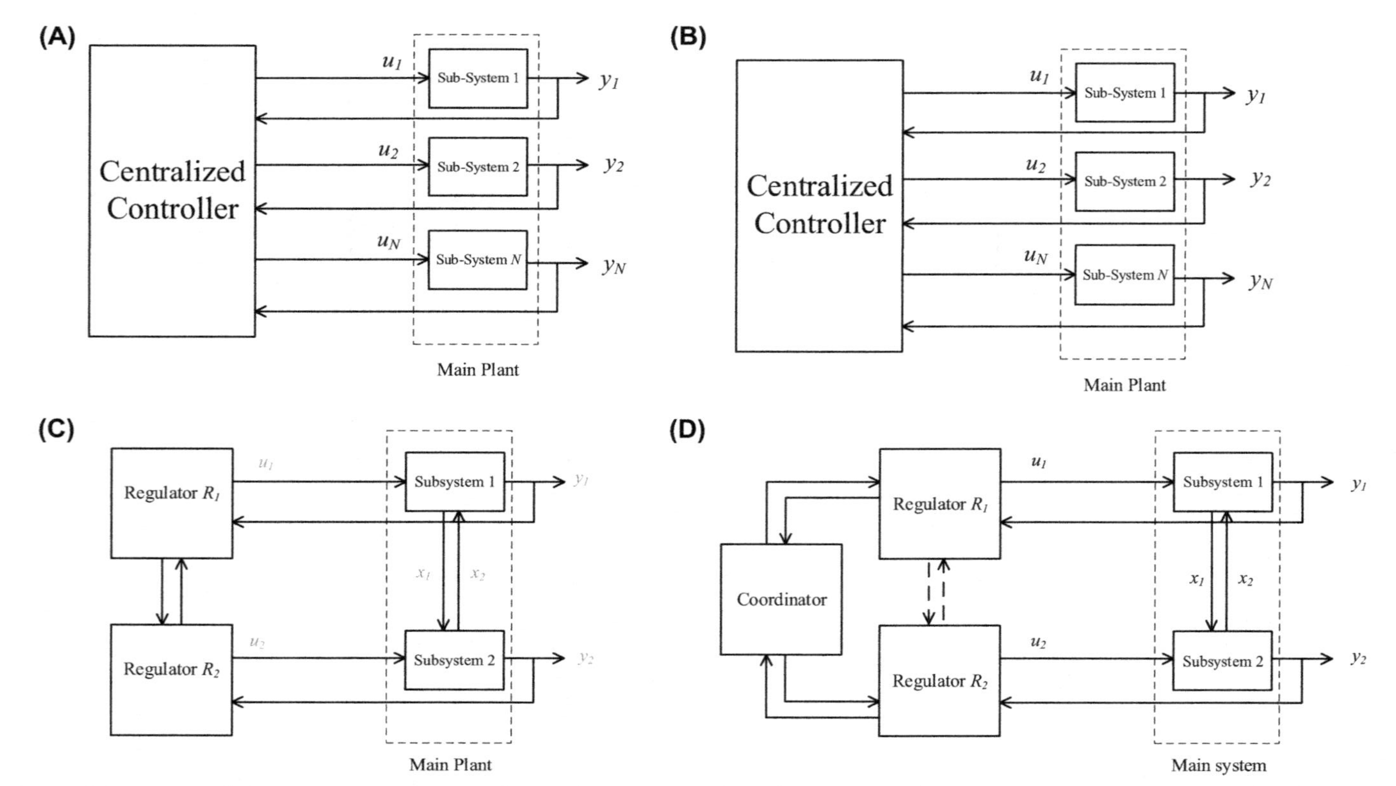

Figure 2.2 Basic block diagram representation of control architectures based on theory of large-scale systems [29]: (A) centralized control, (B) decentralized control, (C) distributed control, and (D) hierarchical control.

Table 2.1 Comparison of the characteristics of different control strategies of microgrids.

Control structure	References	Communication	Implementation	Reliability	Advantages	Disadvantages
Centralized	[24,25]	Fast, two-way, HBC	High	Low	Guarantee of stability, optimality	Inefficiency and no applicability in large-scale systems, needs fast two-way HBC, low reliability, and slow performance
Decentralized	[26,27]	LBC	Reasonable	High	High reliability, reasonable implementation cost, needs only LBC	Stability, operation in case of strong interaction between subsystems
Distributed	[30–32]	LBC	Low	High	High reliability, needs LBC, global optimality low computation burden	Concerns for stability and robustness
Hierarchical	[6,9,17,28]	LBC	Reasonable	High	Easy implementation simple and easy operation and maintenance, fast data transmission between all layers low computation burden	Concerns for stability and robustness

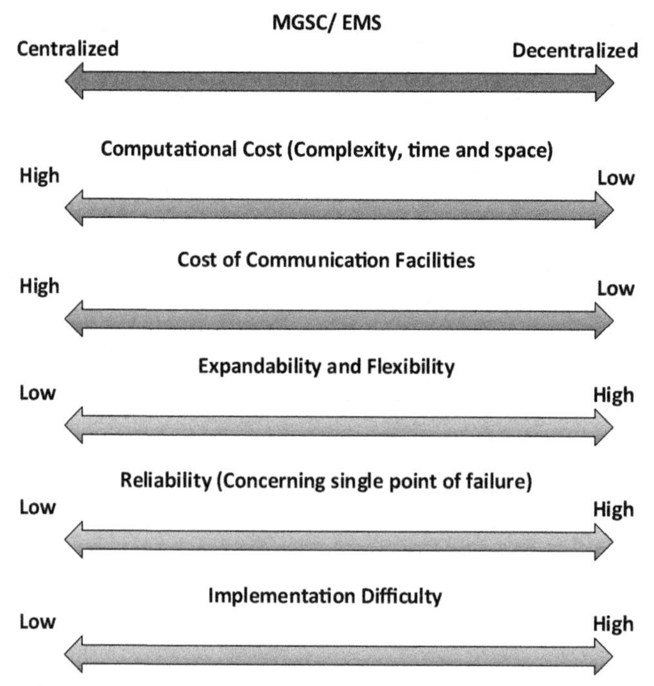

Figure 2.3 Characteristics of microgrid control system in different levels of decentralization [14].

As presented in Fig. 2.2, the MGSC/EMS can have centralized or decentralized structures. The capabilities of local controllers (LCs), to execute commands of upper level or make their own decisions, determine the level of decentralization. The characteristics, which determine the different levels of decentralization, are compared in Fig. 2.3. The centralized and decentralized solutions have advantages and disadvantages, and their selection depends on the type of MG, which can be residential, industrial, commercial, or military. Also, the legal and physical features of MG, i.e., its ownership, location, size, and topology, affect the selected solution.

MGSC/EMS must use the data collected from MG elements and main grid. Using these data, MG scheduling and optimization are carried out to determine efficient and economic operating point. The real-time monitoring of the MG can be considered as the advantages of centralized MGSC/EMS. This results in a strong supervision, protection, and control of the entire MG elements. It must be mentioned that the confidential and private data must be protected within MGSC/EMS center. These requirements show that this center must be highly powerful to be able to process and manage huge amount of data and make efficient decisions. Also, to timely exchange data with MG elements, high bandwidth communication system must be used. The main drawback of the centralized solutions is its low reliability, because a fault in a centralized MGSC/EMS will cause the outage of the MG. Another demerit of this approach is its low flexibility/expandability. Therefore, centralized MGSC/EMS can be suggested for the following MGs [14]:

- MGs with limited geographical dimensions or small-scale MGs. In these systems, the communication system can be implemented with low cost, and its computational burden is low as well.
- MGs, whose elements belong to one owner or DNO. In this case, the MGSC/EMS can provide optimal and economic operation for the MG.
- Military MGs, where high confidentiality is crucial.
- MGs with fixed topology and configuration. They do not need high flexibility/expandability.

For instance, a hybrid distributed design has been presented in Ref. [33] for power control. In each DG unit, a distributed power controller has been used to guarantee tracking of optimal set points determined by the central MGSC/EMS. The EMS has used the measured average power to determine each unit share using a low bandwidth communication link, which its structure has a good tolerance to communication system delays. Also, master–slave method [34], centralized load sharing approach [35], average load sharing system [36–38], current limitation controller [39], and circulating chain control method [40] have been presented and studied, which propose a good voltage regulation and power sharing [40,41]. For example, in Ref. [42], the output voltage is near to its nominal value. These solutions can be applied to adjacent DG units or the ones that are connected to the same bus. In an MG with DGs distributed in the whole system, the communication system must have high bandwidth system and cover all the MG elements, which can be costly and, as mentioned before, has a low system reliability.

3.2 Decentralized control

In the decentralized control structure, all subsystems of the plant are separately controlled by their own LCs. There may be some interactions between the subsystems so that the less interactions may lead to be closer to the suboptimal condition (Fig. 2.3B). In decentralized scheme, the role of all controllable agents and their LCs are the same, and none of them has more important role. Compared with the centralized control, this scheme has higher reliability because it has not a centralized controller with excessive communication. On the other hand, the decentralized scheme provides highest independence for controllable agent so that each agent has its own LC, which operates based on its local measurements' feedbacks. Thus, there is no need for fast two-way high bandwidth communication system and centralized controller, which were vital in the previous control scheme. This means that failure in each of the controllable agents or LCs will not result in collapse of whole plant. When a failure occurs, without redesigning of controller, one (or more) controllable agent(s) and the related LC(s) can be added to or removed from the plant [15,43,44]. Successful usage of communication technologies such as Wi-Fi and Zigbee [45] and effective algorithms for information exchange such as Peer-To-Peer, Gossip, and Consensus [46–49] result in practical application of decentralized control and management. Therefore, it is possible to implement frequency and voltage control, DERs power sharing, and their energy management in a decentralized approach. As shown in Fig. 2.3, the decentralization level may vary from centralized to fully decentralized [50]. Decentralized control can be realized using multiagent systems (MASs). They can improve the intelligence level of LCs by transferring decision-making capability to the elements in the local side. Here, the communication system plays an important role since the local decision-making is based on data received from the environment of each element and its adjacent devices. The decision making is locally processed, and energy management system must perform information sharing/coordination. Therefore, the decentralized approach decreases the computational burden. Also, after loss of MGSC/EMS, the system can operate further, which results in higher reliability than the centralized one. Another privilege is that it can provide a suitable framework for plug-and-play functionality, which is important for the application of plug-in hybrid electric vehicles (PHEVS). This capability improves MG flexibility/expandability. However, synchronization among the MG elements and information security in its communication systems are important challenges in decentralized mode of operation. In the following MGs, a decentralized control system can be considered as a good solution [51–53].

- The MGs, which have a large size, or their DGs, ESSs, and loads are widely dispersed. In this case, data acquisition can be difficult or costly.
- In case of frequent reconfigurations in MG due to need for plug-and-play operation mode of devices such as PHEVs and RERs, a decentralized control system is used.

- The MGs, whose DERs belong to different owners and should be considered as different players in MG. They have their own operation objectives and decision-making procedures.

The decentralized control method has been realized using MAS [30,54], sliding-mode control [55,56], and droop control [25,31,32,57–61]. The DERs, in these methods, have access to their local measured signals and operate independently. Therefore, their plug-and-play operation is possible with less problems in comparison with centralized methods. In the droop control method, the voltage and frequency of DGs are controlled considering their active and reactive power exchange with the MG without using any communication links [62–65]. Therefore, higher reliability and better flexibility can be achieved [17,32,40], and as a result, many researchers have suggested the application of the droop control method [66] and its variants. Also, almost all the experimentally implemented MGs are based on this method [67].

3.3 Distributed control

Distributed control is in fact a subset of decentralized control in which the information can be exchanged among LCs of the subsystems and thus has higher reliability in contrast with the centralized control scheme (Fig. 2.3C). This scheme is discriminated from decentralized one based on three features, namely, communication, cooperation, and coordination. The communication means a weak interaction between LCs. In other words, LCs only exchange the information of subsystems; however, these exchanges will not generally and necessarily result in optimality, stability, and robustness of the whole system. The coordination is close to cooperation with a difference that in coordination, the subproblems are coordinated by an observer for the sake of achieving global optimality.

3.4 Hierarchical control

Hierarchical control of LSSs is basically a controller consisting of decentralized and centralized control strategies. In this control method, the upstream control layers produce the required set points of lower control levels, which are closer to the physical layer that has faster dynamics and less operation time. Fig. 2.3D illustrates the basic block diagram of the hierarchical control system. The upper layers have slower dynamics. The control variables are calculated in the upper layers and transmitted to the lower layers as reference signal (Fig. 2.3D). Considering difference between dynamics and operation time of different layers, hierarchical scheme needs a

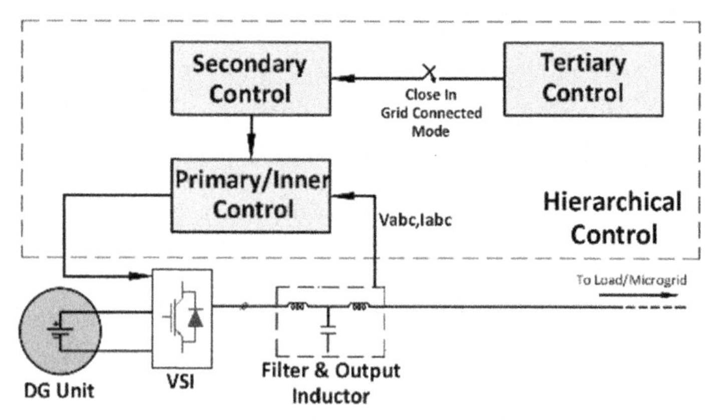

Figure 2.4 Hierarchical control overall scheme.

stable and robust platform for stabilization of plant. This scheme is simple, and it can easily be implemented. Its operation and maintenance are easy and have fast data transmission among all the layers. Also, it can quickly be modified/fixed in case of any change/failure. Also, considering its distributed computational structure, it does not demand massive computational burden. A hierarchical control method has been suggested as an effective solution for MGs control and management [3,15,17,18], to handle problems such as market participation, load power sharing, short- and long-term scheduling, voltage and frequency control, and power quality regulation. It is a complex multiobjective control system.

The terminology used in this chapter has some differences with recent researches. For example, in Ref. [59], the energy management is in the secondary control level, whereas its tertiary level is specified for multiple MGs coordination. Here, the definitions presented in Refs. [15,17,18] have been used. As shown in Figs. 2.4 and 2.5, the hierarchical control scheme contains the following three levels:

(1) ***Primary control level***: The local power, voltage, and current are controlled in this level. This level receives the set points from second-level controllers and applies them to interface PCs.

(2) ***Secondary control level***: This control level provides commands for the primary control level. It should handle issues, such as voltage and frequency restoration, voltage unbalance, and harmonic compensation. Also, MG power exchange and synchronization with the main grid must be adjusted in this level.

(3) ***Tertiary control level***: This control level should optimize the MG performance considering MG efficiency and its economics. Therefore,

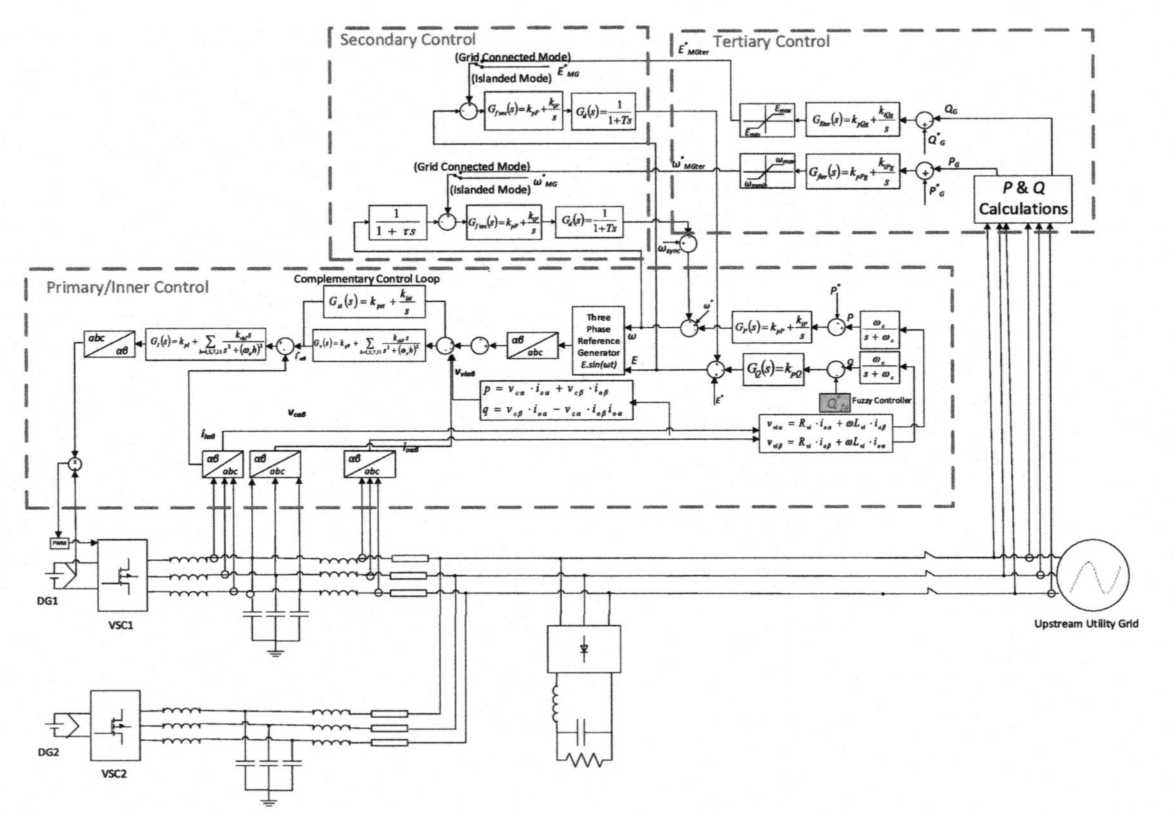

Figure 2.5 Block diagram of hierarchical control for a microgrid consisting two DER units and a common load. *DER*, distributed energy resource.

it must have data transfer capability with both MG and main grid to carry out the optimization functions and take measures using decision-making algorithms.

In this structure and framework, tertiary and secondary control levels are fulfilled in MGSC/EMS unit. In higher control levels, the reactions are slower. In the droop control method, the primary and secondary control levels have a response speed of 1–10 and 100 ms^{-1} s, respectively, whereas the tertiary control level makes decisions in discrete time steps ranging from a few seconds to hours. Therefore, the different control level bandwidths are separated by at least an order of magnitude. This means that different levels have different dynamics and can be decoupled during their modeling and analysis.

4. Current challenges and future trends in control and management of microgrids

In isolated MGs, the control system should aim at MG voltage and frequency control considering active and reactive power sharing [17,19,68–71]. In grid-connected operation mode, it must adjust DERs active and reactive powers, and the MG voltage and frequency depend on the main grid voltage and frequency. The MGs are subjected to various small-signal and large-signal disturbances in both modes. In MGs, nonlinear dynamics of DERs and loads, frequent topological changes, uncertainties of RERs and loads, and communication system reliability issues can be found. A resilient and stable MG should have robust performance against these problems [72,73]. The stability of MGs versus various disturbances has been well studied in Refs. [4,6,9,10,12,19,22]. Also, novel approaches have been presented in Refs. [6,12,68–70,74,75], considering nonlinear and unbalanced loads and nonlinearities of DERs. In Refs. [2,9,69–71,76–82], MGs control and robust power flow considering their topological, parametric, and communication system uncertainties, including time delays, have been investigated. Based on the reviewed literature, the control and management of MGs will face the following challenges:

- Operation and power quality issues,
- Design of protection system considering the effect of PC control system
- Improving fault ride through (FRT) capability of RERs
- Improving power/current sharing of DER units considering stability
- Robust power flow analysis
- Uncertainty handling techniques of demand and supply side

- Effect of communication latencies such as noise, time delay, etc.
- Integration of hug number of charging stations for electric vehicles
- Demand response programs and transactive energy realization
- Improvement of online measurement systems
- Improving resiliency, security, and reliability

In the coming sections, some examples of commonly used hierarchical droop-based and distributed control strategies are described, including their problem formulation, controller design related strategies, and simulation results.

5. Case study 1: design of an improved hierarchical control structure

As mentioned before, in islanded mode of operation of MGs, the voltage and frequency will change after a change in the load, and they will have new values. Therefore, a mechanism must be used to restore the system frequency and voltage to its nominal values [83–85]. In power systems, this function is known as secondary control and is slower than the primary control. Also, to enhance the performance of the droop control method, virtual impedance control loop has been suggested in Ref. [17]. In some investigations, the MG has been studied under normal conditions or versus small-signal disturbances without considering the MG stability due to large-signal disturbances, its fault ride through (FRT) capability, and desirable power sharing among DERs. In the following, a new hierarchical control scheme is presented, and the reactive power reference will be adjusted by a fuzzy logic controller. The simulations are carried out in the MATLAB/Simulink environment, and OPAL-RT real-time digital simulator is used to indicate the efficacy of the suggested scheme in enhancement of small and large signal stability, improvement of FRT capability and modification of power sharing in MG.

5.1 Design procedure of control structure

As mentioned before, the hierarchical control system of MG consists of primary, secondary, and tertiary control levels. In the first control level, the voltage source converters (VSCs) inner control loops and power distribution primary points must be adjusted. The secondary control level must restore the frequency and voltage deviations. In the tertiary control level, the power sharing between the MG and the main grid has to be regulated (Fig. 2.4).

5.1.1 *Primary control level*

Fig. 2.1B shows an MG, which has two DERs supplying a nonlinear load.

5.1.1.1 Droop control

The proposed primary control for parallel VSCs is the droop control, which should control the frequency and voltage (v_{ref}) using the following characteristics [86]:

$$\begin{aligned} \omega &= \omega^* - G_P(s)\cdot\left(P - P^*\right) \\ E &= E^* - G_Q(s)\cdot\left(Q - Q^*\right) \end{aligned} \tag{2.1}$$

where P and Q are real and reactive powers, respectively, ω is output voltage angular frequency, E is output voltage amplitude, ω^* is reference value of ω, E^* is reference value of E, and P^* and Q^* are reference values of P and Q. The transfer functions $G_P(s)$ and $G_Q(s)$ are droop control transfer functions given by the following equations:

$$G_P(s) = k_{pP} + \frac{k_{iP}}{s} \tag{2.2}$$

$$G_P(s) = k_{pP} + \frac{k_{iP}}{s} \tag{2.3}$$

where k_{iP} and k_{pQ} are coefficients of static droop and k_{pP} is system virtual inertia or transient droop term. The values of these coefficients are selected to have k_{iP} equal to the maximum frequency deviation divided by the active power nominal value, and k_{pQ} is equal to the maximum value of the voltage deviation divided by the reactive power nominal value [8]. As shown in Fig. 2.5, the instantaneous powers, p and q, can be determined using Clark's transformation [87] and after transforming the three-phase values into the $\alpha\beta$ coordinate, as follows:

$$\begin{aligned} p &= v_{c\alpha}\cdot i_{o\alpha} + v_{c\beta}\cdot i_{o\beta} \\ q &= v_{c\beta}\cdot i_{o\alpha} - v_{c\alpha}\cdot i_{o\beta} \end{aligned} \tag{2.4}$$

where $v_{c\alpha\beta}$ and $i_{o\alpha\beta}$ are capacitor voltage and output current in $\alpha\beta$ coordinate, respectively. Then, the active and reactive powers are determined by the following equations after passing through a low-pass filter with cutoff frequency of ω_c [65].

$$(P, Q) = \frac{\omega_c}{s + \omega_c}\cdot(p, q) \tag{2.5}$$

As shown in Fig. 2.5, the primary control level contains virtual impedance loops, which should guarantee inductive behavior of system at the power frequency [28], and its inductance and resistance are L_{vi} and R_{vi}, respectively, and we have [83]:

$$\begin{aligned} v_{vi\alpha} &= R_{vi} \cdot i_{o\alpha} + \omega L_{vi} \cdot i_{o\beta} \\ v_{vi\beta} &= R_{vi} \cdot i_{o\beta} + \omega L_{vi} \cdot i_{o\alpha} \end{aligned} \tag{2.6}$$

5.1.1.2 Inner control loops

The internal control loops model has been discussed in Refs. [25,65,83]. The presented model in Refs. [25,65] has control loops of voltage and current with feedback and feedforward terms. A proportional integral (PI) controller obtains the output voltage and current. The following equations present the state equations:

$$\frac{d\varphi_\alpha}{dt} = v_\alpha^* - v_{c\alpha}, \quad \frac{d\varphi_\beta}{dt} = v_\beta^* - v_{c\beta} \tag{2.7}$$

$$\frac{d\gamma_\alpha}{dt} = i_\alpha^* - i_{l\alpha}, \quad \frac{d\gamma_\beta}{dt} = i_\beta^* - i_{l\beta} \tag{2.8}$$

A linearized state-space small-signal representation can be obtained by differential algebraic equations (DAEs). In Ref. [83], using proportional resonant (PR) controller, the control loops of voltage and current have been obtained, and also, in the primary control, a virtual impedance has been used. The current control loop, voltage control loop, and an additional novel controller can be seen for each VSC, all of them based on the $\alpha\beta$ reference frame. The control loops of voltage and current have PR terms to regulate the fundamental and 5, 7, and 11 harmonics. To mitigate the voltage harmonics generated due to nonlinear loads, both the current and voltage control loops should be involved. As can be seen in Fig. 2.5, an LC filter is used after each VSC. The voltage controller $G_v(s)$ and current controller $G_i(s)$ transfer functions are expressed by the following equations:

$$G_v(s) = k_{pV} + \sum_{h=1,5,7,11} \frac{k_{rhV} s}{s^2 + (\omega_o h)^2} \tag{2.9}$$

$$G_I(s) = k_{pI} + \sum_{h=1,5,7,11} \frac{k_{rhI} s}{s^2 + (\omega_o h)^2} \tag{2.10}$$

where k_{pV} and k_{pI} are proportional term coefficients, k_{rhV} and k_{rhI} are resonant term coefficients, and $\omega_o = 2\pi f$ and h is the order of harmonic [8]. The suggested complementary novel control loop, which has a feedback from the control loop of voltage, has a PI term. This controller transfer function $G_{st}(s)$ can be seen in Fig. 2.5, and is given by the following equation:

$$G_{st}(s) = k_{pst} + \frac{k_{ist}}{s} \tag{2.11}$$

where k_{pst} and k_{ist} are coefficients of proportional and integral terms, respectively. This new controller not only reduces the resonance effects of resonant terms of the voltage controller but also preserves the effectiveness of these terms. The simulation results demonstrate that the stability margin is enhanced versus small-signal and large-signal disturbances.

5.1.2 Secondary control level

After any load or generation change in MG, the secondary control level should try to reduce the deviations of the voltage and frequency to zero. Here, the following PI controllers must compensate these deviations.

$$\omega_{rest} = \left(k_{pF} + \frac{k_{iF}}{s}\right) \cdot \left(\omega_{MG}^{*} - \omega_{MG}\right) \tag{2.12}$$

$$E_{rest} = \left(k_{pE} + \frac{k_{iE}}{s}\right) \cdot \left(E_{MG}^{*} - E_{MG}\right) \tag{2.13}$$

where k_{pF}, k_{iF}, k_{pE}, and k_{iE} are PI controller coefficients, and ω_{rest} and E_{rest} are frequency and voltage deviations, respectively, which must be in their rated limits [28]. To adjust the parameters of the secondary controller parameters and analyze the MG stability, a model has been suggested in Ref. [83].

5.1.3 Tertiary control level

In on-grid operation mode, the power sharing among DERs can be adjusted by controlling the frequency and voltage of DERs. The active and reactive power exchange at the point of common coupling (PCC), i.e., P_G and Q_G, must be compared with their desired values, i.e., P_G^* and Q_G^*, as follows [20]:

$$\omega_{MGter}^{*} = \left(k_{pPg} + \frac{k_{iPg}}{s}\right) \cdot \left(P_G^{*} - P_G\right) \tag{2.14}$$

$$E_{MGter}^{*} = \left(k_{pQg} + \frac{k_{iQg}}{s}\right) \cdot \left(Q_G^{*} - Q_G\right) \tag{2.15}$$

As shown in Fig. 2.5, this control level is inactive in off-grid operation mode, and the frequency and voltage references in the secondary control level are selected. In on-grid operation mode, the tertiary controller generates these references.

5.1.4 Control of reactive power reference using fuzzy logic control

In case of heavy-load motor starting in an MG, the voltage drop may be out of the acceptable range. In this section, the proposed control system must handle this problem by injecting reactive power and by adjusting the reactive power reference, i.e., Q using the fuzzy logic. The fuzzy logic controllers are nonlinear controllers, which provide more flexibility compared with the conventional PI controllers.

Fuzzy set theory expresses vague data using membership functions for logical reasoning. After determination of input and output of a fuzzy logic controller (FLC), it can be designed. In the MG, *Ei* (the voltages of the MG) are the nonfuzzy inputs of the FLC. As shown in Fig. 2.6A, the membership function of a triangular number *μA(X)* is defined by the following equation [88]:

$$\mu_A(X) = \begin{cases} 0 & x \leq a \\ \dfrac{x-a}{b-a} & a \leq x \leq b \\ \dfrac{c-x}{c-b} & b \leq x \leq c \\ 0 & c \leq x \end{cases} \tag{2.16}$$

where *μA(X)* is the membership function of a triangular number. Eq. (2.16) can be rewritten in the following form using the minimum and maximum relations.

$$\mu_A(X) = \max\left(\min\left(\frac{x-a}{b-a}, \frac{c-x}{c-b}\right), 0\right) \tag{2.17}$$

The triangular membership functions are shown in Fig. 2.6B for different voltages indicating a linguistic variable. Using the input and output membership functions, the input and output values can easily be fuzzified. After fuzzification of the inputs, the linguistic variables must be combined through the rules. They must be applied to a fuzzy inference system (FIS). The rules in FIS are on this basis that whatever reduces the voltage, the

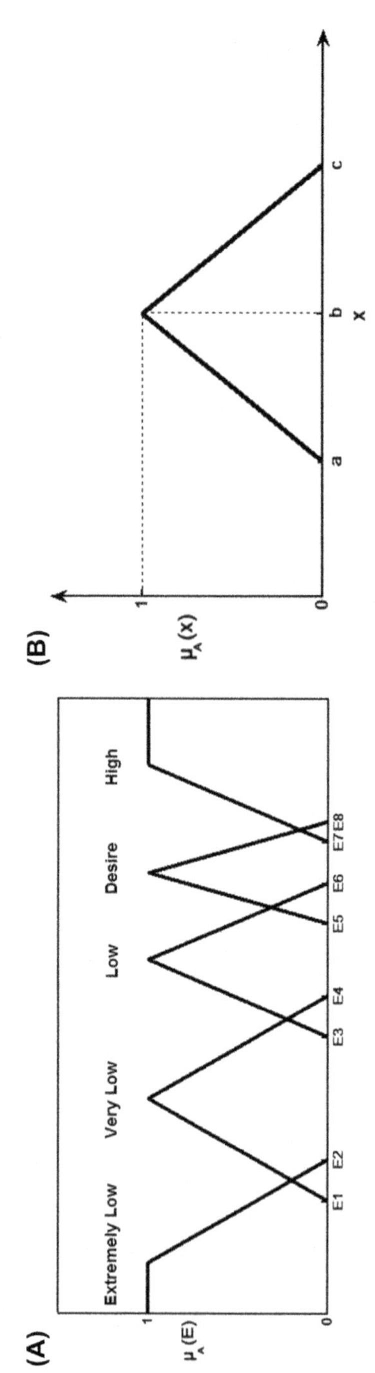

Figure 2.6 Membership function of (A) a triangular number μA(X) and (B) triangular number μA(E) for different voltages.

weight of the generated fuzzy output must be increased. There is a direct relation between the reactive power injection and the output fuzzy weight. After defuzzification, the nonfuzzy output Δq is obtained in the last step. The reactive power after applying a delay in Δq is ΔQ, which must be added to the initial reference value of the reactive power, i.e., Q^*, as follows:

$$\begin{aligned} Q^*_{fu} &= Q^* + \Delta Q \\ \Delta Q &= \Delta q \cdot \left(\frac{1}{1 + T_d s}\right) \end{aligned} \tag{2.18}$$

where Q^*fu is the new reactive power reference, which must be applied to the voltage droop equation by substituting Q^*fu in Eq. (2.1).

The output surface of the FIS and the block diagram of the Q^* fuzzy control scheme are shown in Fig. 2.3.

5.2 Simulation results

For simulations, an MG is selected, which has four renewable energy–based DERs connected to the main grid at PPC as shown in Fig. 2.8. Each DER has a local load. The MG and its hierarchical control parameters are listed in Table 2.2. The MG is simulated offline in MATLAB/Simulink environment, and the OPAL-RT real-time digital simulator is used for software–hardware in the loop (SHIL) simulations [89] to verify the results (Fig. 2.9). The control system and the power system are separately run on two CPUs, i.e., CPU1 and CPU2, respectively (Fig. 2.7).

5.2.1 Islanded operation with step load change

The power sharing of local loads in the islanded MG is shown in Fig. 2.10A. Its voltage and frequency can be seen in Fig. 2.10B and C, respectively. In Fig. 2.10D, the effect of the step load change at $t = 5$ s can be seen. In this instant, a load of 4 and 3 kW has been added to DG1 and DG2. It can be seen that these loads have successfully been supplied by the DGs.

5.2.2 Motor starting

To study motor starting effect, a three-phase 1.1-kW, four-pole, 400-V, 50-Hz induction motor is simulated at bus DG3 and started at $t = 4$ s under nominal load with direct online starting method and then terminated at $t = 8$ s. Three different cases are studied. In the first case, the control schemes of previous works (without using the new control loop and Q^* FLC) are used. In the second case, the new control loop is added to the case

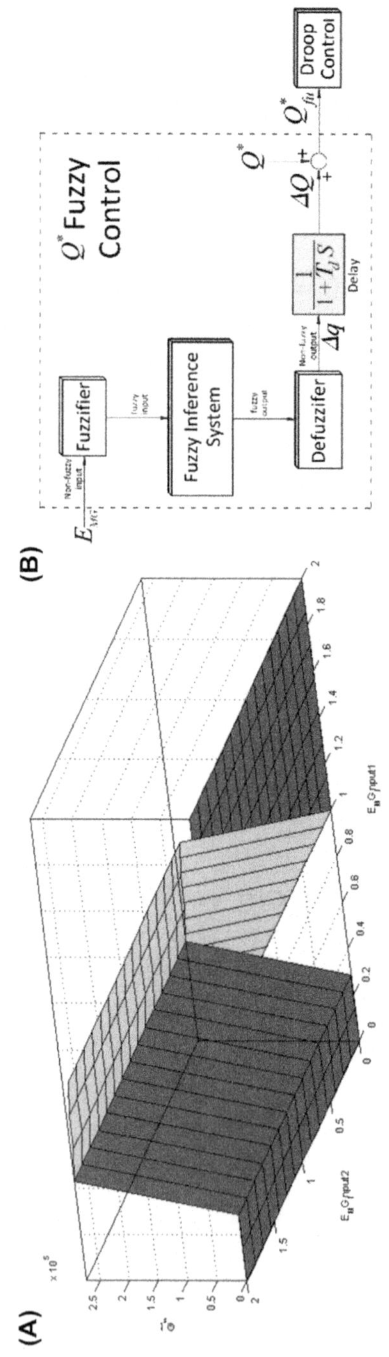

Figure 2.7 (A) Output surface of FIS and (B) Q* fuzzy control scheme block diagram. *FIS*, fuzzy inference system.

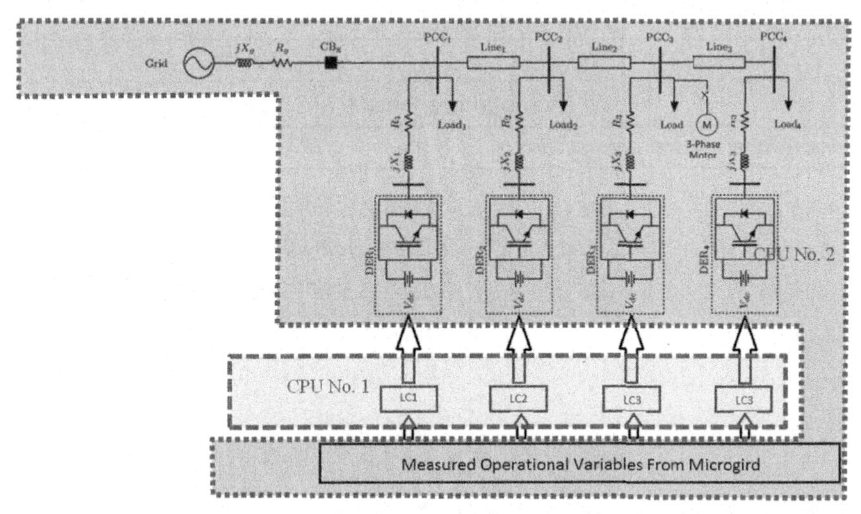

Figure 2.8 Single line diagram of multi-DER MG. *DER*, distributed energy resource; *MG*, microgrid.

Table 2.2 Parameters of hierarchical control.

Items		Values	
		Microgrid parameters	
DG 1	14.7 *kVA*	*Load* 1	11 $kW + j8.1kVAr$
DG 2	23.5 *kVA*	*Load* 2	17.2 $kW + j12.6kVAr$
DG 3	13.6 *kVA*	*Load* 3	9 $kW + j6.5kVAr$
DG 4	28.4 *kVA*	*Load* 4	20 $kW + j16.1\ kVAr$
System voltage (L-L), f		415 V, 50 Hz	
V_{dc}		1500 *Vdc*	
Z_{Line1}		$0.3756+j0.1936\Omega$	
Z_{Line2}		$0.1878+j0.0968\Omega$	
Z_{Line3}		$0.1935+j0.1118\Omega$	
L, C		75.8 mH, 50 μF	
L_{o1}, Lo_2, Lo_3, Lo_4		10.8, 6.94, 12.96, 5.9 mH	

Continued

Table 2.2 Parameters of hierarchical control.—cont'd

Items	Values
Microgrid parameters	
Primary/inner control parameters	
k_{pP1}, k_{iP1}	0.25e-3, 1.25e-6
k_{pP2}, k_{iP2}	0.161e-3, 0.8036e-6
k_{pP3}, k_{iP3}	0.3e-6, 1.5e-6
k_{pP4}, k_{iP4}	0.136e-3, 0.682e-6
k_{pQ1}, k_{pQ2}, k_{pQ3}, k_{pQ4}	0.725, 0.466, 0.87, 0.395
R_{vi}, L_{vi}	1 Ω,4e-4 mH
k_{pV}, k_{r1V}, k_{r5V}, k_{r7V}, kr_{11V}	0.35, 400, 4, 20, 11
k_{pI}, k_{r1I}, k_{r5I}, k_{r7I}, kr_{11I}	0.7, 100, 30, 30, 30
New control loop parameters	
k_{pst}, k_{ist}	0.23, 0.11
Secondary control parameters	
k_{pF}, k_{iF}	0.5e-3, 0.1
k_{pE}, k_{iE}	0.1e-3, 0.11
T	50 ms
Tertiary control parameters	
k_{pPg}, k_{iPg}, k_{pQg}, k_{iQg}	0.7e-3, 9.8, 0.25e-3, 13
Q* fuzzy control parameter	
T_d	23.5 ms

Figure 2.9 The OPAL-RT eMEGAsim real-time digital simulator in the lab.

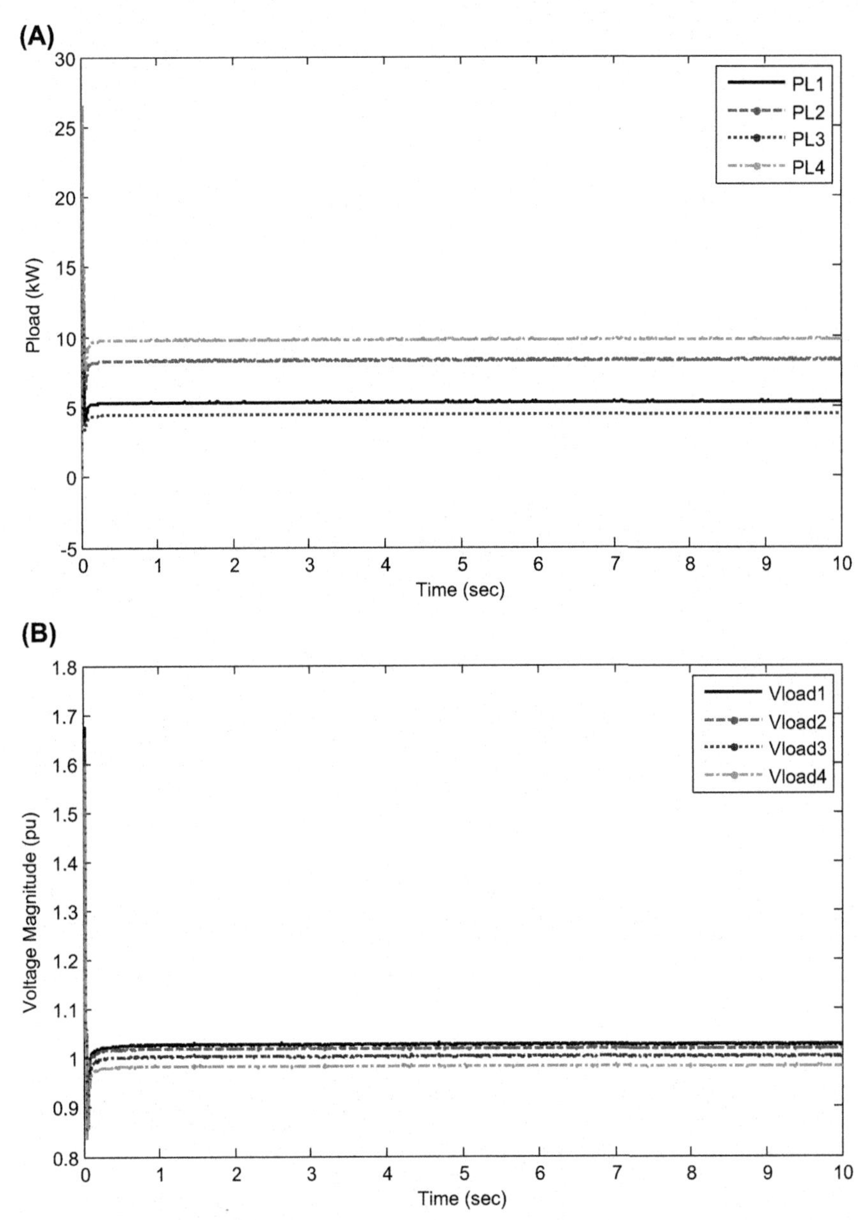

Figure 2.10 Microgrid in islanded mode: (A) steady-state power sharing, (B) voltage, (C) frequency, and (D) power sharing in step-load change.

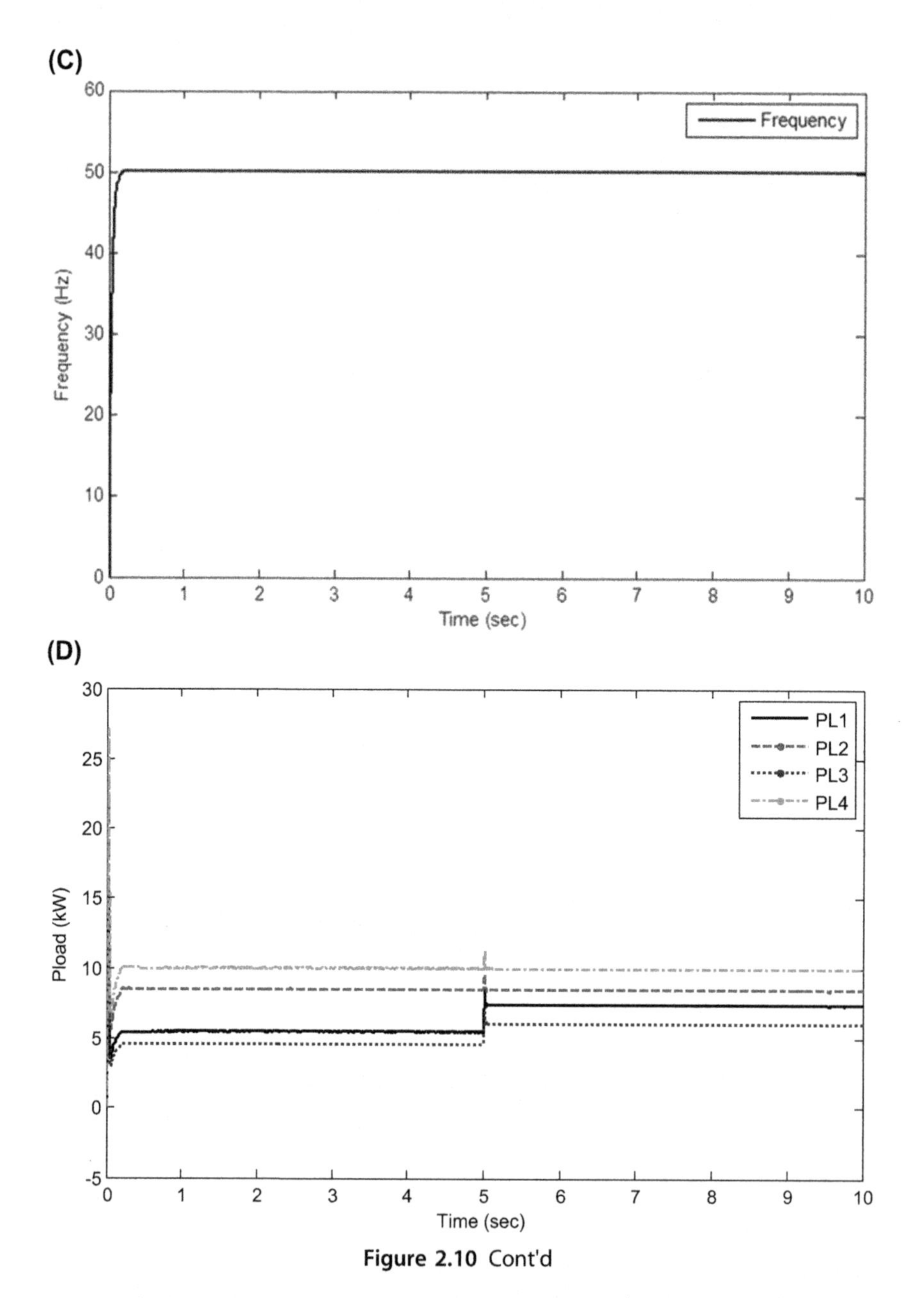

Figure 2.10 Cont'd

one. In the last case, the new control loop and the Q* FLC are added to the case 1. Fig. 2.11A shows the effect of the motor starting on the MG voltage. A heavy voltage drop can be seen. In the second case, the voltage drop is improved as shown in Fig. 2.11B, but the voltage level is out of the

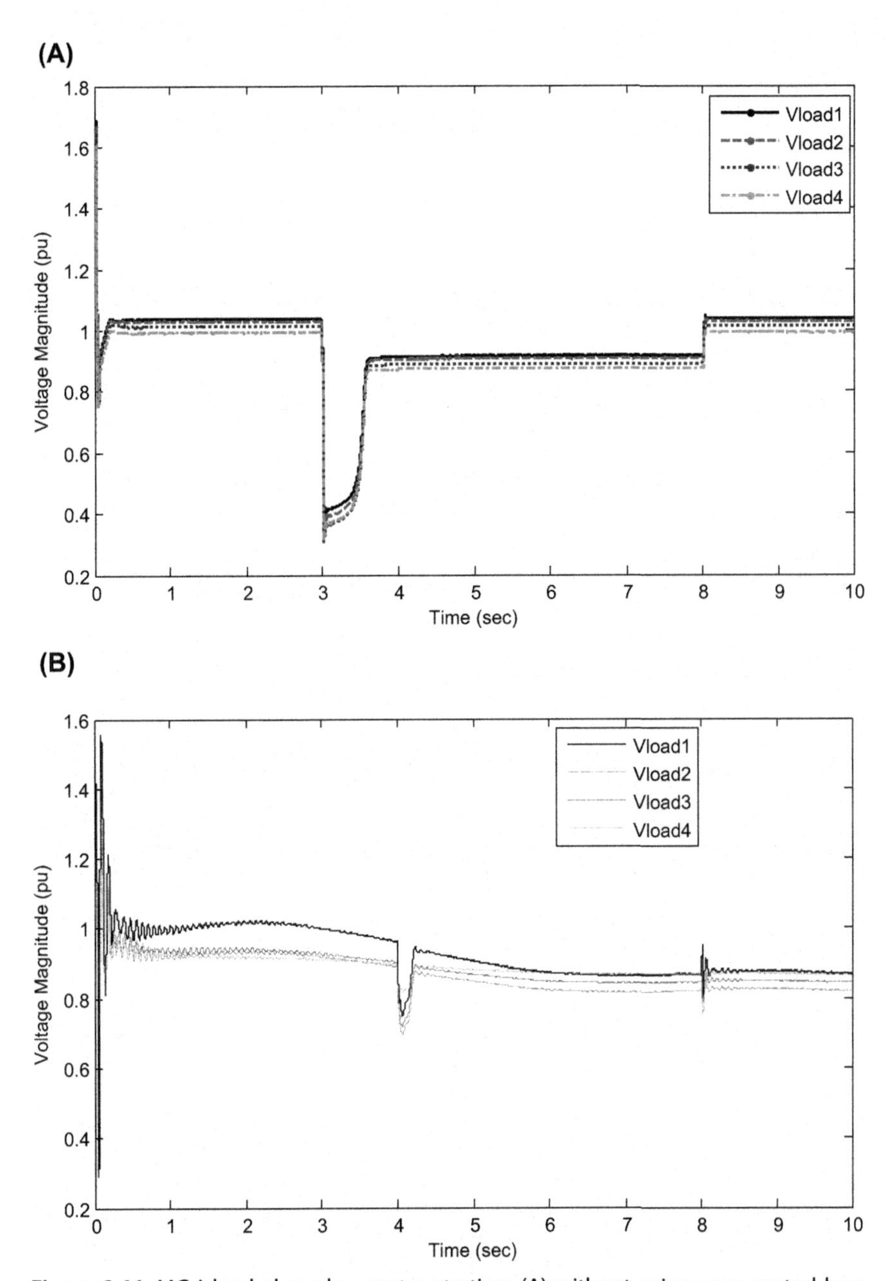

Figure 2.11 MG islanded mode—motor starting: (A) without using new control loop and the Q* FLC, (B) with new control loop but without using Q* FLC, (C) Effect of new control loop and the Q* FLC, and (D) effect of using new control loop and the Q* FLC on power sharing. *FLC*, fuzzy logic controller; *MG*, microgrid.

(C)

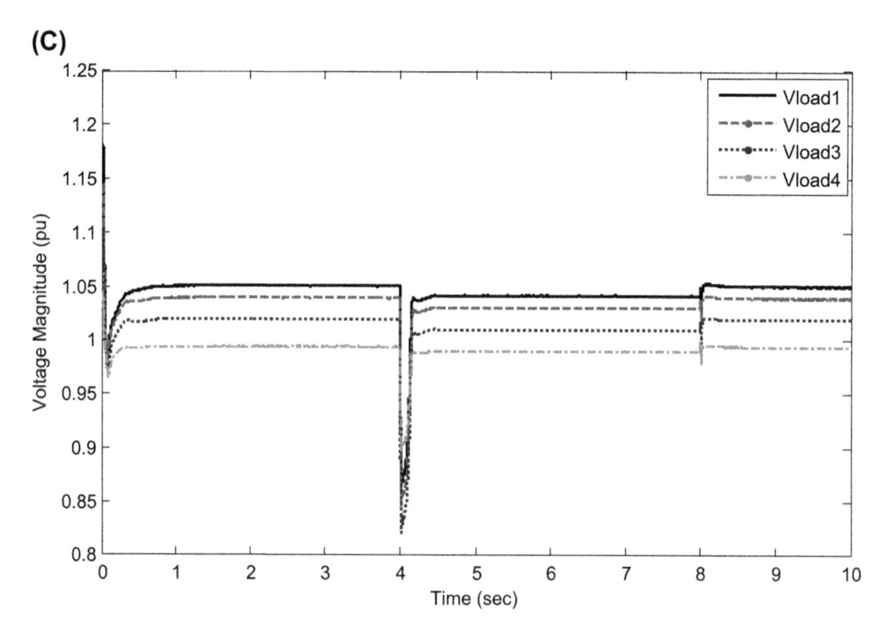

(D)

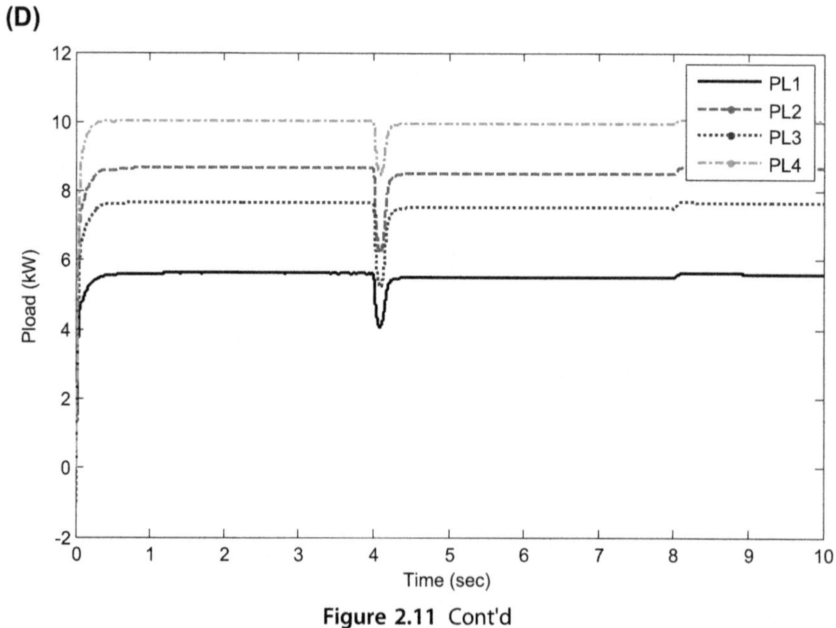

Figure 2.11 Cont'd

standard range and not acceptable. In the third case, the proposed hierarchical controller, including the new control loop and Q* FLC, is used. Fig. 2.11C shows the MG voltage. It can be seen that after a voltage sag, the voltage is immediately recovered to its nominal value. The power sharing in the third case can be seen in Fig. 2.11D.

5.2.3 Grid-connected operation mode

In this mode, the hierarchical control has the tertiary control level. The power sharing among the MG, loads, and the main grid can be seen in Fig. 2.12A. In this case, the MG supplies the local loads and also injects 9 kW to the main grid. It is assumed that an islanding occurs at $t = 5$ s. The MG voltages at different buses can be seen in Fig. 2.12B.

To study the fault effect on this operation mode, another scenario is simulated. A single-phase to ground fault in the line connecting DG2 and DG3 is simulated at $t = 4$ s for six cycles, and then, the circuit breaker opens this line. Therefore, DG1 and DG2 are isolated. Fig. 2.12C shows the power sharing, and Fig. 2.12D indicates the voltage before, during, and after the fault occurrence. The voltages are stable, and the power sharing is acceptable.

5.2.4 Three-phase to ground fault in islanded mode and fault ride through capability, real-time verification

In this section, a three-phase fault at $t = 4$ s is considered in the middle of the line connecting DG1 and DG2, and it is cleared after six cycles. The OPAL-RT real-time simulator is used to verify the results [89]. In Fig. 2.8A–D, it can be seen that the proposed new hierarchical controller results in stable voltage and an acceptable power sharing after fault clearing, while the voltages and power flows become unstable and the power sharing is undesirable in case of using the conventional solution. The MG voltage is properly restored from the fault presenting a good FRT capability (Fig. 2.13).

5.3 Discussions

In this case study, the basic concepts and necessary requirements for control structure and MGSC/EMS have been reviewed. A novel hierarchical control system has been proposed for an MG. The control system has a new control loop in the internal control loops of the VSC and a Q* FLC. The performance of the suggested solution has been validated using off-line simulations in MATLAB environment and verified by real-time simulations of OPAL-RT simulator. In the islanded and gird-connected modes, the MG has maintained its stability versus step load change (small-signal) and three-phase and single-phase to ground faults (large-signal). By using the suggested Q* FLC, the voltage drop caused by the motor starting has been improved. The performance of the MG control for different cases is summarized in Table 2.3.

It seems that although there are so many reported researches in this area, there are lots of issues that open a wide area of future research for the experts and researchers.

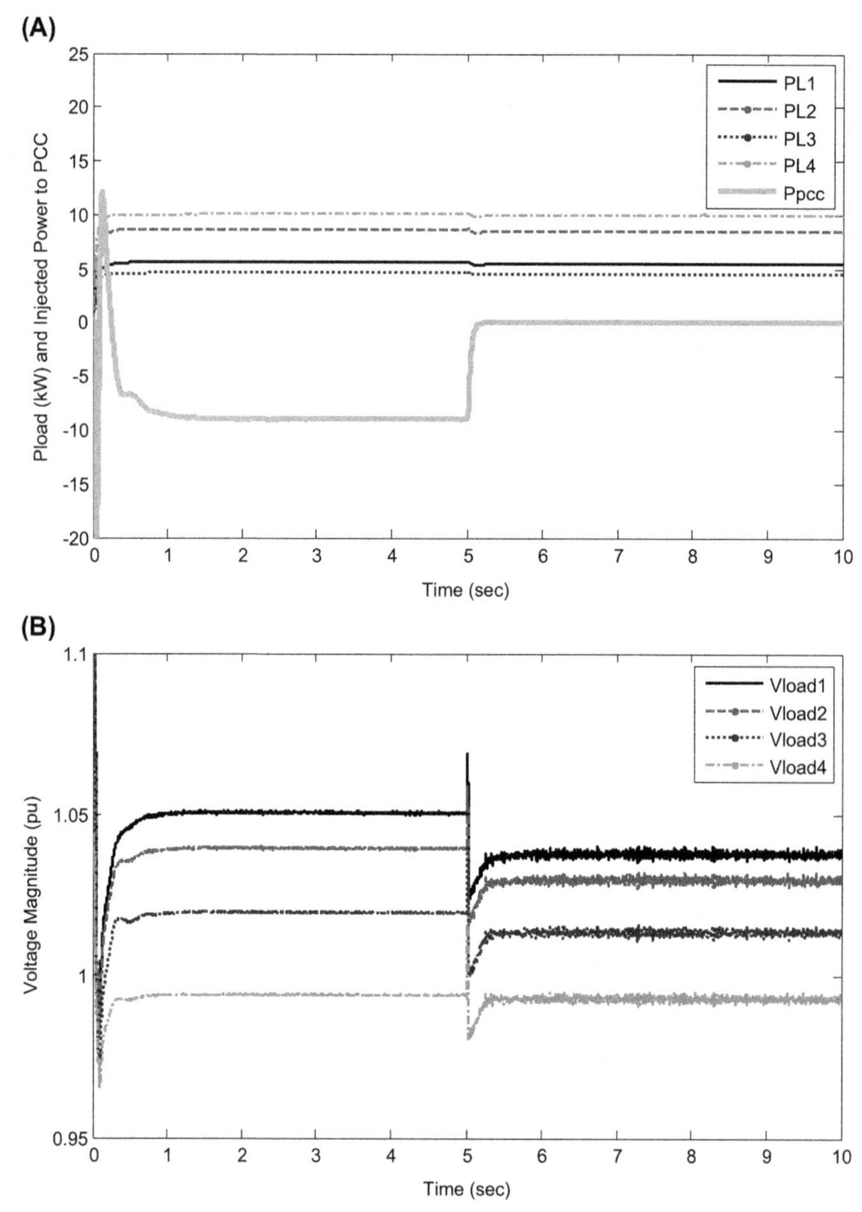

Figure 2.12 MG-connected mode: (A) power sharing, (B) voltages, (C) power sharing after single-phase to ground fault, and (D) voltage after single-phase to ground fault. *MG*, microgrid.

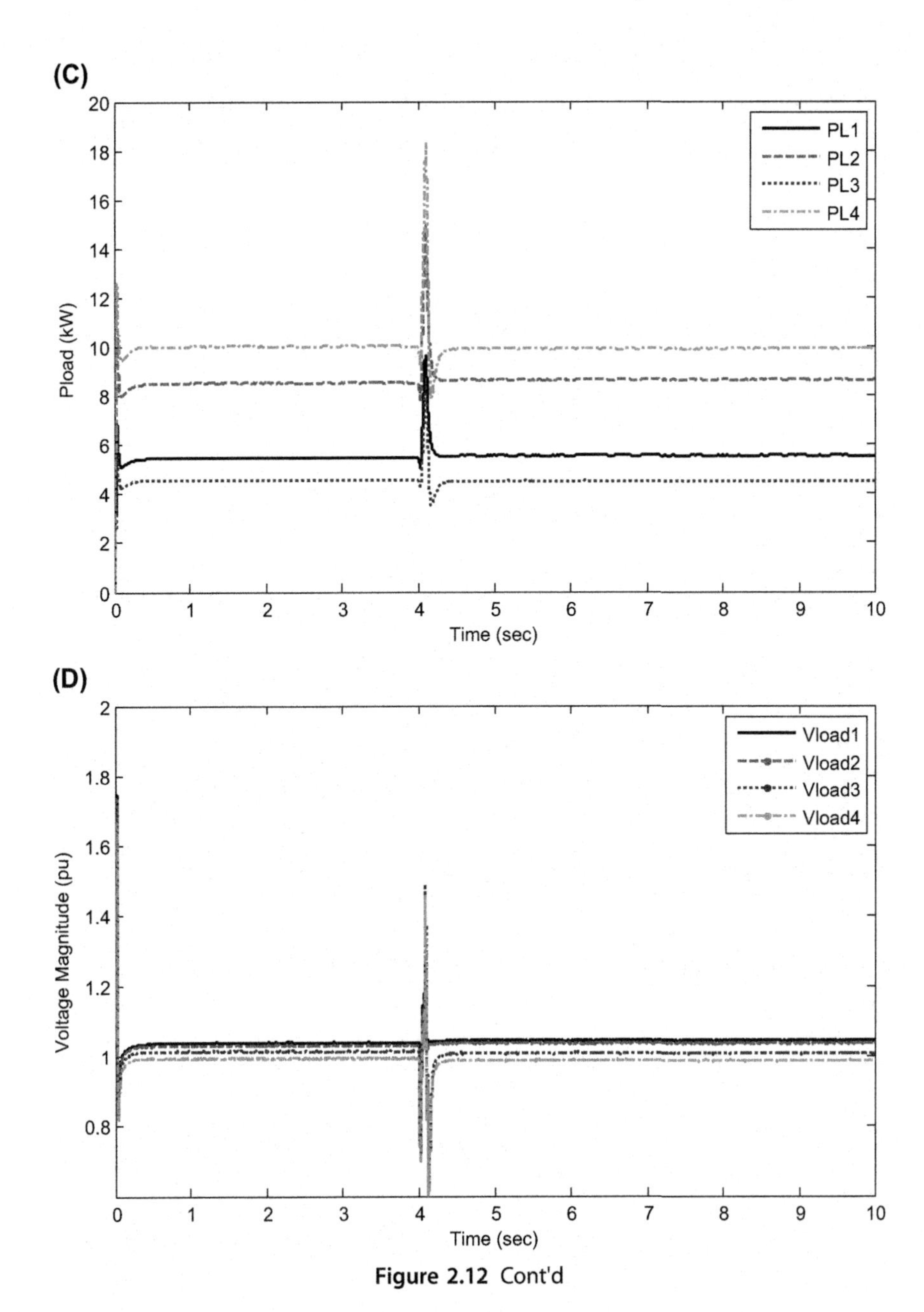

Figure 2.12 Cont'd

Table 2.3 Comparison of microgrid performance for different control schemes.

Control scheme	Disturbance description		
	Small-signal events: Step load change	Large-signal disturbance: Short circuit fault	Large-signal disturbance: Heavy motor starting
Hierarchical control:	Good performance	Unstable, bad performance	Unstable, bad performance
Hierarchical control + new control loop	Good performance	Good performance	Stable, high voltage drop, weak performance
Hierarchical control + new control loop + Q^* fuzzy controller	Good performance	Good performance	Good performance

6. Case study 2: improvement of hierarchical control performance for unbalanced and nonlinear loads

In this case study, an MG is studied, which has nonlinear and unbalanced loads. An improved hierarchical control is suggested for unbalanced harmonic voltage compensation and power sharing among DERs. The proposed hierarchical control has the same control levels of the conventional one [17,28,32]. However, to enhance the MG stability, it uses a new unbalance/harmonic virtual impedance (UHVI) and has a supplementary control loop [2,19]. For positive and negative sequences (PS/NS), the line current and voltage at PCC adjust the virtual impedance at fundamental and harmonic frequencies, respectively. To solve harmonic power flow (HPF) and determine the voltage harmonics and active and reactive powers, a novel power calculation method is suggested for MG balanced and unbalanced modes of operation, which uses the nonlinear mapping capability of radial basis function neural networks (RBFNNs). To validate the effectiveness of the proposed control system, off-line simulations in MATLAB environment and real-time simulations using OPAL-RT simulator [89] are used.

In Ref. [71], a fundamental frequency deterministic power flow calculation method using RBFNNs has been proposed to enhance the power sharing of droop-based hierarchical control system. The suggested method has several advantages compared with other ones such as (1) unlike the methods presented in Refs. [17,28,65], it uses a new virtual impedance which is a combination of fundamental and unbalanced/harmonics virtual impedances and (2) it generates a signal, which is added to the three-phase reference generator output to adjust the voltage controller input, as shown in Fig. 2.14 [71]. To generate this signal, the real value of the voltage and current total harmonic distortion (THD) at PCC calculated by RBFNN has been compared with their maximum values according to IEEE Standards 519 and 1159 [90,91].

The main function of RBFNN is solving the HPF as will be presented in Chapter 3. In case of communication system failure, the RBFNN can approximate the harmonics and estimate set points for the proposed control system. This function can be considered as the second one. Actually, the RBFNN not only determines voltage harmonics, THD, and active and reactive powers for tertiary control level, but also it is a low-bandwidth communication system backup. It should be mentioned that the suggested structure may be implemented either decentralized or distributed in the secondary control level considering the design criteria for communication system.

6.1 Basic structure and control levels

The structure of the proposed control in this section is the same as the one discussed in Section 5. To improve the performance of this hierarchical structure for nonlinear and unbalanced loads, a self-tuning filter, a harmonic virtual impedance, and a harmonic compensation scheme are added.

6.1.1 Self-tuning filter

In Ref. [92], in the synchronous reference frame (SRF), it has been shown that we have

$$V_{xy} = e^{j\omega t} \int e^{-j\omega t} U_{xy}(t)\mathrm{dt} \tag{2.19}$$

where U_{xy} and V_{xy} are instantaneous signals before and after integration in SRF, respectively. Using Laplace transform, Eq. (2.20) can be expressed in the frequency domain as follows:

$$H(s) = \frac{V_{xy}(s)}{U_{xy}(s)} = \frac{s + j\omega}{s^2 + \omega^2} \tag{2.20}$$

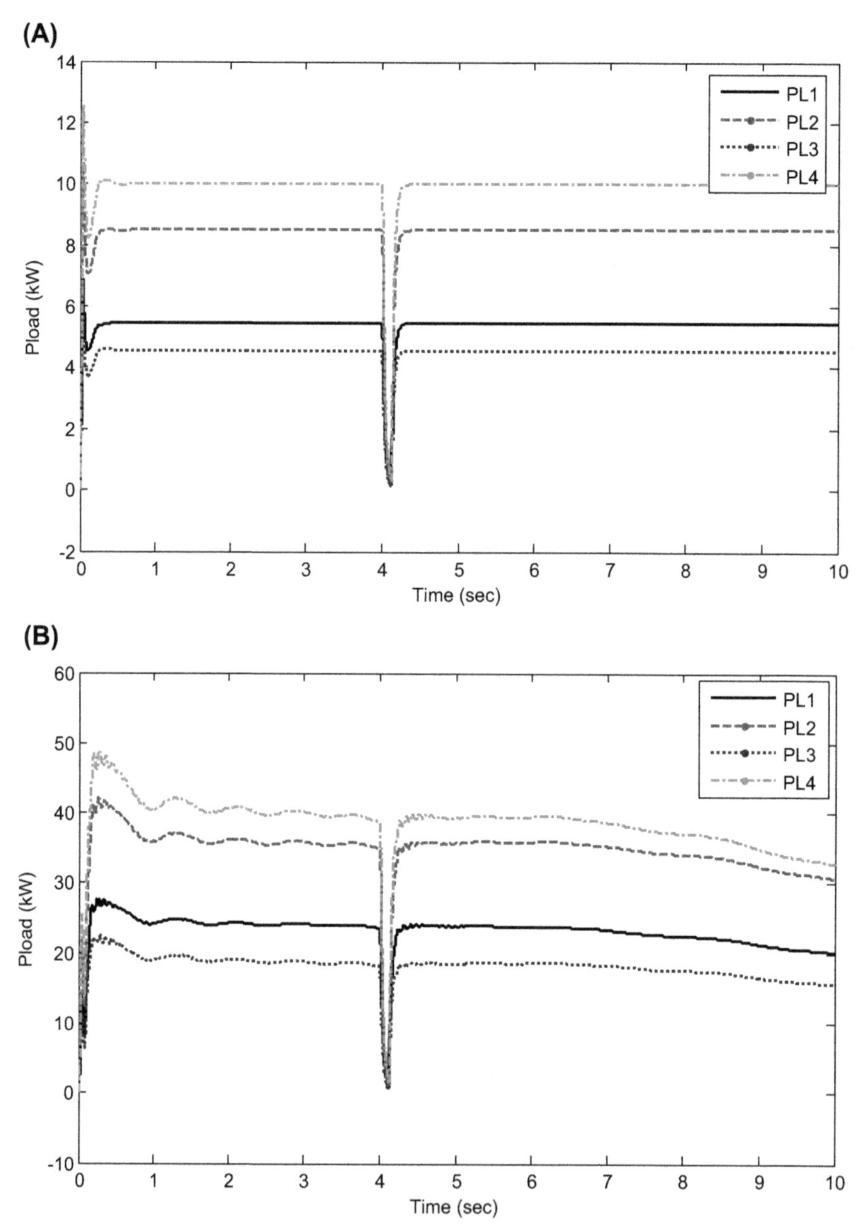

Figure 2.13 MG in islanded mode—symmetric three-phase fault condition: (A) the proposed control effect on power sharing, (B) power sharing without using new control loop, (C) the proposed control effect on voltages, and (D) voltages (without using new control loop). *MG*, microgrid.

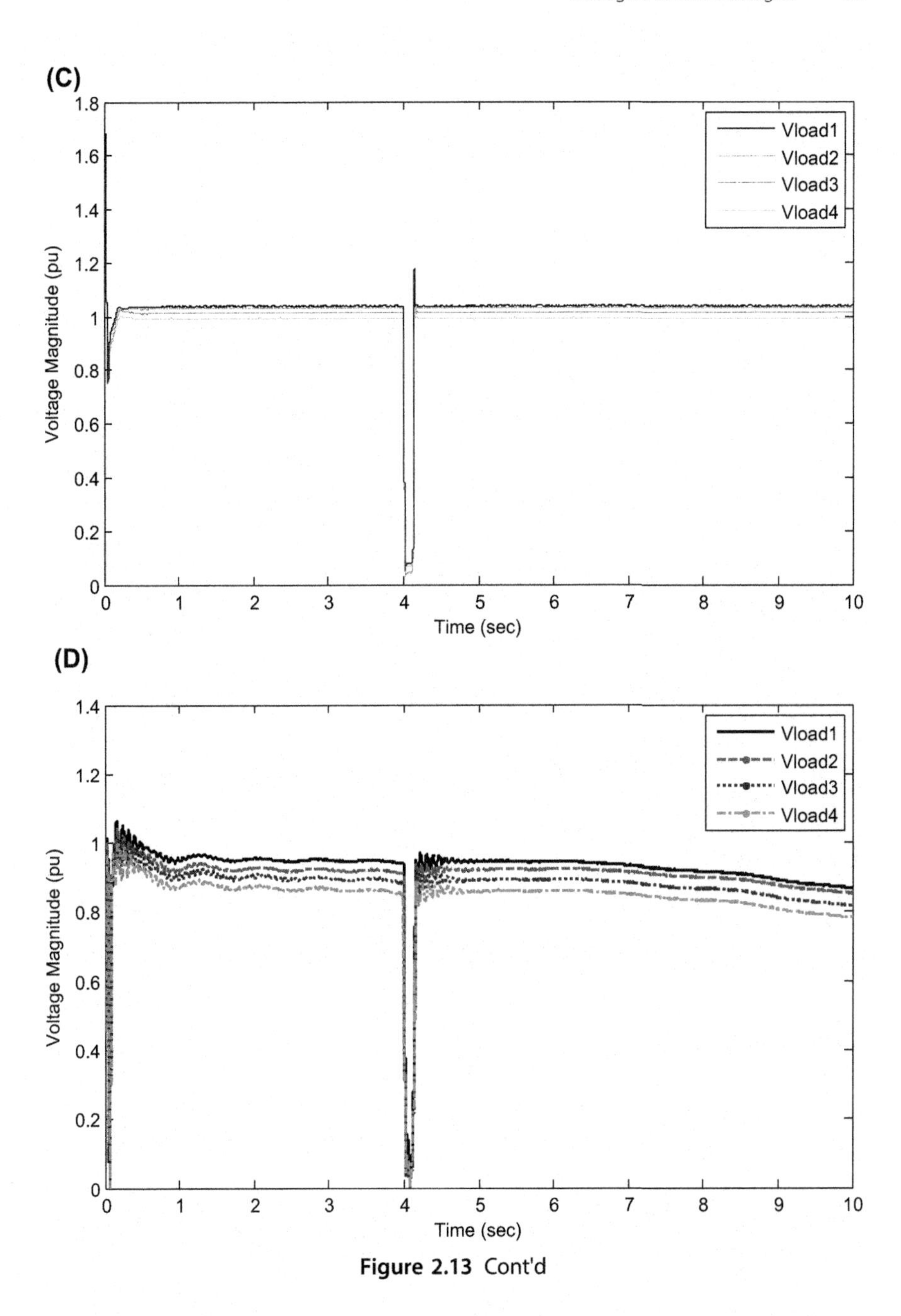

Figure 2.13 Cont'd

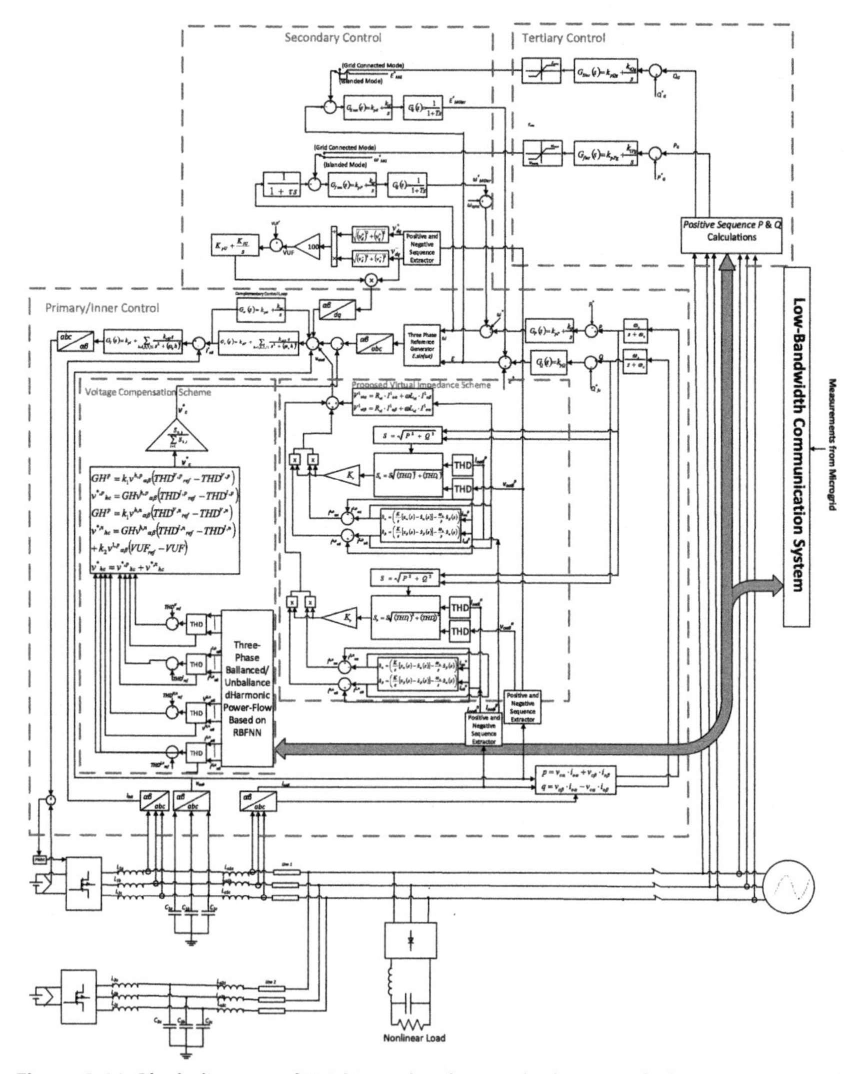

Figure 2.14 Block diagram of MG hierarchical control scheme including two DERs with common load. *DER*, distributed energy resource; *MG*, microgrid.

To suggest a self-tuning filter (STF) with cutoff frequency of ω_n, a fixed parameter (K) has been presented in Ref. [92] based on transfer function $H(s)$, as follows [92,93]:

$$H(s) = \frac{V_{xy}(s)}{U_{xy}(s)} = K\frac{(s+K)+j\omega}{(s+K)^2+\omega^2} \tag{2.21}$$

After, respectively, replacing the input and output signals U_{xy} and V_{xy} with $x_{\alpha\beta}(s)$ and $\widehat{x}_{\alpha\beta}(s)$, and after simplifying (2.21), we have

$$\widehat{x}_{\alpha} = \left(\frac{K}{s}[x_{\alpha}(s) - \widehat{x}_{\alpha}(s)] - \frac{\omega_n}{s}\widehat{x}_{\beta}(s)\right) \tag{2.22}$$

$$\widehat{x}_{\beta} = \left(\frac{K}{s}[x_{\beta}(s) - \widehat{x}_{\beta}(s)] - \frac{\omega_n}{s}\widehat{x}_{\alpha}(s)\right) \tag{2.23}$$

where $x_{\alpha\beta}(s)$ and $\widehat{x}_{\alpha\beta}(s)$ can be either voltage or current signals before and after filtering in STF, respectively, and ω_n is the desired frequency at the output. The lower value of K results in higher accuracy of extracting desired frequency components. The suggested STF, used in the proposed hierarchical control system, filters the current and voltage to extract the desired frequency components without any delay and reduction in stability margin [92]. It can be shown that at $f = 50$ Hz, the $H(s)$ magnitude and phase angle is 1 and 0, respectively. Consequently, the current and voltage reference can properly be tracked [92].

6.1.2 Virtual impedance

To design a virtual impedance, there are many options [32,76,94,95]. The one suggested in Refs. [19,71] can guarantee its inductive behavior at the power frequency. In this section, the presented one consists of fundamental and harmonic virtual impedances for PS and NS components, respectively, as follows (Fig. 2.14).

$$\begin{aligned} V_{vi\alpha}^{1,p} &= R_{vi}^{p} \cdot I_{o\alpha}^{1,p} + \omega L_{vi} \cdot I_{o\beta}^{1,p} \\ V_{vi\beta}^{1,p} &= R_{vi}^{p} \cdot I_{o\beta}^{1,p} + \omega L_{vi} \cdot I_{o\alpha}^{1,p} \end{aligned} \tag{2.24}$$

$$\begin{aligned} V_{vi\alpha}^{1,n} &= R_{vi}^{n} \cdot I_{o\alpha}^{1,p} + \omega L_{vi} \cdot I_{o\beta}^{1,n} \\ V_{vi\beta}^{1,n} &= R_{vi}^{n} \cdot I_{o\beta}^{1,p} + \omega L_{vi} \cdot I_{o\alpha}^{1,n} \end{aligned} \tag{2.25}$$

$$S^{p} = \sqrt{(P^{p})^2 + (Q^{p})^2} \tag{2.26}$$

$$S^{n} = \sqrt{(P^{n})^2 + (Q^{n})^2} \tag{2.27}$$

$$S_{n}^{P} = S\sqrt{(THD_{I}^{p})^2 + (THD_{V}^{p})^2} \tag{2.28}$$

$$S_{n}^{n} = S\sqrt{(THD_{I}^{n})^2 + (THD_{V}^{n})^2} \tag{2.29}$$

$$R^{p}_{v,harm} = K^{p}_{v} S_{n} \tag{2.30}$$

$$R^{n}_{v,harm} = K^{n}_{v} S_{n} \tag{2.31}$$

where $v_{c\alpha\beta}$ and $i_{o\alpha\beta}$ are capacitor output voltage and current in $\alpha\beta$ coordinate, $V^{1}{}_{vi\alpha\beta}$ and $I^{1}{}_{o\alpha\beta}$ are fundamental frequency components of $v_{vi\alpha\beta}$ and $i_{o\alpha\beta}$, L_{vi} and R_{vi} are inductance and resistance of the virtual impedance, THD_{v}, and THD_{i} are output voltage and current THD of DERs, K_{v} is small positive constant determined based on nominal power of DERs, $R_{v,harm}$ is harmonic virtual resistance, $v^{h}{}_{o}$ and $Z^{h}{}_{inv}$ are VSC output voltage and impedance at harmonic order of h in the absence of virtual impedance, $v^{h}{}_{out}$ and $v^{h}{}_{PCC}$ are, respectively, DER and PCC harmonic voltage in presence of the harmonic virtual impedance, and THD^{v} and THD^{i} are voltage and current THD, respectively. The superscripts p and n indicate for PS and NS variables. S_{n} is calculated based on IEEE Standard 1459 [96] using Eqs. (2.28) and (2.29). As S_{n} increases, the value of $R_{v,harm}$ also increases for a specific value of $R_{v,harm}$ considering Eqs. (2.30) and (2.31), which is a limiting factor for S_{n}. For a better clarification, an equivalent circuit is shown in Fig. 2.15 [97,98]. It is obvious that the DER output voltage THD, and as a result, PCC voltage may increase on account of voltage drop across the $R_{v,harm}$. The other harmonic extraction methods uses Fourier transform and communication systems [99,100], or low-pass (LPF) or band-pass filters (BPF) [84,85,101–103], but the proposed solution uses the HPF nonlinear equation set (NLES) solved by RBFNN.

6.1.3 Harmonic compensation

As mentioned and shown in Fig. 2.14, the RBFNN is used to carry out the HPF and extract the harmonics for the voltage compensation, and then THD^{v} and THD^{i} are determined and compared with their reference values, i.e., $THD^{v}{}_{ref}$ and $THD^{i}{}_{ref}$, to generate the supplementary signal for voltage compensation $v^{*}{}_{c}$ using the following equations:

$$GH^{p} = k_{1} v^{h,p}_{\alpha\beta}\left(THD^{V,p}_{ref} - THD^{V,p}\right) \tag{2.32}$$

$$GH^{n} = k_{1} v^{h,n}_{\alpha\beta}\left(THD^{V,n}_{ref} - THD^{V,n}\right) \tag{2.33}$$

$$v^{*,p}_{hc} = GHv^{h,p}_{\alpha\beta}\left(THD^{I,p}_{ref} - THD^{I,p}\right) \tag{2.34}$$

$$v^{*,n}_{hc} = GHv^{h,n}_{\alpha\beta}\left(THD^{I,n}_{ref} - THD^{I,n}\right) + k_{2} v^{h,n}_{\alpha\beta}\left(VUF_{ref} - VUF\right) \tag{2.35}$$

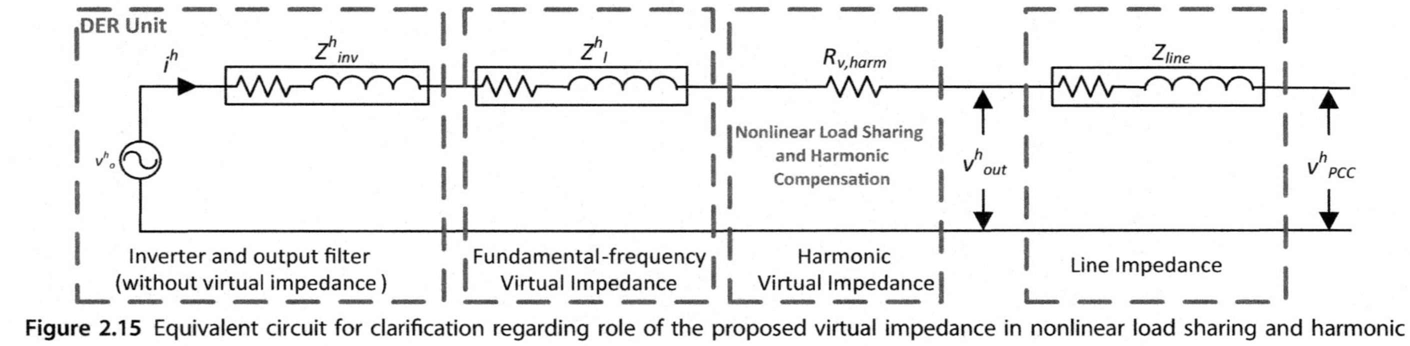

Figure 2.15 Equivalent circuit for clarification regarding role of the proposed virtual impedance in nonlinear load sharing and harmonic compensation.

$$v_c^* = \frac{S_{o,k}}{\sum_{i=1} S_{0,i}} v_{hc}^* \tag{2.36}$$

$$v_{hc}^* = \frac{S_{o,k}}{\sum_{i=1} S_{0,i}} \left(v_{hc}^{*,p} + v_{hc}^{*,n} \right) \tag{2.37}$$

where GH is voltage harmonics compensation gain, $v^h{}_{\alpha\beta}$ is voltage harmonics of PCC in $\alpha\beta$ coordinate, $v^*{}_{hc}$ and $v^*{}_c$ are voltage harmonics compensation reference and compensation reference, respectively, and $S_{O,i}$ is VSC nominal power for ith DER. In addition, an unbalanced voltage compensation plan is used to the control system. Firstly, the voltage unbalance factor (VUF) must be determined using the following equation [84,85,90,91]:

$$VUF = 100 \times \frac{\sqrt{\left(v_d^-\right)^2 + \left(v_q^-\right)^2}}{\sqrt{\left(v_d^+\right)^2 + \left(v_q^+\right)^2}} \tag{2.38}$$

The VUF is compared with its reference value, and the error is compensated by a PI controller. Then, the output is added to the inner control loops of the primary control level as shown in Fig. 2.14. To reduce THD at PCC, GH must be a positive constant determined considering harmonic pollution at PCC; the negative sign is used for the injection of $v^*{}_c$ so that the harmonic voltage is generated with 180° phase shift with respect to $v^h{}_{\alpha\beta}$.

6.1.4 Algorithm of harmonic power flow

The initial conditions of the proposed hierarchical control scheme and the set points of the control loops are determined using an improved harmonic power flow that will be further explained in Chapter 3 of this book.

6.2 Simulation results

To study the proposed control scheme, the same MG discussed in Section 5 and shown in Fig. 2.8 is simulated in MATLAB/Simulink environment with the same parameters listed in Table 2.2. It should be noted that real-time verifications can be also performed like Section 5.2.4. As mentioned before, each DER has an LC as shown in Fig. 2.14.

6.2.1 Nonlinear load effect

A highly nonlinear load, namely, a three-phase six-pulse diode bridge rectifier is simulated at bus DER3. This rectifier supplies 5-hp, 400-V, 1750-rpm shunt DC motor, which starts at $t = 5$ s as shown in Fig. 2.16A. The nonlinear load is completely modeled without using ANN, which was used in Ref. [104]. To study the effect of the proposed virtual impedance and harmonic compensation on the MG performance, three intervals are discussed and shown in Fig. 2.16B.

- Interval 1 ($0 \text{ s} \leq t \leq 2$ s): In this time interval, the fundamental frequency virtual impedance is active, but there is no voltage compensation.
- Interval 2 ($2 \text{ s} \leq t \leq 3.5$ s): The harmonic virtual impedance is active in this period, but there is no voltage compensation as well.
- Interval 3 ($3.5 \text{ s} \leq t \leq 5$ s): The proposed virtual impedance and voltage compensation are active.

Although the input currents of the rectifier have high THD about 31%, it can be seen in Fig. 2.16A and B that the proposed method results in high-quality voltages and their THD satisfies the IEEE Standards 519 and 11,159 [90,91]. The bus 3 and 2 voltage and DER3 and DER2 output current are shown in Fig. 2.17 and Fig. 2.18, respectively. As can be seen in Fig. 2.17, the current sharing is not acceptable in case of using the conventional hierarchical droop control, as shown in Fig. 2.18, the proposed method can solve this problem.

6.2.2 Unbalanced load effect

To study the unbalanced load effect on the performance of the system, a 230 Ω resistive load is simulated at bus 3 between phase b and phase c. The bus 3 and bus 2 voltage and DER3 and DER2 output current are shown in Fig. 2.19 and Fig. 2.20, respectively. It can be seen in Fig. 2.19 that the conventional hierarchical droop control leads to a current sharing, which is not acceptable, but the suggested solution can overcome this problem as shown in Fig. 2.20.

6.2.3 Distributed energy resource outage and plug-and-play operation

In this section, it is assumed that the nonlinear and unbalanced loads are supplied by the MG. Under this condition, a step load change (as a small signal disturbance) and a DER outage (as a large signal disturbance) are

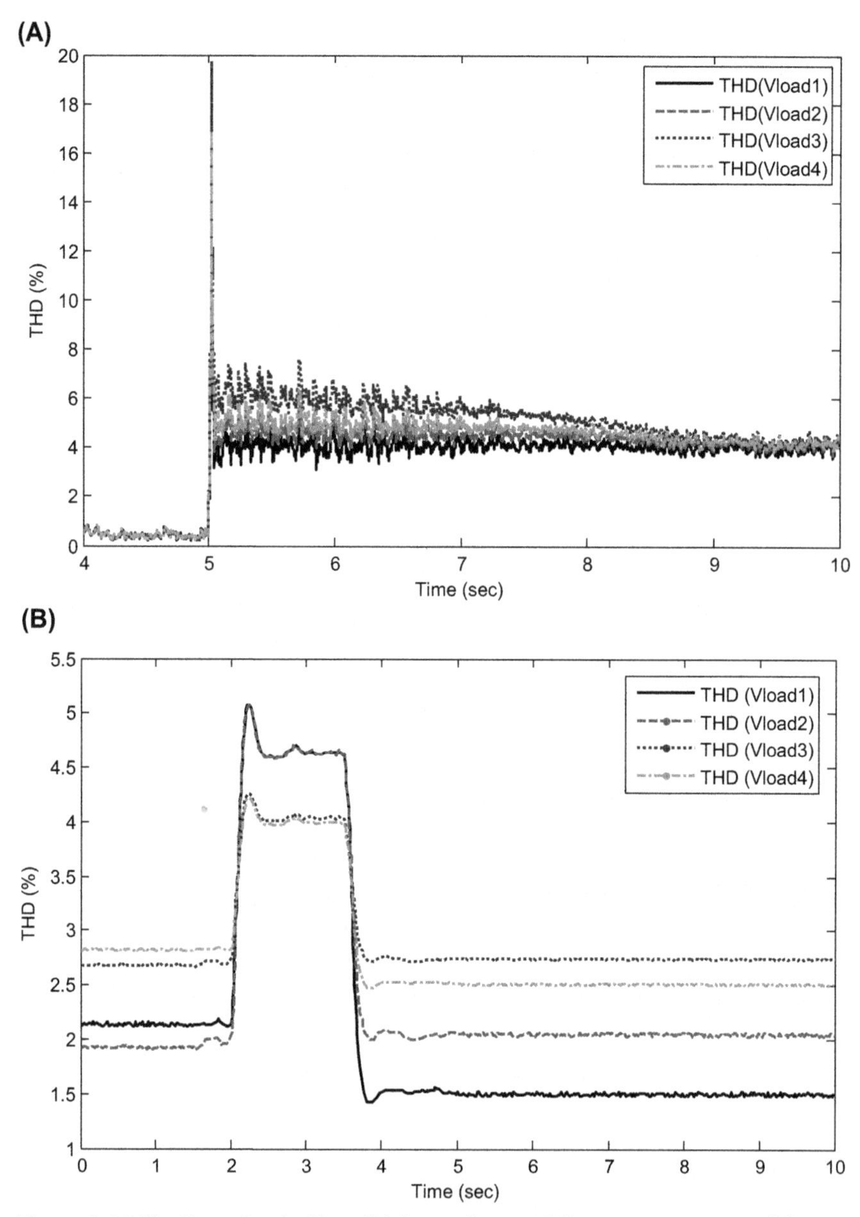

Figure 2.16 Nonlinear load effect: (A) bus voltages THD using conventional hierarchical droop-based control [17,28] and (B) bus voltages THD using the proposed control. *THD*, total harmonic distortion.

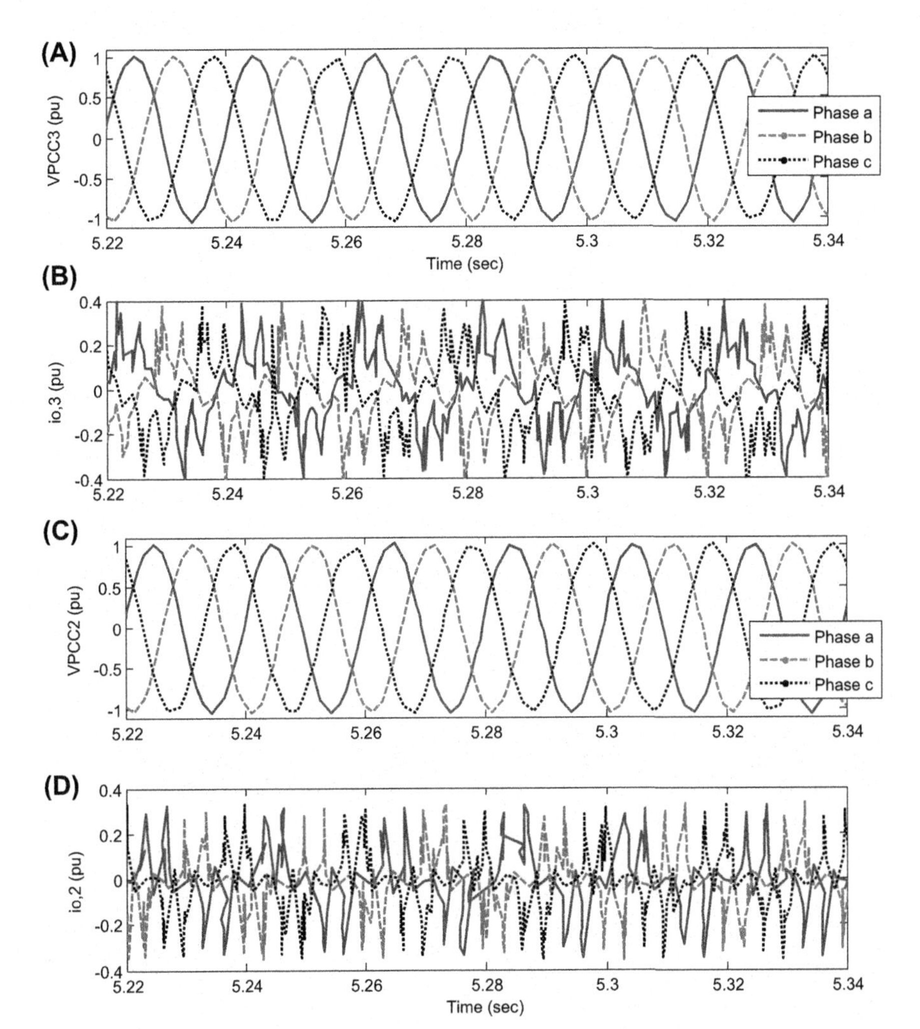

Figure 2.17 Nonlinear load effect: (A and B) bus voltage and output current of DER3, respectively, and (C and D) bus voltage and output current of DER2, respectively, using conventional droop-based control [17,28]. *DER*, distributed energy resource.

applied to the MG, and the MG stability and plug-and-play operation are studied. The MG simulations are presented in the following three periods:

- Period 1 ($0 \text{ s} \leq t \leq 1 \text{ s}$): In this period, the MG is in normal operation mode.
- Period 2 ($1 \text{ s} \leq t \leq 2 \text{ s}$): A step load change is applied to the MG under the conditions presented in Section 4.1.

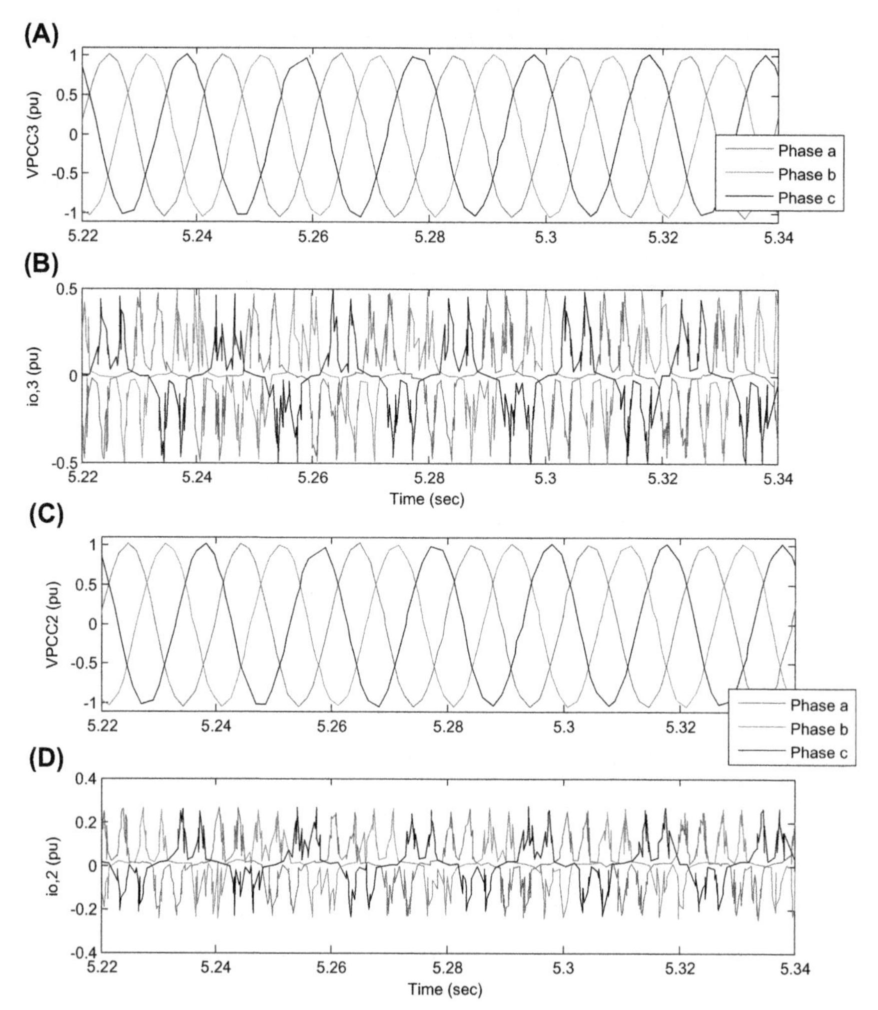

Figure 2.18 Nonlinear load effect: (A and B) bus voltage and output current of DER3, respectively, and (C and D) bus voltage and output current of DER2, respectively, using the proposed control scheme. *DER*, distributed energy resource.

- Period 3 (2 s $\leq t \leq$ 3 s): In this period, the DER4 outage is simulated under the conventional hierarchical droop-based control [17,28] and the proposed control system.

The results can be seen in Fig. 2.21A–D. It is obvious that the performance of the MG with nonlinear and unbalance loads is enhanced using the suggested control system. Also, plug-and-play operation mode and small and large signal stability are improved.

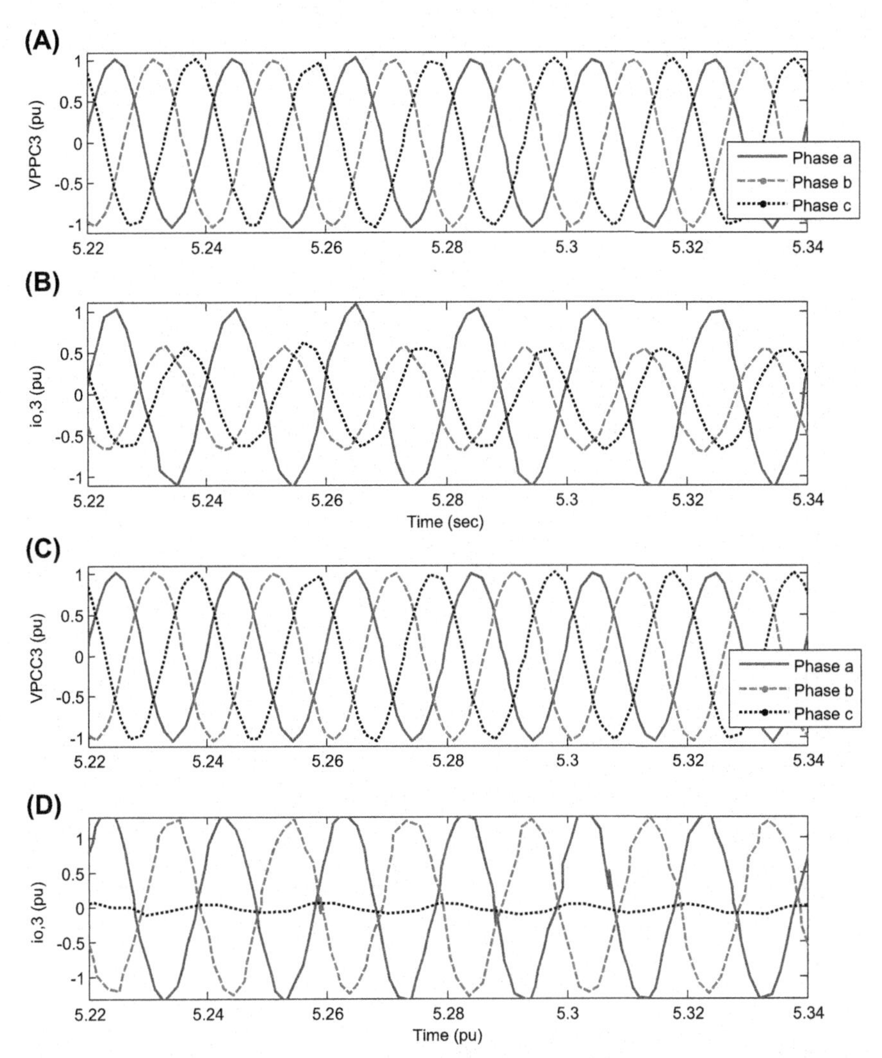

Figure 2.19 Unbalanced load effect: (A and B) 3 voltage and DER3 output current and (C and D) bus 2 voltage and DER2 output current using conventional droop-based control [17,28]. *DER*, distributed energy resource.

7. Summary

In this chapter, basic infrastructures for control and power/energy management in MGs was introduced and classified. The MG basic control architectures are centralized, decentralized, distributed, and hierarchical. Also, current challenges and future trends in control and management of

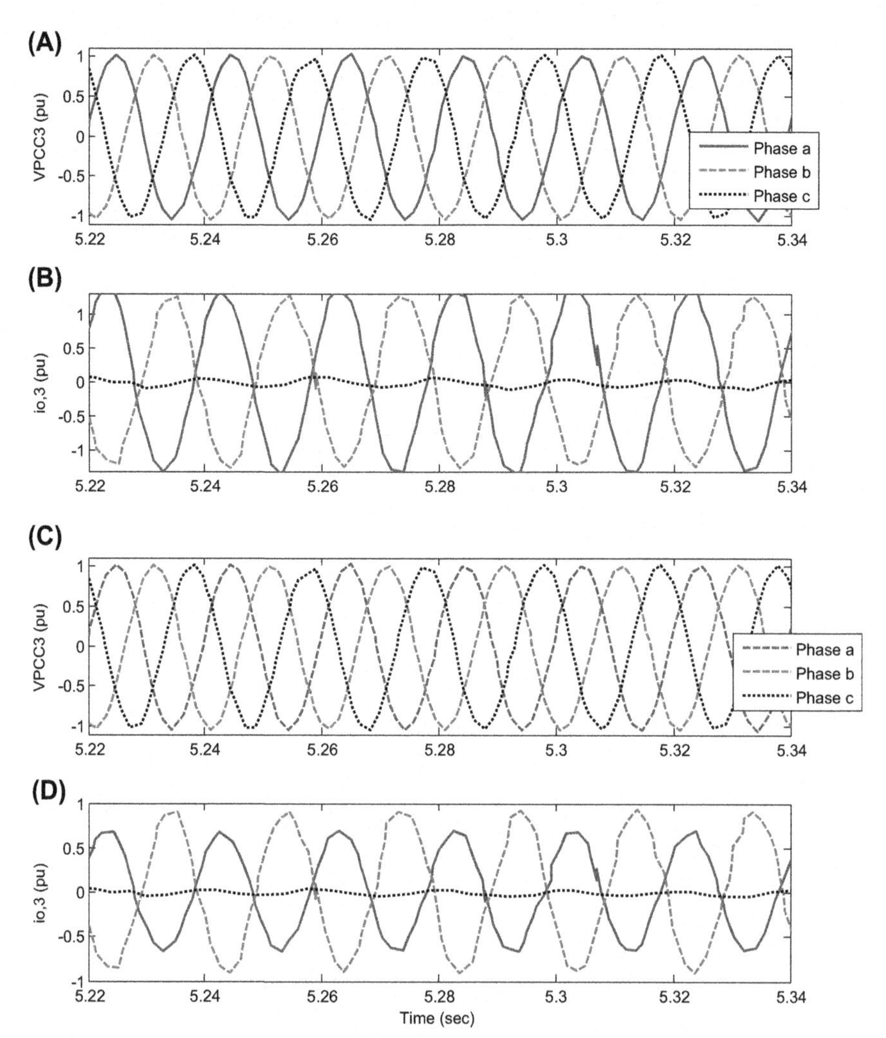

Figure 2.20 Unbalanced load effect: (A and B) 3 voltage and DER3 output current, (C and D) bus 2 voltage and DER2 output current using the proposed control method. *DER*, distributed energy resource.

MGs were reviewed. Finally, two case studies were studied for design and simulation of MGs with the proposed enhanced hierarchical control system.

In the first case study, a new control loop was added to the internal control loops of the VSC and a Q^* FLC. It was shown that the instability due to large signal disturbances can be overcome. Also, in off-grid and on-grid operation modes, the MG can preserve its stability vs. small signal

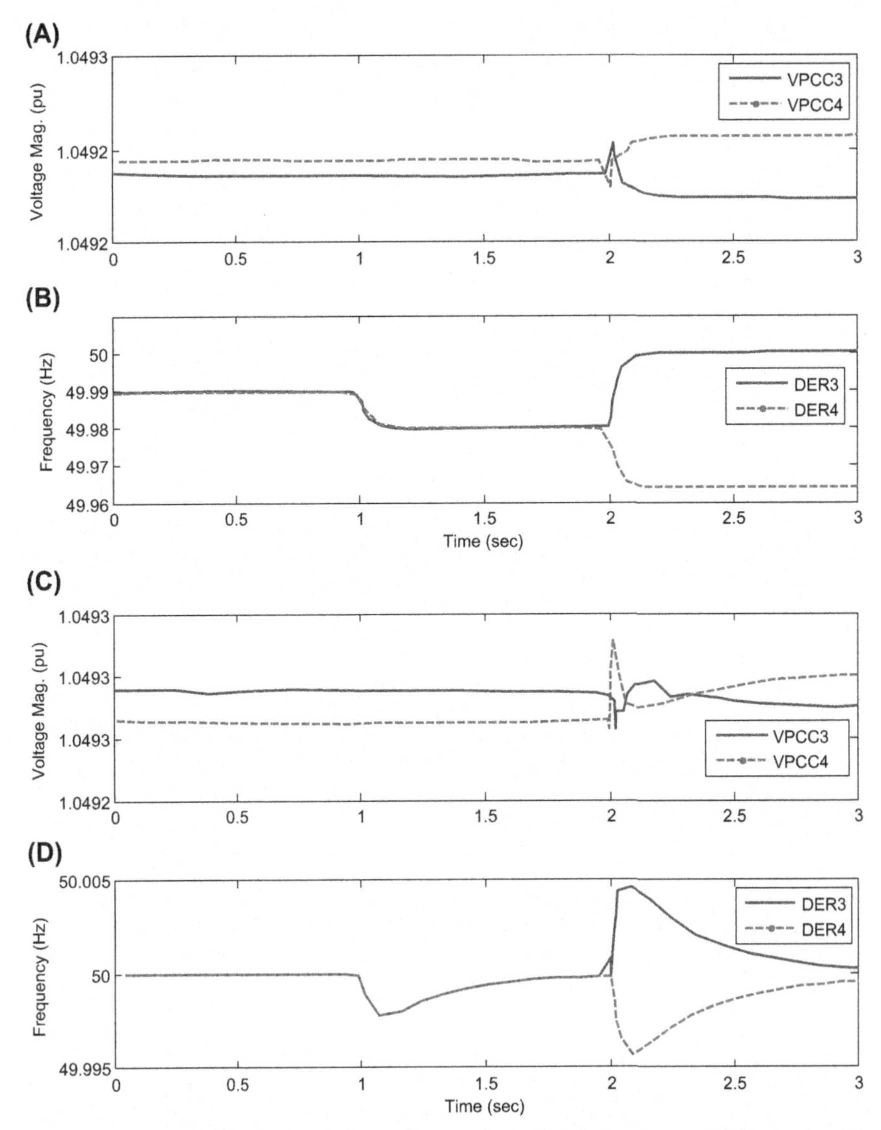

Figure 2.21 MG bus 3 and 4 voltage due to a step load change and DER outage: (A and B) using conventional droop-based control [17,28] and (C and D) using the proposed control system. *DER*, distributed energy resource; MG, microgrid.

disturbances such as a step load change, and large signal changes such as three-phase and single-phase to ground faults. In addition, the MG voltage drop problem due to motor starting was overcome by using an FLC to change the Q^*.

A supplementary stabilizing control loop, a fundamental frequency and harmonics frequencies virtual impedance, voltage and unbalance compensation, and HPF using RBFNNs for harmonic load sharing and voltage compensation were studied using a new control system for LCs of DERs in the second case study. The suggested control system overcomes the demerits of the conventional droop-based control systems such as large signal instability and undesirable load sharing under nonlinear and unbalanced loads. Also, the simulations showed that the MG can obtain the plug-and-play operation and is robust vs. DER outages.

References

[1] C. Li, X. Liu, K. Sun, Y. Cao, F. Ma, B. Zhou, A hybrid control strategy to support voltage in industrial active distribution networks, IEEE Trans. Power Deliv. 99 (July 2018) 1–11.

[2] H.R. Baghaee, M. Mirsalim, G.B. Gharehpetian, H.A. Talebi, Eigenvalue, robustness and time delay analysis of hierarchical control scheme in multi-DER microgrid to enhance small/large-signal stability using complementary loop and fuzzy logic controller, J. Circ. Syst. Comput. 26 (06) (Jun. 2017) 1–26.

[3] D.E. Olivares, A. Mehrizi-Sani, A.H. Etemadi, C.A. Canizares, R. Iravani, M. Kazerani, A.H. Hajimiragha, O. Gomis-Bellmunt, M. Saeedifard, R. Palma-Behnke, G.A. Jimenez-Estevez, N.D. Hatziargyriou, Trends in microgrid control, IEEE Trans. Smart Grid 5 (4) (July 2014) 1905–1919.

[4] H. Karimi, E.J. Davison, R. Iravani, Multivariable servomechanism controller for autonomous operation of a distributed generation unit: design and performance evaluation, IEEE Trans. Power Syst. 25 (2) (May 2010) 853–865.

[5] P. Moutis, N.D. Hatziargyriou, Decision trees-aided active power reduction of a virtual power plant for power system over-frequency mitigation, IEEE Trans. Ind. Inf. 11 (1) (February 2015) 251–261.

[6] H.R. Baghaee, M. Mirsalim, G.B. Gharehpetian, H.A. Talebi, Decentralized sliding mode control of WG/PV/FC microgrids under unbalanced and nonlinear load conditions for on- and off-grid modes, IEEE Syst. J. 12 (4) (December 2018) 3108–3119.

[7] H.R. Baghaee, M. Mirsalim, G.B. Gharehpetian, H.A. Talebi, Fuzzy unscented transform for uncertainty quantification of correlated wind/PV microgrids: possibilistic–probabilistic power flow based on RBFNNs, IET Renew. Power Gener. 11 (6) (May 2017) 867–877.

[8] H.R. Baghaee, M. Mirsalim, G.B. Gharehpetian, H.A. Talebi, Reliability/cost-based multi-objective Pareto optimal design of stand-alone wind/PV/FC generation microgrid system, Energy 115 (November 2016) 1022–1041.

[9] H.R. Baghaee, M. Mirsalim, G.B. Gharehpetian, H.A. Talebi, A generalized descriptor-system robust h1 control of autonomous microgrids to improve small and large signal stability considering communication delays and load nonlinearities, Int. J. Electr. Power Energy Syst. 92 (November 2017) 63–82.

[10] M.B. Delghavi, A. Yazdani, Sliding-mode control of AC voltages and currents of dispatchable distributed energy resources in master-slave-organized inverter-based microgrids, IEEE Trans. Smart Grid 99 (September 2017) 1–12.

[11] H.R. Baghaee, M. Mirsalim, G.B. Gharehpetian, H.A. Talebi, A new current limiting strategy and fault model to improve fault ridethrough capability of inverter interfaced DERs in autonomous microgrids, Sustain. Energy Technol. Assess. 24 (December 2017) 71–81.
[12] A.H. Etemadi, R. Iravani, Overcurrent and overload protection of directly voltage-controlled distributed resources in a microgrid, IEEE Trans. Ind. Electron. 60 (12) (December 2013) 5629–5638.
[13] S. Kar, S.R. Samantaray, M.D. Zadeh, Data-mining model based intelligent differential microgrid protection scheme, IEEE Syst. J. 11 (2) (June 2017) 1161–1169.
[14] L. Meng, E.R. Sanseverino, A. Luna, T. Dragicevic, J.C. Vasquez, J.M. Guerrero, Microgrid supervisory controllers and energy management systems: a literature review, Renew. Sustain. Energy Rev. 60 (July 2016) 1263–1273.
[15] F. Katiraei, R. Iravani, N. Hatziargyriou, A. Dimeas, Microgrids management, IEEE Power Energy Mag. 6 (3) (May 2008) 54–65.
[16] P. Piagi, R. Lasseter, Autonomous control of microgrids, in: IEEE Power Engineering Society General Meeting, IEEE, Montreal, Que., Canada, June 2006, pp. 1–8.
[17] J.M. Guerrero, J.C. Vasquez, J. Matas, L.G. de Vicuna, M. Castilla, "Hierarchical control of droop-controlled AC and DC microgrids—a general approach toward standardization, IEEE Trans. Ind. Electron. 58 (1) (January 2011) 158–172.
[18] A. Bidram, A. Davoudi, Hierarchical structure of microgrids control system, IEEE Trans. Smart Grid 3 (4) (December 2012) 1963–1976.
[19] H.R. Baghaee, M. Mirsalim, G.B. Gharehpetian, Real-time verification of new controller to improve small/large-signal stability and fault ride-through capability of multi-DER microgrids, IET Gener. Trans. Distrib. 10 (12) (September 2016) 3068–3084.
[20] J. Rocabert, A. Luna, F. Blaabjerg, P. Rodriguez, Control of power converters in AC microgrids, IEEE Trans. Power Electron. 27 (11) (November 2012) 4734–4749.
[21] F. Blaabjerg, R. Teodorescu, M. Liserre, A. Timbus, Overview of control and grid synchronization for distributed power generation systems, IEEE Trans. Ind. Electron. 53 (5) (October 2006) 1398–1409.
[22] M. Babazadeh, H. Karimi, A robust two-degree-of-freedom control strategy for an islanded microgrid, IEEE Trans. Power Deliv. 28 (3) (July 2013) 1339–1347.
[23] H. Han, X. Hou, J. Yang, J. Wu, M. Su, J.M. Guerrero, Review of power sharing control strategies for islanding operation of AC microgrids, IEEE Trans. Smart Grid 7 (1) (January 2016) 200–215.
[24] A.G. Tsikalakis, N.D. Hatziargyriou, Centralized control for optimizing microgrids operation, IEEE Trans. Energy Convers. 23 (1) (March 2008) 241–248.
[25] N. Pogaku, M. Prodanovic, T.C. Green, Modeling, analysis and testing of autonomous operation of an inverter-based microgrid, IEEE Trans. Power Electron. 22 (2) (March 2007) 613–625.
[26] Y. Mohamed, E. El-Saadany, Adaptive decentralized droop controller to preserve power sharing stability of paralleled inverters in distributed generation microgrids, IEEE Trans. Power Electron. 23 (6) (November 2008) 2806–2816.
[27] Y.-R. Mohamed, E. El-Saadany, A control method of grid connected PWM voltage source inverters to mitigate fast voltage disturbances, IEEE Trans. Power Syst. 24 (1) (February 2009) 489–491.
[28] J.M. Guerrero, M. Chandorkar, T.-L. Lee, P.C. Loh, "Advanced control architectures for intelligent microgrids—part I: decentralized and hierarchical control, IEEE Trans. Ind. Electron. 60 (4) (April 2013) 1254–1262.
[29] M. Jamshidi (Ed.), Large-Scale Systems: Modeling, Control and Fuzzy Logic, first ed., Prentice Hall PTR, Upper Saddle River, New Jersey, USA, 2001.

[30] H. Cai, G. Hu, F.L. Lewis, A. Davoudi, A distributed feedforward approach to cooperative control of AC microgrids, IEEE Trans. Power Syst. 31 (5) (September 2016) 4057–4067.
[31] J. Guerrero, J. Matas, L.G.D.V.D. Vicuna, M. Castilla, J. Miret, Wireless-control strategy for parallel operation of distributed generation inverters, IEEE Trans. Ind. Electron. 53 (5) (October 2006) 1461–1470.
[32] J.M. Guerrero, J. Matas, L.G. de Vicuna, M. Castilla, J. Miret, Decentralized control for parallel operation of distributed generation inverters using resistive output impedance, IEEE Trans. Ind. Electron. 54 (2) (April 2007) 994–1004.
[33] A. Kahrobaeian, Y.A.-R.I. Mohamed, Networked-based hybrid distributed power sharing and control for islanded microgrid systems, IEEE Trans. Power Electron. 30 (2) (February 2015) 603–617.
[34] B. Zhao, X. Zhang, J. Chen, Integrated microgrid laboratory system, in: 2013 IEEE Power & Energy Society General Meeting, IEEE, 2013.
[35] T. Iwade, S. Komiyama, Y. Tanimura, M. Yamanaka, M. Sakane, K. Hirachi, A novel small-scale ups using a parallel redundant operation system, in: 25th International Telecommunications Energy Conference, 2003, INTELEC '03., Yokohama, Japan, October 2003, pp. 480–483.
[36] X. Sun, L.-K. Wong, Y.-S. Lee, D. Xu, Design and analysis of an optimal controller for parallel multi-inverter systems, IEEE Trans. Circuits Syst. II: Express Briefs 53 (1) (January 2006) 56–61.
[37] Z. He, Y. Xing, Y. Hu, Low cost compound current sharing control for inverters in parallel operation, in: IEEE 35th Annual Power Electronics Specialists Conference (IEEE Cat. No. 04CH37551), IEEE, Aachen, Germany, June 2004, pp. 222–227.
[38] H.-M. Hsieh, T.-F. Wu, Y.-E. Wu, H.-S. Nien, Y.-E. Wu, Y.-K. Chen, "A compensation strategy for parallel inverters to achieve precise weighting current distribution, in: Fourteenth IAS Annual Meeting. Conference Record of the 2005 Industry Applications Conference, vol. 2, IEEE, Kowloon, Hong Kong, China, October 2005, pp. 945–960.
[39] S. Chiang, C. Lin, C. Yen, Current limitation control technique for parallel operation of UPS inverters, in 35th Annual Power Electronics Specialists Conference (IEEE Cat. No.04CH37551), vol. 3, IEEE, Aachen, Germany, June, pp. 1922–1926.
[40] J.M. Guerrero, L. Hang, J. Uceda, Control of distributed uninterruptible power supply systems, IEEE Trans. Ind. Electron. 55 (8) (August 2008) 2845–2859.
[41] A. Tuladhar, H. Jin, T. Unger, K. Mauch, Control of parallel inverters in distributed AC power systems with consideration of line impedance effect, IEEE Trans. Ind. Appl. 36 (1) (January 2000) 131–138.
[42] T. Vandoorn, J.D. Kooning, B. Meersman, L. Vandevelde, Review of primary control strategies for islanded microgrids with power electronic interfaces, Renew. Sustain. Energy Rev. 19 (1) (March 2013) 613–628.
[43] S.D.J. McArthur, E.M. Davidson, V.M. Catterson, A.L. Dimeas, N.D. Hatziargyriou, F. Ponci, T. Funabashi, "Multi-agent systems for power engineering applications—part I: concepts, approaches, and technical challenges, IEEE Trans. Power Syst. 22 (4) (November 2007) 1743–1752.
[44] S.D.J. McArthur, E.M. Davidson, V.M. Catterson, A.L. Dimeas, N.D. Hatziargyriou, F. Ponci, T. Funabashi, "Multi-agent systems for power engineering applications—part II: technologies, standards, and tools for building multi-agent systems, IEEE Trans. Power Syst. 22 (4) (November 2007) 1753–1759.
[45] D. Niyato, L. Xiao, P. Wang, Machine-to-machine communications for home energy management system in smart grid, IEEE Commun. Mag. 49 (4) (April 2011) 53–59.

[46] R.C. Qiu, Z. Hu, Z. Chen, N. Guo, R. Ranganathan, S. Hou, G. Zheng, Cognitive radio network for the smart grid: experimental system architecture, control algorithms, security, and microgrid testbed, IEEE Trans. Smart Grid 2 (4) (December 2011) 724–740.
[47] Z. Fan, P. Kulkarni, S. Gormus, C. Efthymiou, G. Kalogridis, M. Sooriyabandara, Z. Zhu, S. Lambotharan, W.H. Chin, Smart grid communications: overview of research challenges, solutions, and standardization activities, IEEE Commun. Surveys Tutorials 15 (1) (2013) 21–38.
[48] K. Iniewski, Smart Grid Infrastructure & Networking, first ed., Mc Graw Hill, New York, NY, USA, 2013.
[49] V.C. Gungor, D. Sahin, T. Kocak, S. Ergut, C. Buccella, C. Cecati, G.P. Hancke, Smart grid technologies: communication technologies and standards, IEEE Trans. Ind. Inf. 7 (4) (November 2011) 529–539.
[50] A. Dimeas, N. Hatziargyriou, Operation of a multiagent system for microgrid control, IEEE Trans. Power Syst. 20 (3) (August 2005) 1447–1455.
[51] W. Liu, W. Gu, W. Sheng, X. Meng, Z. Wu, W. Chen, Decentralized multi-agent system-based cooperative frequency control for autonomous microgrids with communication constraints, IEEE Trans. Sustain. Energy 5 (2) (April 2014) 446–456.
[52] Y. Xu, W. Liu, Novel multiagent based load restoration algorithm for microgrids, IEEE Trans. Smart Grid 2 (1) (March 2011) 152–161.
[53] J. He, Y.W. Li, An enhanced microgrid load demand sharing strategy, IEEE Trans. Power Electron. 27 (9) (September 2012) 3984–3995.
[54] Q. Li, F. Chen, M. Chen, J.M. Guerrero, D. Abbott, Agent-based decentralized control method for islanded microgrids, IEEE Trans. Smart Grid 7 (2) (March 2016) 637–649.
[55] Z. Chen, A. Luo, H. Wang, Y. Chen, M. Li, Y. Huang, Adaptive sliding-mode voltage control for inverter operating in islanded mode in microgrid, Int. J. Electr. Power Energy Syst. 66 (March 2015) 133–143.
[56] N.M. Dehkordi, N. Sadati, M. Hamzeh, A back-stepping high-order sliding mode voltage control strategy for an islanded microgrid with harmonic/interharmonic loads, Contr. Eng. Pract. 58 (January 2017) 150–160.
[57] J. Guerrero, J. Vasquez, J. Matas, M. Castilla, L. de Vicuna, Control strategy for flexible microgrid based on parallel line-interactive UPS systems, IEEE Trans. Ind. Electron. 56 (3) (March 2009) 726–736.
[58] E. Barklund, N. Pogaku, M. Prodanovic, C. Hernandez-Aramburo, T. Green, Energy management in autonomous microgrid using stability-constrained droop control of inverters, IEEE Trans. Power Electron. 23 (5) (September 2008) 2346–2352.
[59] M.B. Delghavi, A. Yazdani, A unified control strategy for electronically interfaced distributed energy resources, IEEE Trans. Power Deliv. 27 (2) (April 2012) 803–812.
[60] Z. Zhang, X. Huang, J. Jiang, B. Wu, A load-sharing control scheme for a microgrid with a fixed frequency inverter, Elec. Power Syst. Res. 80 (3) (March 2010) 311–317.
[61] T.L. Vandoorn, B. Meersman, J.D.M.D. Kooning, L. Vandevelde, Analogy between conventional grid control and islanded microgrid control based on a global DC-link voltage droop, IEEE Trans. Power Deliv. 27 (3) (July 2012) 1405–1414.
[62] T.L. Vandoorn, B. Renders, L. Degroote, B. Meersman, L. Vandevelde, Active load control in islanded microgrids based on the grid voltage, IEEE Trans. on Smart Grid 2 (1) (March 2011) 139–151.
[63] T.L. Vandoorn, B. Meersman, L. Degroote, B. Renders, L. Vandevelde, A control strategy for islanded microgrids with DC-link voltage control, IEEE Trans. Power Deliv. 26 (2) (April 2011) 703–713.

[64] X. Huang, X. Jin, T. Ma, Y. Tong, A voltage and frequency droop control method for microsources, in: 2011 International Conference on Electrical Machines and Systems, IEEE, Beijing, China, August 2011, pp. 1–5.
[65] P.H. Divshali, A. Alimardani, S.H. Hosseinian, M. Abedi, Decentralized cooperative control strategy of microsources for stabilizing autonomous VSC-based microgrids, IEEE Trans. Power Syst. 27 (4) (November 2012) 1949–1959.
[66] R. Zamora, A.K. Srivastava, Controls for microgrids with storage: review, challenges, and research needs, Renew. Sustain. Energy Rev. 14 (7) (September 2010) 2009–2018.
[67] N. Lidula, A. Rajapakse, Microgrids research: a review of experimental microgrids and test systems, Renew. Sustain. Energy Rev. 15 (1) (January 2011) 186–202.
[68] H.R. Baghaee, M. Mirsalim, G.B. Gharehpetian, H.A. Talebi, "A decentralized robust mixed H_2/H_∞ voltage control scheme to improve small/large-signal stability and FRT capability of islanded multi-DER microgrid considering load disturbances", IEEE Syst. J. 12 (3) (April 2017) 2610–2621.
[69] H.R. Baghaee, M. Mirsalim, G.B. Gharehpetian, H.A. Talebi, Nonlinear load sharing and voltage compensation of microgrids based on harmonic power-flow calculations using radial basis function neural networks, IEEE Syst. J. 12 (3) (September 2018) 2749–2759, https://doi.org/10.1109/JSYST.2016.2645165.
[70] H.R. Baghaee, M. Mirsalim, G.B. Gharehpetian, H.A. Talebi, Unbalanced harmonic power sharing and voltage compensation of microgrids using radial basis function neural network-based harmonic power-flow calculations for distributed and decentralized control structures, IET Gener., Transm. Distrib. 12 (7) (April 2018) 1518–1530, https://doi.org/10.1049/iet-gtd.2016.1277.
[71] H.R. Baghaee, M. Mirsalim, G.B. Gharehpetian, Power calculation using RBF neural networks to improve power sharing of hierarchical control scheme in multi-DER microgrids, IEEE J. Emerg. Select. Topics Power Electron. 4 (4) (December 2016) 1217–1225, https://doi.org/10.1109/JESTPE.2016.2581762.
[72] Z. Li, M. Shahidehpour, F. Aminifar, A. Alabdulwahab, Y. Al-Turki, Networked microgrids for enhancing the power system resilience, Proc. IEEE 105 (7) (July 2017) 1289–13010.
[73] H.R. Baghaee, Control of Microgrids to Respond to Large-Signal Disturbances and Nonlinearities and its Real-Time Implementation (Ph.D. dissertation), Department of Electrical and Computer Engineering, Amirkabir University of Technology, Tehran, Iran, 2017.
[74] H.R. Baghaee, M. Mirsalim, G.B. Gharehpetian, Performance improvement of multi-DER microgrid for small and large-signal disturbances and nonlinear loads: novel complementary control loop and fuzzy controller in a hierarchical droop-based control scheme, IEEE Syst. J. 12 (1) (March 2018) 444–451, https://doi.org/10.1109/JSYST.2016.2580617.
[75] H.R. Baghaee, M. Mirsalim, G.B. Gharehpetian, H.A. Talebi, A decentralized power management and sliding mode control strategy for hybrid AC/DC microgrids including renewable energy resources, IEEE Trans. Ind. Inf. 99 (March 2017) 1–10, https://doi.org/10.1109/TII.2017.2677943.
[76] H.R. Baghaee, M. Mirsalim, G.B. Gharehpetian, H.A. Talebi, Three phase AC/DC power-flow for balanced/unbalanced microgrids including wind/solar, droop-controlled and electronically-coupled distributed energy resources using RBF neural networks, IET Power Electron. 10 (3) (March 2017) 313–328, https://doi.org/10.1049/iet-pel.2016.0010.
[77] H.R. Baghaee, M. Mirsalim, G.B. Gharehpetian, H. Talebi, Generalized three phase robust load-flow for radial and meshed power systems with and without uncertainty in energy resources using dynamic radial basis functions neural networks, J. Clean.

Prod. 174 (part C) (February 2018) 96–113, https://doi.org/10.1016/j.jclepro.2017.10.316.
[78] A.H. Etemadi, E.J. Davison, R. Iravani, A generalized decentralized robust control of islanded microgrids, IEEE Trans. Power Syst. 29 (6) (October 2014) 3102–3113.
[79] M.S. Sadabadi, A. Karimi, H. Karimi, Fixed-order decentralized/distributed control of islanded inverter-interfaced microgrids, Contr. Eng. Pract. 45 (December 2015) 174–193.
[80] X. Lu, X. Yu, J. Lai, J.M. Guerrero, H. Zhou, Distributed secondary voltage and frequency control for islanded microgrids with uncertain communication links, IEEE Trans. Ind. Inf. 99 (August 2016) 1–13.
[81] N. Mahdian-Dehkordi, H.R. Baghaee, N. Sadati, J.M. Guerrero, Distributed noise-resilient secondary voltage and frequency control for islanded microgrids, IEEE Trans. Smart Grid 10 (99) (2018) 1–11, https://doi.org/10.1109/TSG.2018.2834951.
[82] J. Lai, H. Zhou, X. Lu, X. Yu, W. Hu, Droop-based distributed cooperative control for microgrids with time-varying delays, IEEE Trans. Smart Grid 7 (4) (July 2016) 1775–1789.
[83] J.C. Vasquez, M. Guerrero, M. Savaghebi, J. Eloy-Garcia, R. Teodorescu, Modeling, analysis, and design of stationary-reference-frame droop-controlled parallel three-phase voltage source inverters, IEEE Trans. Ind. Electron. 60 (4) (April 2013) 1271–1280.
[84] M. Savaghebi, A. Jalilian, J.C. Vasquez, J.M. Guerrero, Secondary control scheme for voltage unbalance compensation in an islanded droopcontrolled microgrid, IEEE Trans. Smart Grid 3 (2) (June 2012) 797–807.
[85] M. Savaghebi, A. Jalilian, J.C. Vasquez, J.M. Guerrero, Secondary control for voltage quality enhancement in microgrid, IEEE Trans. Smart Grid 3 (4) (December 2012) 1893–1902.
[86] M.C. Chandorkar, D.M. Divan, R. Adapa, Control of parallel connected inverters in standalone ac supply systems, IEEE Trans. Ind. Appl. 29 (1) (January/February 1993) 136–143.
[87] P. Krause, O. Wasynczuk, S. Sudhoff, Analysis of Electric Machinery and Drive Systems, second ed., Wiley, Hoboken, NJ, 2002.
[88] Mathwork Inc., Fuzzy Logic Toolbox, User's Guide, The MathWorks, Inc., Natick, MA, USA, 2015.
[89] J.-F. Cecile, L. Schoen, V. Lapointe, A. Abreu, J.B. Elanger [Online], www.opal-rt.com, 2006, http://www.opal-rt.com/technical-document/distributed-real-time-framework-dynamic-management-heterogeneousco-simulations.
[90] IEEE recommended practice for monitoring electric power quality, IEEE Standard 1159 (June 2009).
[91] IEEE recommended practices and requirements for harmonic control in electrical power systems, IEEE Standard 519 (June 2014).
[92] M. Abdusalam, P. Poure, S. Karimi, S. Saadate, New digital reference current generation for shunt active power filter under distorted voltage conditions, Elec. Power Syst. Res. 79 (1) (2009) 759–765.
[93] R. Ghanizadeh, M. Ebadian, G.B. Gharehpetian, Non-linear load sharing and voltage harmonics compensation in islanded microgrids with converter interfaced units, Int. Trans. Elect. Energy Syst. 27 (1) (2017) 1–20.
[94] Interconnecting Committee, IEEE standard for interconnecting distributed resources with electric power systems, IEEE Standard 1547 (2003) 1–28.
[95] W. Yao, M. Chen, J. Matas, J. Guerrero, Z.M. Qian, Design and analysis of the droop control method for parallel inverters considering the impact of the complex impedance on the power sharing, IEEE Trans. Ind. Electron. 58 (2) (February 2011) 576–588.

[96] IEEE standard definitions for the measurement of electric power quantities under sinusoidal, non-sinusoidal, balanced or unbalanced conditions, IEEE Standard 1459 (2010).
[97] P. Sreekumar, V. Khadkikar, A new virtual harmonic impedance scheme for harmonic power sharing in an islanded microgrid, IEEE Trans. Power Deliv. 31 (3) (June 2016) 936–945.
[98] P. Sreekumar, V. Khadkikar, Nonlinear load sharing in low voltage microgrid using negative virtual harmonic impedance, in: Proc. 41st Annu. Conf. IEEE Ind. Electron. Soc., Yokosuka, Japan, November 2015, pp. 3353–3358.
[99] J. He, Y.W. Li, J.M. Guerrero, F. Blaabjerg, J.C. Vasquez, An islanding microgrid power sharing approach using enhanced virtual impedance control scheme, IEEE Trans. Power Electron. 28 (11) (November 2013) 5272–5282.
[100] J. He, Y.W. Li, F. Blaabjerg, An enhanced islanding microgrid reactive power, imbalance power, and harmonic power sharing scheme, IEEE Trans. Power Electron. 30 (6) (June 2015) 3389–3401.
[101] M. Savaghebi, J.M. Guerrero, A. Jalilian1, J.C. Vasquez, T.L. Lee, Hierarchical control scheme for voltage harmonics compensation in an islanded droop controlled microgrid, in: Proc. IEEE 9th Int. Conf. Power Electron. Drive Syst., Singapore, December 2011, pp. 89–94.
[102] M. Savaghebi, J.C. Vasquez, A. Jalilian1, J.M. Guerrero, T.L. Lee, Selective harmonic virtual impedance for voltage source inverters with LCL filter in microgrids, in: Proc. IEEE Energy Convers. Congr. Expo., Raleigh, NC, USA, September 2012, pp. 1960–1965.
[103] X. Wang, F. Blaabjerg, Z. Chen, Autonomous control of inverter interfaced distributed generation units for harmonic current filtering and resonance damping in an islanded microgrid, IEEE Trans. Ind. Appl. 50 (1) (January/February 2014) 452–461.
[104] M.A. Moreno, J. Usaola, A new balanced harmonic load flow including nonlinear loads modeled with RBF networks, IEEE Trans. Power Deliv. 19 (2) (Apr. 2004) 686–693.

CHAPTER 3

Power-flow analysis of microgrids

1. Fundamental-frequency deterministic power flow

1.1 Introduction

Microgrid (MG) design, control and protectiont, and its analysis and optimal operation under high penetration level of distributed energy resources (DERs) need powerful software. To determine MG steady-state condition, the power flow analysis (PFA) must be used. The focus of this chapter is on the tools for PFA in steady-state conditions [1–3].

There are well-known algorithms for PFA such as Newton–Raphson (NR) and Gauss–Seidel (GS) methods [4]. Also, there are many approaches in the literature for PFA [1–12]. To overcome the limitations of these classical solutions, metaheuristic intelligent algorithms [4–6], and probabilistic techniques [7–9] have been suggested. Due to the radial topology of distribution system and also its lines' high resistance to reactance (R/X) ratio, the PFA in distribution systems and MGs is an ill-conditioned problem, and the NR-based algorithms cannot be used. In this case, the backward–forward sweep method is a well-known solution [10–12].

The methods used for MG PFA are time-consuming because they should inverse the Jacobian matrix (JM). Also, their main challenges are modeling of nonlinear loads and components such as electronically coupled distributed energy resources (ECDERs). Therefore, there is a need for methods capable of finding multiple solutions to the PFA. The PFA has approximately been solved in Ref. [13] using radial basis function neural network (called RBFNN). The problem inputs and the results, obtained from the NR method, have been used to train the RBFNN and estimate the solutions under other operating conditions. In Ref. [14], a multilayer perceptron neural network (MLPNN) has been used for the same purpose. The challenges of MGs and their ECDER unit modeling in the PFA have been discussed in many researches [15–19]. In Ref. [1], the ECDER model

Microgrids and Methods of Analysis
ISBN 978-0-12-816172-2
https://doi.org/10.1016/B978-0-12-816172-2.00003-1

has been suggested for different strategies under balanced and unbalanced conditions and considering voltage source converter (VSC) limitations [15]. Symmetrical components have been used in Ref. [16] to study a three-phase PFA for droop-controlled isolated MGs. The solution was based on the optimization of a nonlinear equation set (NLES). The same approach can be seen in Ref. [17] for solving probabilistic load flow in an MG with a photovoltaic (PV) and wind generation (WG) DERs. The improved NR method has been presented in Refs. [18,19] for off-grid MGs, which are based on partial derivatives of power flow problem (PFP) equations. The two-step power flow method, like the one suggested in Ref. [1], has three main problems:

- They should calculate the JM and partial derivatives of power flow equations.
- They need more robustness versus the high load multiplier (LM) and R/X ratio of MGs.
- DER-based MG problem has unequal numbers of equations and unknowns. One solution can be the selection of some variables arbitrarily [1].

To have a fast, robust, and efficient algorithm for MGs, including WG, PV, and other ECDERs, a new artificial neural network (ANN)–based method is proposed in this chapter to map PFP inputs to its outputs (solutions) without inverting JM and calculating partial derivatives. Unlike Ref. [5], the suggested method is able to solve PFA NLES even with unequal number of equations and operational variables. The efficiency, precision, and solution speed of the suggested approach is shown using simulation results of different MGs, which have ECDERs such as doubly fed induction generator (DFIG) WG system and PV panels.

The nomenclature of all variables in this chapter is presented as follows:

Symbol	Definition
V	RMS voltage of the bus
θ	The angle of the bus voltage
f_n	System frequency
P_g	The real power of the generator
Q_g	Reactive power of the generator
v_g	Internal voltage of instantaneous generator
θ_g	Voltage angle of the generator
i_g	The current of the instantaneous generator
f_g	Frequency of generator output
R_g	The internal resistance of the generator
X_g	The internal reactance of the generator

Symbol	Definition
v_i	The input voltage of the instantaneous converter
v_o	The output voltage of the instantaneous converter
i	The output current of the instantaneous converter
V_{dc}	The voltage of converter DC link
A_m	Converter magnitude modulation index
α_m	Converter angle modulation index
R_f	Resistance of filter
X_f	Reactance of filter
P_{Li}, Q_{Li}	Active and reactive power of bus i (under nominal operating voltage)
$\|V_o\|$	Magnitude of nominal voltage
ω_o	Nominal angular frequency
ω	System angular frequency
α, β	Exponents of active and reactive powers
K_{pf}, K_{qf}	Load frequency sensitivity parameters
P_G, Q_G	DER active and reactive powers
m_p, n_q	Coefficients of frequency and voltage droop
P_o, Q_o	Setpoints of active and reactive powers
P_m	Rotor mechanical power
A	Area
ρ	Air density
U	Speed of wind
C_p	Coefficient of power
λ	Tip speed ratio
$C_1 \cdots C_6$	Constants
β	Pitch angle of blade
ω_t, ω_r	Angular speed of turbine and rotor
R_t	Turbine radius
K_g	Coefficient of gear box
$[P]_{abc}$, $[Q]_{abc}$	Real and reactive powers of three-phase WG
$[V]_{abc}$	Magnitude of three-phase voltage at PCC
$[\delta]_{abc}$	Phase angle of three-phase voltage at PCC
$(P_{s0}, P_{s1}, P_{s2})/P_{r0}, P_{r1}, P_{r2}$	Stator/rotor powers (symmetrical components)
$(\bar{I}_{s0}, \bar{I}_{s1}, \bar{I}_{s2})/(\bar{I}_{r0}, \bar{I}_{r1}, \bar{I}_{r2})$	Stator/rotor winding current phasors (symmetrical components)
$(Z_{s0}, Z_{s1}, Z_{s2})/(Z_{r0}, Z_{r1}, Z_{r2})$	Stator/rotor winding impedance (symmetrical components)
$(\bar{V}_{s0}, \bar{V}_{s1}, \bar{V}_{s2})/(\bar{V}_{r0}, \bar{V}_{r1}, \bar{V}_{r2})$	Stator/rotor winding voltages (symmetrical components)

Continued

Symbol	Definition
$[P]_{g,abc}$, $[Q]_{g,abc}$	Real and reactive powers (form grid-side converter to PCC)
$\overline{E}_0, \overline{E}_1, \overline{E}_2$	Induced emf (in symmetrical component networks)
$s_0, s_1, s_2, [s]_{012}$	Converted symmetrical component slip
Z_{m0}, Z_{m1}, Z_{m2}, $[Z]_{m012}$	Symmetrical component magnetizing reactance of stator
PL_0, PL_1, PL_2	Stator and rotor Cu losses
I_{inv}	PV inverter current injection
S_{inv}, P_{inv}	Apparent and real power of inverter
$V_{Ph\text{-}N}$	PV inverter phase to neutral voltage
$*$	Complex conjugate
P_{DC}	PV modules generated DC power
η_{inv}	Inverter efficiency
η_m	PV modules mismatch
η_d	Dirt effects
V_{DC}, I_{DC}	Voltage and current of PV module
t	Ambient temperature
G	Ambient irradiance level
P	PV module electrical parameters vector
φ	Function determining PV module voltage
ψ	Function determining PV module current
P_i, Q_i	Net injected active and reactive powers to ith bus
V_i, ™$_i$	Magnitude and phase angle of voltage at bus i
Y_{ij}, $\backslash_{ij}$	Magnitude and argument of ijth element of Ybus matrix
PG, PD	Bus active power (generation and demand)
QG, QD	Bus reactive power (generation and demand)
NAC, NDC	AC and DC buses numbers
$YBabc$	Elementary matrices of three-phase Ybus
YB	Elements of Ybus (three DC buses)
$PGDC$, $PDDC$	Active power generation and demand (DC bus)
H	RBFNN hidden neurons number
W	RBFNN weight matrix
σ_h	RBFNN distance scaling parameter
μ_h	Center of kernel unit of RBFNN
$\|\|.\|\|$	Euclidian norm operator
X	Input matrix (consisting of *KN*-dimensional vectors)
Y	Output matrix (consisting of *KM*-dimensional vectors)

Symbol	Definition
a_{hk}	Activation of h-th unit in hidden layer given input x_k.
μ_h	N-dimensional position of h-th radial unit center (in input space)
σ_h	Parameter of distance scaling
$\|\|.\|\|$	Euclidian norm
y_{mk}	Output of m-th neuron of output layer to k-th input vector (X_k)
w_{mh}	Weighting factor (hth neuron of hidden layer to mth neuron of output layer)
T_k	T target output to kth input vector
A	$H \times K$ matrix composed of a_{hk} as its components
T	Input matrix target output
W	Matrix of output layer's weighting factors
μ, σ	Mean value and standard deviation (normal distribution)
PL	Load power, pu
v	Wind speed
α, ρ	Shape and scale parameter of Weibull PDF
$PWT(v)$	WT generated power at v
Pr, WT	WT active power
$Vcut\text{-}in$, $Vcut\text{-}out$	WT low and high cutoff speeds
vr	WT rated speed
NWT, NPV	Number of WT units in a wind farm and number of PV units
$\alpha\beta$, $\beta\beta$	Shape factor of beta PDF
r	Solar irradiation
Prs	PV unit rated power
Rc	Radiation usually set to 150 W/m^2
$RSTD$	Standard radiation (1000 W/m^2)
k	Coefficient of maximum power temperature
PPV	PV module output power at GING
$GING$	Sun irradiance
$GSTC$	Standard test condition irradiance
$PSTC$	PV rated power at $GSTC$
Tr	PV cell temperature, °C
$\varepsilon\ell j$	Error index of random variable j for statistical characteristic ℓ
$\overline{Y}_i^k$	PQ loads and generators harmonic admittance
$\overline{I}_j^k$	NLD injected current

Continued

Symbol	Definition
Bij, *βij*	Generic NLD data and parameters
d, *l*	Special nonlinear functions (for NLD behavior and its injected current)

1.2 Microgrid dynamic and steady-state modeling

1.2.1 Voltage source converter–based distributed energy resources

Fig. 3.1 indicates the MG single line diagram, including two converter-based DERs. The steady-state model of DG2 in *dqo* rotating reference frame [20] is present here [1]:

$$R_{g2}I_{dg2} - L_{g2}\omega_{g2}I_{qg2} - V_{mg2} = \frac{-1}{2}A_{mi2}V_{dc}\cos\left(\alpha_{mi2} - \theta_{g2}\right) \tag{3.1}$$

$$R_{g2}I_{qg2} + L_{g2}\omega_{g2}I_{dg2} = \frac{-1}{2}A_{mi2}V_{dc}\sin\left(\alpha_{mi2} - \theta_{g2}\right) \tag{3.2}$$

$$R_{f2}I_{d2} - L_{f2}\omega_n I_{q2} + V_{m2} = \frac{1}{2}A_{mo2}V_{dc}\cos(\alpha_{mo2} - \theta_2) \tag{3.3}$$

$$R_{f2}I_{q2} + L_{f2}\omega_n I_{d2} = \frac{1}{2}A_{mo2}V_{dc}\sin(\alpha_{mo2} - \theta_2) \tag{3.4}$$

$$P_2 = V_{m2}I_{d2} \tag{3.5}$$

$$Q_2 = -V_{m2}I_{q2} \tag{3.6}$$

$$P_{g2} = V_{m2g}I_{dg2} \tag{3.7}$$

$$Q_{g2} = -V_{mg2}I_{qg2} \tag{3.8}$$

Eq. (3.9) presents the power balance between bus 2 and DG2.

$$P_{g2} = P_2 + R_{f2}\left(I_{d2}^2 + I_{q2}^2\right) + R_{g2}\left(I_{dg2}^2 + I_{qg2}^2\right) \tag{3.9}$$

The DG3 steady-state model can be expressed as follows [1]:

$$R_{f3}I_{d3} - L_{f3}\omega_n I_{q3} + V_{m3} = V_{mo3}\cos(\alpha_{mo3} - \theta_3) = V_{do3} \tag{3.10}$$

$$R_{f3}I_{q3} + L_{f3}\omega_n I_{d3} = V_{mo3}\sin(\alpha_{o3} - \theta_3) = V_{qo3} \tag{3.11}$$

$$R_{g3}I_{dg3} - L_{g3}\omega_{g3}I_{qg3} - V_{mg3} = -V_{mi3}\cos\left(\alpha_{mi3} - \theta_{g3}\right) = -V_{di3} \tag{3.12}$$

$$R_{g3}I_{qg3} + L_{g3}\omega_{g3}I_{dg3} = -V_{mi3}\sin\left(\alpha_{mi3} - \theta_{g3}\right) = -V_{qi3} \tag{3.13}$$

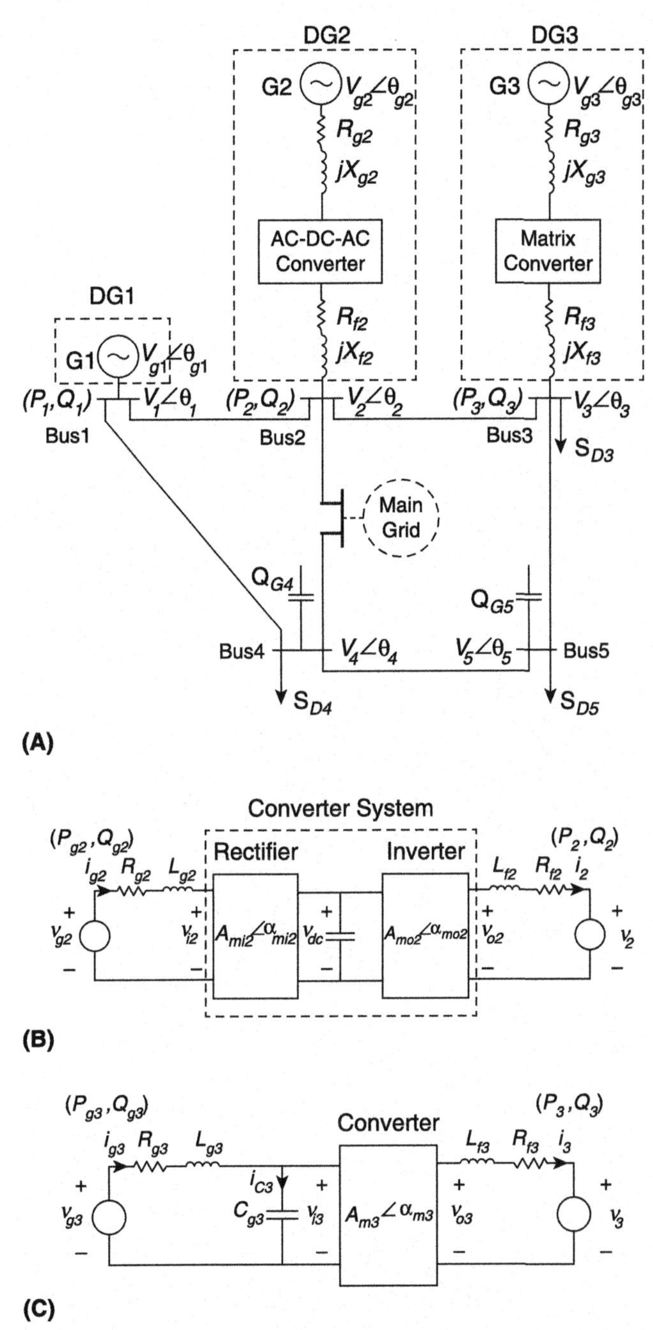

Figure 3.1 Case study: (A) MG, (B) DG2, and (C) DG3 [1]. *DG*, distributed generation; *MG*, microgrid.

$$I_{dC3} = -C_{g3}\omega_{g3}V_{qi3} \tag{3.14}$$

$$I_{qC3} = C_{g3}\omega_{g3}V_{di3} \tag{3.15}$$

$$P_3 = V_{m3}I_{d3} \tag{3.16}$$

$$Q_3 = -V_{m3}I_{q3} \tag{3.17}$$

$$P_{g3} = V_{m3g}I_{dg3} \tag{3.18}$$

$$Q_{g3} = -V_{mg3}I_{qg3} \tag{3.19}$$

Eq. (3.20) represents balance of active power between bus 3 and DG3.

$$P_{g3} = P_3 + R_{f3}\left(I_{d3}^2 + I_{q3}^2\right) + R_{g3}\left(I_{dg3}^2 + I_{qg3}^2\right) \tag{3.20}$$

1.2.2 Load modeling

The power of a static load is modeled by the following equations [16,19]:

$$P_{Li} = P_{Lio}\left(\frac{|V_i|}{|V_o|}\right)^{\alpha}\left(1 + K_{pf}(\omega - \omega_0)\right) \tag{3.21}$$

$$Q_{Li} = Q_{Lio}\left(\frac{|V_i|}{|V_o|}\right)^{\beta}\left(1 + K_{qf}(\omega - \omega_0)\right) \tag{3.22}$$

where α and β are defined for different types of load [21,22]. K_{pf} and K_{qf} can vary between 0 to 3 and -2 to 0, respectively, considering geographical and seasonal conditions [22,23].

1.2.3 Droop-controlled distributed energy resources

In the case of having inductive output impedance, the droop characteristics of each DER can be expressed by the following equations.

$$\omega = \omega_0 - m_p(P_G - P_o) \tag{3.23}$$

$$|V| = |V_o| - n_q(Q_G - Q) \tag{3.24}$$

In case of having resistive output impedance or high R/X ratio in distribution lines, for an ECDER, the active and reactive powers change their role in droop equations, as follows [24]:

$$\omega = \omega_o + m_p(Q_G - Q_o) \tag{3.25}$$

$$|V| = |V_o| - n_q(P_G - P_o) \tag{3.26}$$

Here, considering the IEEE standard 1547.7, Eqs. (3.23) and (3.24) are applied to the studies [25]. However, it must be noted that P and Q are not decoupled and they are dependent on f and V. In this status, we have the following forms of droop equations [26].

$$\omega = \omega_o - m_p(P_G - Q_G) \tag{3.27}$$

$$|V| = |V_o| - n_q(P_G + Q_G) \tag{3.28}$$

1.2.4 Nonlinear model of type-3 doubly fed induction generator–based wind generation

It is well known that mechanical power of a turbine is written as follows [27–30]:

$$P_m = \frac{1}{2} A \rho U^3 C_p \tag{3.29}$$

where we have

$$C_p(\lambda, \beta) = C_1\left(\frac{C_2}{\lambda} - C_3 \cdot \beta - C_4\right) \cdot e^{\left(\frac{-C_5}{\lambda}\right)} + C_6 \cdot \lambda \tag{3.30}$$

Also, the turbine and rotor angular speed is given by the following equation:

$$\overline{\omega}_t = \frac{\lambda \cdot U}{R_t} \tag{3.31}$$

$$\overline{\omega}_r = \overline{\omega}_t \cdot K_g \tag{3.32}$$

The power of the grid-side converter (GSC) is injected to the point of common coupling (PCC). Its active and reactive powers can be expressed by the following equations:

$$[P]_{g,abc} = real\left\{[V\angle\delta]_{abc} \cdot \left(\frac{[V\angle\delta]_{abc} - [V\angle\delta]_{gabc}}{[Z]_{gabc}}\right)^*\right\} \tag{3.33}$$

$$[Q]_{g,abc} = image\left\{[V\angle\delta]_{abc} \cdot \left(\frac{[V\angle\delta]_{abc} - [V\angle\delta]_{gabc}}{[Z]_{gabc}}\right)^*\right\} \tag{3.34}$$

The induction machine slip in the frame of symmetrical components is as follows:

$$[s]_{012} = \begin{bmatrix} 0 \\ s \\ 2 - s \end{bmatrix} \tag{3.35}$$

For the sequence network impedances, we have

$$Z_{s\varepsilon} = Z_s; \varepsilon = 0, 1, 2 \tag{3.36}$$

$$Z_{r0} = \infty \tag{3.37}$$

$$Z_{r1} = \frac{real(Z_r)}{s_1} + image(Z_r) \tag{3.38}$$

$$Z_{r2} = \frac{real(Z_r)}{s_2} + image(Z_r) \tag{3.39}$$

$$Z_{m\varepsilon} = Z_m; \varepsilon = 0, 1, 2 \tag{3.40}$$

So, stator and rotor induced electromotive force (EMF), and stator currents are written as follows:

$$\overline{E}_\varepsilon = \frac{\left(\overline{V}_{s\varepsilon} / Z_{s\varepsilon}\right) + \left(\left(\overline{V}_{r\varepsilon} / S_\varepsilon\right) / Z_{r\varepsilon}\right)}{1/Z_{s\varepsilon} + 1/Z_{r\varepsilon} + 1/Z_{m\varepsilon}}; \varepsilon = 1, 2 \tag{3.41}$$

$$\begin{aligned} \overline{I}_{s\varepsilon} &= \left(\overline{V}_{s\varepsilon} - \overline{E}_\varepsilon\right) / Z_{s\varepsilon} \\ \overline{I}_{r\varepsilon} &= \left(\overline{V}_{r\varepsilon} - \overline{E}_\varepsilon\right) / Z_{r\varepsilon}; \varepsilon = 1, 2 \\ \overline{I}_{r0} &= 0 \end{aligned} \tag{3.42}$$

The balance of the active power is represented using the following equations:

$$
\left.
\begin{aligned}
PL_\varepsilon &= \left|\overline{I}_{s\varepsilon}\right|^2 \cdot real(Z_{s\varepsilon}) + \left|\overline{I}_{r\varepsilon}\right|^2 \cdot real(Z_{r\varepsilon}) \\
P_{s\varepsilon} &= real\left(\overline{V}_{s\varepsilon} \cdot \overline{I}_{s\varepsilon}^{*}\right) \\
P_{r\varepsilon} &= real\left(\overline{V}_{r\varepsilon} \cdot \overline{I}_{r\varepsilon}^{*}\right) \\
\sum_{\varepsilon}(P_{s\varepsilon} + P_{r\varepsilon}) &= P_m/3 + \sum_{\varepsilon} PL_\varepsilon
\end{aligned}
\right\} ; \varepsilon = 0, 1, 2 \qquad (3.43)
$$

Now, the WG active and reactive powers can be written as follows:

$$
\begin{aligned}
[P]_{abc} &= [P]_{s,abc} + [P]_{g,abc} \\
[Q]_{abc} &= [Q]_{s,abc} + [Q]_{g,abc} = 0
\end{aligned} \qquad (3.44)
$$

The set of Eqs. (3.29)–(3.44) are the DFIG-based WG NLES combined with PFA equations.

$$
f(U, [V]_{abc}, [V]_{abc}, [\delta]_{abc}, [P]_{abc}, [Q]_{abc}) = 0 \qquad (3.45)
$$

1.2.5 Solar photovoltaic nonlinear model

The single or three-phase converters have been used to connect PV units to the grid. In case of using a single-phase converter, the PV injects active power into one phase of the three phases [31–33], i.e., it is located between phase a, b, or c and neutral conductor, and can its current be calculated, as follows:

$$
I_{inv} = \left(\frac{P_{inv} + j\sqrt{S_{inv}^2 - P_{inv}^2}}{V_{Ph-N}}\right)^{*} \qquad (3.46)
$$

where P_{inv} and S_{inv} are real and apparent powers generated by the inverter, respectively. The DC power of the PV module (P_{DC}) determines P_{inv} as follows:

$$
P_{inv} = \eta_{inv} \times \eta_m \times \eta_d \times P_{DC} \qquad (3.47)
$$

P_{DC} can be calculated using the PV module V–I characteristic as follows:

$$P_{inv} = \max(V_{DC} \times I_{DC}) \tag{3.48}$$

In this equation, max(.) denotes the function MPPT (maximum power point tracking) in the P–V curve. The ambient temperature (t) and irradiation (G) affect the V_{DC} and I_{DC} as given in the following equations [31–33]:

$$V_{DC} = \phi(t, P) \tag{3.49}$$

$$I_{DC} = \psi(t, G, \mathrm{P}, V_{DC}, I_{DC}) \tag{3.50}$$

Therefore, a numerical technique must be applied to solve them [33]. The RBFNN can also be used to solve PFA equations of an MG, including PV modules. It must be mentioned that further details about P, φ, and ψ can be found in Refs. [30,31].

1.3 Power flow problem

The formulation of the conventional AC PFP is formed in terms of the following equations [34]:

$$P_i = [PG_{abc}]_i - [PD_{abc}]_i = \sum_{j=1}^{NAC} |V_i||V_j||Y_{ij}|\cos(\delta_i - \delta_k - \theta_{ij}) = \mathrm{Re}\left\{ [V_{abc}]_i \left(\sum_{j=1}^{NAC} [YB_{abc}]_{ij} [V_{abc}]_j \right)^* \right\} \tag{3.51}$$

$$Q_i = [QG_{abc}]_i - [QD_{abc}]_i = \sum_{j=1}^{NAC} |V_i||V_j||Y_{ij}|\sin(\delta_i - \delta_k - \theta_{ij}) = \mathrm{Im}\left\{ [V_{abc}]_i \left(\sum_{j=1}^{NAC} [YB_{abc}]_{ij} [[V_{abc}]_j] \right)^* \right\} \tag{3.52}$$

The power balance in the DC section of the network is written as follows [35]:

$$P_{DC,i} = [PG_{DC}]_i - [PD_{DC}]_i = V_{DC,i} \sum_{j=1}^{NDC} (YB_{ij} \cdot V_{DC,j}) \tag{3.53}$$

Integration of MG equations in AC and DC networks equations results in the following equation [35]:

$$\begin{bmatrix} (PG-PD)_{AC} \\ (QG-QD)_{AC} \\ \cdots \\ (PG-PD)_{DC} \end{bmatrix} = \begin{bmatrix} V_{AC} \\ V_{AC} \\ \cdots \\ V_{DC} \end{bmatrix} \left(\begin{bmatrix} YBUS_{AC} & \vdots & \\ \cdots & \cdots & \cdots \\ & \vdots & YBUS_{DC} \end{bmatrix} \begin{bmatrix} V_{AC} \\ V_{AC} \\ \cdots \\ V_{DC} \end{bmatrix} \right)^{*} \tag{3.54}$$

2. Radial basis function neural networks

The architecture of an RBFNN is shown in Fig. 3.2 [37]. As described in Ref. [38], the output of the hidden layer, which has H basic nonlinear Gaussian activation functions, is defined as follows:

$$a_{hk} = e^{\left(-\frac{\|x_k-\mu_h\|^2}{\sigma_h^2}\right)} \tag{3.55}$$

The network output is expressed using the following equation:

$$y_{mk} = \sum_{h=1}^{H} w_{mh} a_{hk} \tag{3.56}$$

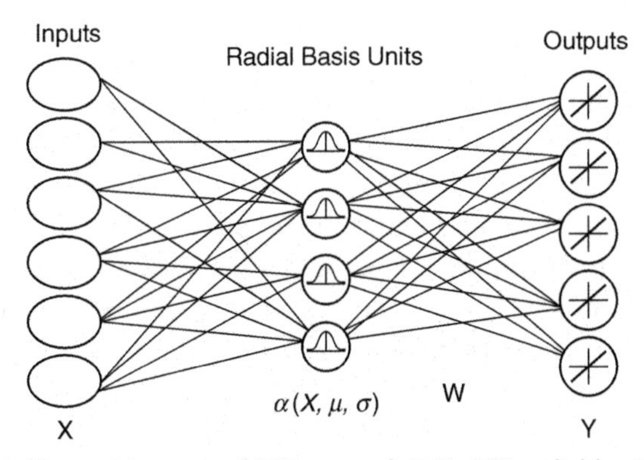

Figure 3.2 The architecture of RBF network [38]. *RBF*, radial basis function.

The training goal is to gain kernels centers and widths, and weighting factors of the output layer, which is obtained after minimization of the following objective function [38]:

$$E(\mu, \sigma, w) = \sum_{k=1}^{K} \|Y_k - T_k\|^2 \tag{3.57}$$

This function has been optimized using the algorithm "orthogonal least squares" [37] and the method "exact fit" (EF) [36]. The NLES of power flow can accurately be solved by the method EF. The hidden layer of training is neglected to significantly raise calculations speed. The network sensitivity can be increased by selecting a low kernel width, resulting in an overfitting problem. In the case of large kernel width, a singularity of matrix A may happen. Therefore, a trade-off is necessary. In the end, the following objective function must be optimized:

$$f = \|T - WA\|^2 \tag{3.58}$$

to find the optimal weighting factor matrix W, as follows [36–39]:

$$E(\mu, \sigma, w) = \sum_{k=1}^{K} \|Y_k - T_k\|^2 \tag{3.59}$$

3. Proposed algorithm

3.1 Power flow algorithm

Assume that the following equation set, which has m unknowns and n equations, must be solved:

$$Y = f(X) \tag{3.60}$$

The NR method can solve this equation set with acceptable speed and accuracy. But this method has two drawbacks. It needs the determination of the JM and then the calculation of its elements, which is a time-consuming task for PFA. The second problem is the singularity of the JM in some cases such as in case of MG studies. Thus, using derivative-free approaches and methods that do not need the inverse of the JM can be a good option. In this subsection, an algorithm is presented for PFP solving, which uses the merit of ANNs nonlinear mapping capability and their accuracy. It can handle NLESs with $n \neq m$ and does not need inverse of JM.

The PFP inputs are load demands and DERs generation (active and reactive powers), which are presented by X in Eq. (3.60). The output, i.e., Y, contains DERs operational parameters, magnitude and phase of voltages, and slack bus real and reactive power generation.

The following six steps form the proposed algorithm:

Step 1: The random vectors X_i, including all the variables, are generated. The magnitude and phase angle of voltages are randomly produced. They are in the range of 0.8−1.1 pu and 0−360°, respectively.

Step 2: X_i is mapped to Y_i by the function f.

Step 3: The RBFNN is trained by the vectors Y_i and X_i as inputs and outputs, respectively. The trained RBFNN should generate X^* in case of applying the input vector Y^*.

Step 4: The Y^* is applied to the ANN, which generates X^0. In this step, the kernel, which generates the least response at its output, is selected as the farthest center of the input vector. The input/output of this center is used in the sixth step as replaced set.

Step 5: After applying f to X^0, we have $Y^0 = f(X^0)$. The difference between Y^0 and Y^* determines the mismatch. In case of having a mismatch less than a prespecified tolerance, the procedure will stop, and the solution is X^0. Otherwise, the procedure will go to step 6.

Step 6: Replace one of the (Y_i, X_i)-s with (Y^0, X^0), and go to step 3. In case of a violation of maximum or minimum operational parameters of the DERs, that parameter has to be fixed at the maximum or minimum value, and the procedure must be repeated in step 5.

The error function, i.e., Eq. (3.57), is minimized using the RBFNN weighting factors determined by Eqs. (3.58) and (3.59). This training procedure is fast and different from orthogonal least squares or linear least squares regression.

It must be noted that in some conditions, the DERs are not able to adjust the voltage of their terminal at the prespecified values because of technical limitations on reactive power generation of PV buses. Therefore, their reactive powers must be checked in each iteration, and in case of a limitation violation, the developed program changes the PV bus to a load bus, and after rearrangement of buses, the procedure will go to step 2.

Unlike the work presented in Ref. [6], the PFP solution is simultaneously determined here, considering the operational equations, i.e., Eqs. 3.1−3.50, and the nonlinear PFP equations. This combination results in computation time (CT) reduction and high accuracy. Also, unlike Ref. [6], it can handle PFP equations even in the case of $n > m$ without using any arbitrary operating variables. In this section, the initial spread, which is in the range of 300–500, decreases exponentially with an increase in the number of iterations. The coefficient of the exponent is 1.02. Also, it must be mentioned that the hidden neurons number is in the range of 150–350 (Fig. 3.3).

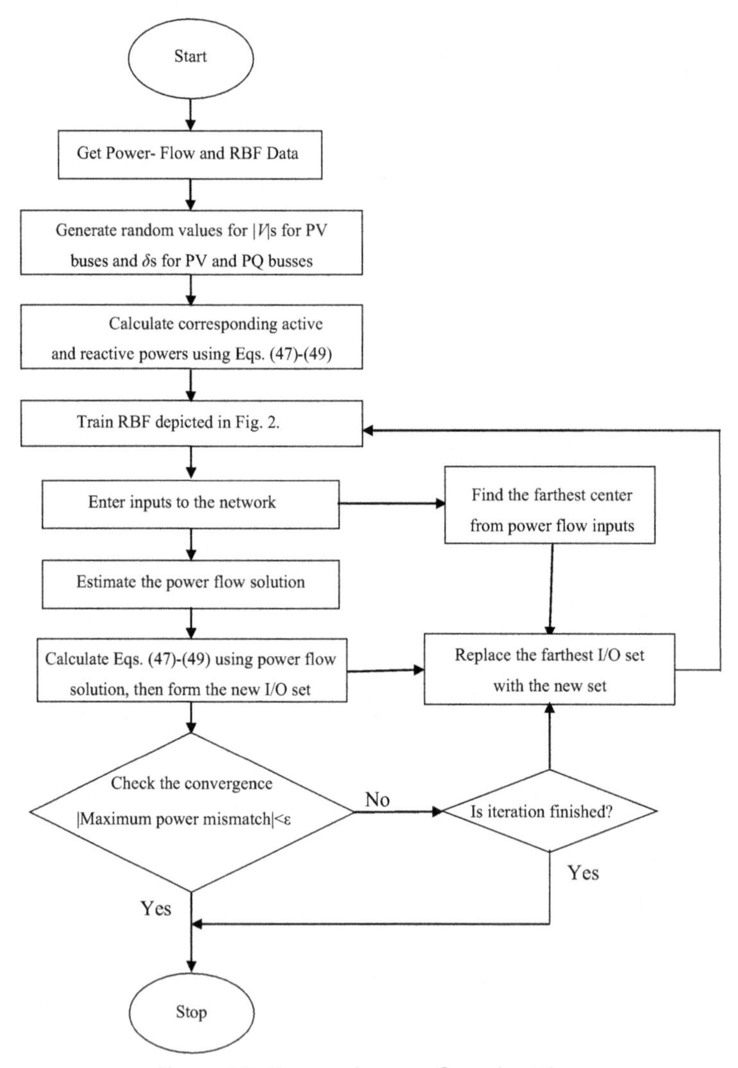

Figure 3.3 Proposed power flow algorithm.

3.2 Simulation results

Different networks have been simulated by the developed program. In all the tested cases, there was a good convergence even in the case of analyzing radial ill-conditioned MGs. The program has been simulated on Pentium IV (CPU 2.9 GHz, RAM 1.96 GB) in environment of version 7.11.0 Matlab.

3.2.1 Case study 1: meshed microgrid

The meshed MG, shown in Fig. 3.1A [1], is used as case study 1, to study a three-phase power flow. Here, the ECDER models given by Eqs. (3.1)−(3.9) and (3.10)−(3.20) are, respectively, used for DG2 and DG3. The parameters of this case study system are given in Ref. [1].

3.2.2 Case study 2: radial microgrid

To have a radial MG, in the MG shown in Fig. 3.1A, bus 4 and the lines, which are connected to this bus, have been removed. In this new MG, called case study 2, bus 5 is renamed to bus 4. The parameters of this case study can be found in Ref. [1]. The chosen and given variables and both systems parameters are listed in Table 3.1. Both case studies have been solved by the suggested algorithm and the conventional NR method. The convergence time and iterations number have been compared in Table 3.2. The proposed method is more accurate and needs less CT because it solves all the equations simultaneously, unlike Ref. [5] that solves the PFP equations in two steps. The used CPU time includes the times required for training and other calculations. Fig. 3.4A and B show the voltage variations of bus 4 versus R/X and also LM for both cases.

3.2.3 Case study 3: microgrid with droop-controlled distributed energy resources

For a three-phase balanced study, the proposed algorithm is applied to 6-bus MG and also the modified IEEE 38-bus system shown in Fig. 3.5A and B, respectively. The DERs are droop-controlled, and their data are given in Refs. [16,19,40]. The other parameters can be found in Ref. [40]. In this subsection, the following three scenarios are studied:

Scenario 1: 3 DGs have $P-\omega$ and $Q-V$ droop controllers.

Scenario 2: The same as the first case, but DGs have resistive output.

Scenario 3: The same as the first case, but DGs output has an impedance.

Table 3.1 PFA results for case studies 1 and 2.

Test case	Scenarios		Slack bus	PV buses Buses 2 and 3	PQ buses Buses 4 and 5	DG 2	DG 3
Study system (1)	Example 1	Selected/ given parameters	$V_1 = 1.00$, $\theta = 0$	$P_2 = 1.00$, $V_2 = 1.00$ $P_3 = 1.00$, $V_3 = 1.00$ $S_{D3} = 0.2 + j0.1$	$S_{D4} = 1.5 + j0.75$ $Q_{G4} = 1.3$ $S_{D5} = 1.25 + j0.65$ $Q_{G5} = 1.5$	–	–
		Computed parameters	$P_1 = 1.07$, $Q_1 = -0.02$	$Q_2 = -0.04$, $\theta_2 = -2.05$ $Q_3 = -0.17$, $\theta_3 = -6.69$	$V_4 = 1.04$, $\theta_4 = -21.84$ $V_5 = -1.11$, $\theta_4 = -27.98$	$P_{g2} = 1.04$, $V_{dc} = 1.32$ $V_{i2} = 1.45$, $V_{o2} = 1$ $\alpha_{i2} = \alpha_{mi2} = -18.12$ $\alpha_{o2} = \alpha_{mo2} = 4.2$ $A_{mi2} = 1$, $A_{mo2} = 0.67$ $V_{g2} = 1$, $\theta_{g2} = 30$, $Q_{g2} = 0.01$	$P_{g3} = 1.02$, $V_{i3} = 1.18$, $\alpha_{mi3} = -25.8$ $V_{o3} = 1.13$, $\alpha_{mo3} = 5.42$ $A_{mi3} = 0.8$, $\alpha_{m3} = -20.17$ $V_{g2} = 1$, $\theta_{g2} = 25.6$, $Q_{g2} = 0.01$

	Example 2	Selected/given parameters	$V_1 = 1.00$, $\theta_1 = 0$	$P_2 = 0.74$, $V_2 = 0.98$ $P_3 = 1.00$, $V_3 = 1.00$ $S_{D3} = 0.2 + j0.1$	$S_{D4} = 2.00 + j1.00$ $Q_{G4} = 1.30$ $S_{D5} = 1.25 + j0.65$ $Q_{G5} = 1.50$	–	–
		Computed parameters	$P_1 = 1.90$, $Q_1 = 0.5$	$Q_2 = 0.2$, $\theta_2 = -11.38$ $Q_3 = 0.06$, $\theta_3 = -17.74$	$V_4 = 0.94$, $\theta_4 = -35.16$ $V_5 = 1.06$ $\theta_4 = -41.27$	$P_{g2} = 0.75$, $V_{dc} = 1.32$ $V_{i2} = 1.25$, $V_{o2} = 1$ $\alpha_{i2} = \alpha_{mi2} = -24.62$ $\alpha_{o2} = \alpha_{mo2} = -8.3$ $A_{mi2} = 1$, $A_{mo2} = 0.76$ $V_{g2} = 0.99$, $\theta_{g2} = 15.9$, $Q_{g2} = 0.2$	$P_{g3} = 1.02$, $V_{i3} = 1.19$, $\alpha_{mi3} = -29.48$ $V_{o3} = 1.15$, $\alpha_{mo3} = 4.98$ $A_{mi3} = 0.81$, $\alpha_{m3} = -23.19$ $V_{g2} = 0.98$, $\theta_{g2} = 17.2$, $Q_{g2} = 0.05$

Continued

Table 3.1 PFA results for case studies 1 and 2.—cont'd

Test case	Scenarios		Slack bus	PV buses Buses 2 and 3	PQ buses Buses 4 and 5	DG 2	DG 3
Radial microgrid (study system (2))	Example 3	Selected/given parameters	$V_1 = 1.00$, $\theta_1 = 0$	$P_2 = 0.85$, $V_2 = 1$ $P_3 = 1.00$, $V_3 = 1.00$ $S_{D3} = 0.2 + j0.1$	$S_{D4} = 1.5 + j0.75$ $Q_{G4} = 1.3$	–	–
		Computed parameters	$P_1 = 0.85$, $Q_1 = 0.43$	$Q_2 = 0.19$, $\theta_2 = -18.84$ $Q_3 = 0.1$, $\theta_3 = -11.7$	$V_4 = 1.05$, $\theta_4 = -28.94$	$P_{g2} = 0.89$, $V_{dc} = 1.38$ $V_{i2} = 1.25$, $V_{o2} = 1.02$ $\alpha_{i2} = \alpha_{mi2} = -20.45$ $\alpha_{o2} = \alpha_{mo2} = -7.03$ $A_{mi2} = 0.96$, $A_{mo2} = 0.81$ $V_{g2} = 0.98$, $\theta_{g2} = 12.56$, $Q_{g2} = 0.23$	$P_{g3} = 1.02$, $V_{i3} = 1.15$, $\alpha_{mi3} = -26.83$ $V_{o3} = 1.13$, $\alpha_{mo3} = 5.21$ $A_{mi3} = 0.89$, $\alpha_{m3} = -19.95$ $V_{g2} = 0.95$, $\theta_{g2} = 14.5$, $Q_{g2} = 0.08$

PFA, power flow analysis; *PV*, photovoltaic.

Table 3.2 Comparison of iterations number, used CPU time, and time used for each iteration.

Scenario		Test cases	
		Case study 1	Case study 2
Iterations no.	NR	5.0	6.0
	RBFNN	3.0	3.0
Used CPU time in (s)	NR	0.9035	0.9983
	RBFNN	0.4221	0.4411
Used time for each iteration (s)	NR	0.1807	0.1664
	RBFNN	0.1407	0.1470

NR, Newton–Raphson; *RBFNN*, radial basis function neural network.

The PFA results using the proposed and modified NR (MNR) methods [19], are respectively, listed in Tables 3.3 and 3.4 for 6-bus MG and 38-bus MG. For three scenarios, the suggested algorithm results in an error less than 1% for V and θ and converges with less computational effort (time and iterations).

For unbalanced power flow studies, the proposed method and the Newton trust region (NTR) method presented in Ref. [16] are applied to the 25-bus network indicated in Fig. 3.5C [16,41]. The parameters and data used for simulations can be found in Ref. [41], and the drop control parameters of Ref. [16] are used for simulations. The results are given and compared in Table 3.5. The suggested RBFNN-based method is able to solve the unbalanced PFP with the error less than 2%.

3.2.4 Case study 4: comparison between the proposed and modified ladder iterative power flow analysis algorithms

Both power flow algorithms, i.e., the proposed one, which is based on RBFNN, and the modified algorithm, which is based on ladder iterative algorithm [27,28], are applied to the reformed IEEE 37-bus network shown in Fig. 3.6A. A WG system, which is based on type-3 DFIG [27], is connected to this system at bus 775. The PFA results, the calculations time, and the iterations number are compared in Table 3.6. Although WG and PV systems optimal sizing can be found in Ref. [42], considering the control strategies presented in Ref. [43], the DFIG-based WG unit parameters presented in Refs. [27,28] are used in this section. The other parameters are available in Ref. [44].

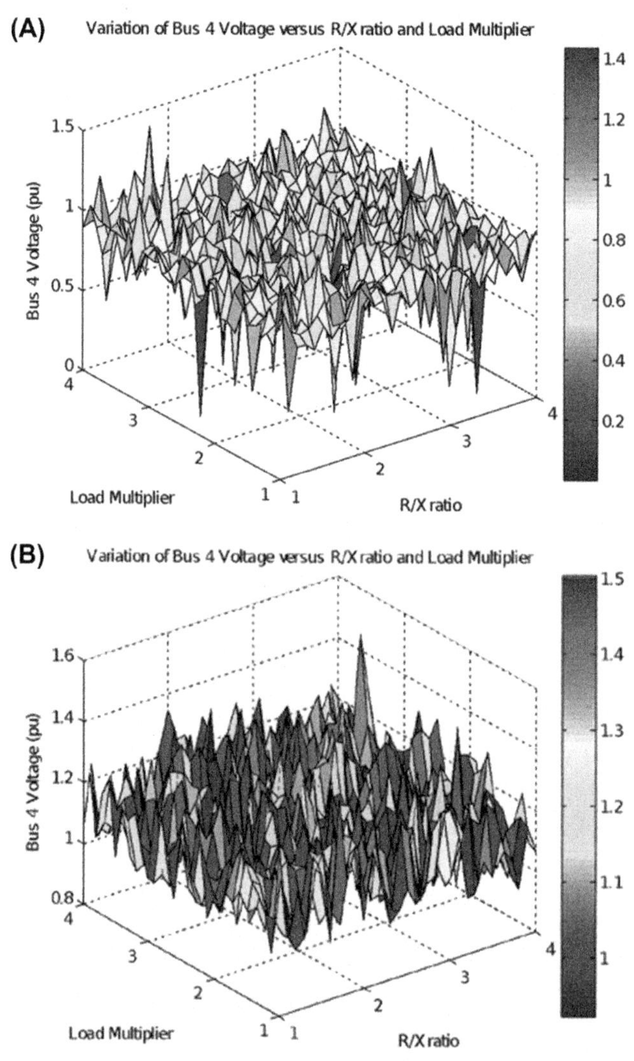

Figure 3.4 Voltage variations of bus 4 vs. R/X and load multiplier for case studies (A) 1 and (B) 2. *R/X*, resistance to reactance.

In Ref. [27], the ANN has been used to obtain an approximate solution for Eq. (3.45). The reformed IEEE 37-bus network, including the WG system, has simultaneously been solved by the proposed algorithm. It has been shown that its calculations time is lower than the modified ladder iterative power flow algorithm (MLIPFA), and also, it had better accuracy and authenticity.

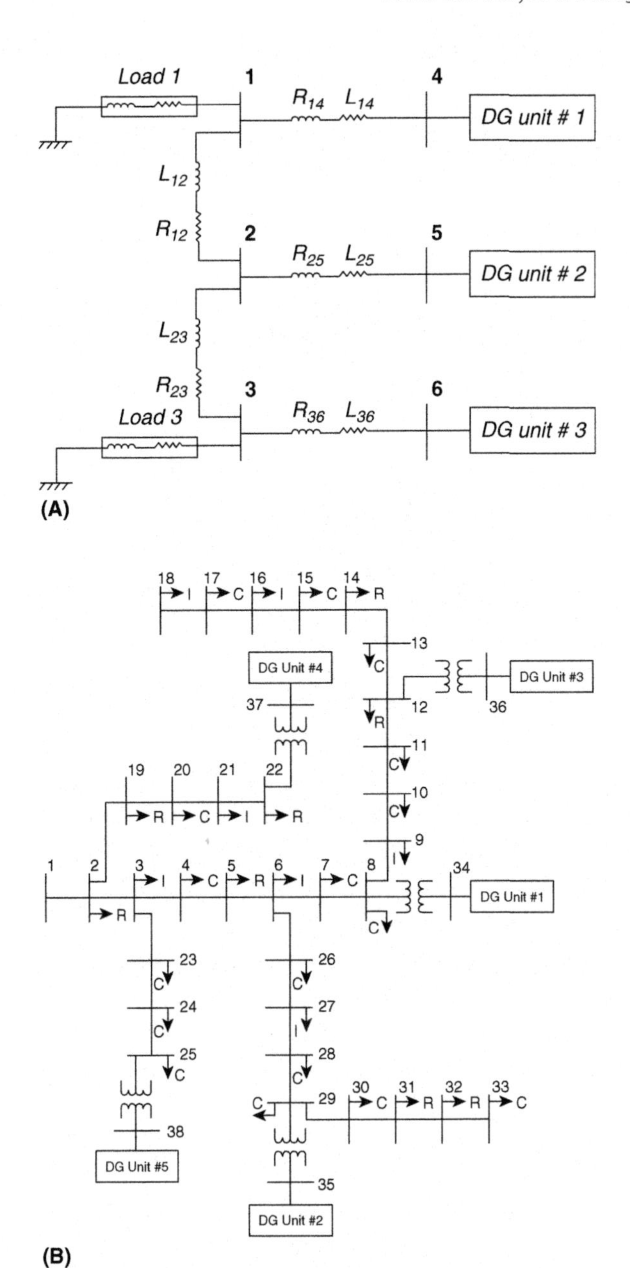

Figure 3.5 (A) 6-bus MG [16], (B) Modified IEEE 38-bus network [19,40] and (C) 25-bus network [16.42]]. *MG*, microgrid.

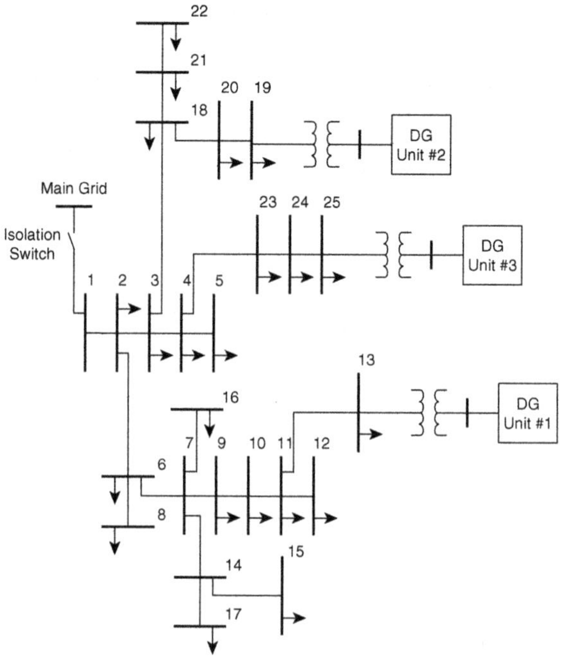

Figure 3.5 Cont'd

3.2.5 Case study 5: microgrid with AC and DC sections

In this section, the suggested RBFNN-based algorithm is used to solve PFP of the modified IEEE 13-bus network indicated in Fig. 3.6A [35]. This system presents an MG with AC and DC sections, which its parameters are available in Ref. [44]. The other components of data can be found in Ref. [35]. To model the three-phase AC/DC VSC, the equivalent π-mode is utilized. The proposed RBFNN-based algorithm and NR method results are compared in Table 3.7. Again, it can be seen a better performance for the proposed algorithm.

3.3 Discussion and conclusion

Table 3.8 summarizes the comparison of different algorithms for five cases. In this table, the convergence of all the algorithms is compared considering the effect of the R/X and LM. The comparison emphasizes the better performance of the suggested RBFNN-based algorithm.

Table 3.3 Results of 6-bus MG [19] for RBFNN-based and MNR methods.

Case 1 (P−ω and Q−V droop functions)	Bus	Voltage							
		$\alpha = 0, \beta = 0$)				($\alpha = 2, \beta = 2$)			
		Magnitude (pu)		θ (deg)		Magnitude (pu)		θ (deg)	
		MNR	RBFNN	MNR	RBFNN	MNR	RBFNN	MNR	RBFNN
	1	0.9566	0.9566	0.0000	0.0000	0.9601	0.9602	0.0000	0.0000
	2	0.9704	0.9703	−0.5597	−0.5595	0.9726	0.9725	−0.5222	−0.5214
	3	0.9611	0.9610	−2.8714	−2.8716	0.9639	0.9638	−2.6708	−2.6710
	4	0.9861	0.9861	−0.0870	−0.0870	0.9873	0.9872	−0.0727	−0.0723
	5	0.9893	0.9893	−0.4769	−0.4769	0.9901	0.9902	−0.4453	−0.4450
	6	0.9670	0.9671	−3.0693	−3.0694	0.9694	0.9693	−2.8535	−2.8532
		f (pu)							
		($\alpha = 0, \beta = 0$)				($\alpha = 2, \beta = 2$)			
		MNR		RBFNN		MNR		RBFNN	
		0.99903		0.99903		0.99911		0.99911	
	Max. Error of V (%)	Mag.		θ		Mag.		θ	
	Absolute	0.0001		0.0002		0.0001		0.0008	
	Relative	0.0104		0.0357		0.0104		0.5502	
		Iterations number, used CPU time, and time used for each iteration							
		MNR		RBFNN		MNR		RBFNN	
	Iterations no.	4		3		4		3	
	Used CPU time (s)	0.9120		0.4430		0.9140		0.4570	
	Used time for each iteration (s)	0.2280		0.1477		0.2285		0.1523	

Continued

Case 2 (P–V and Q–ω droop functions)	Bus	V							
		($\alpha = 0,\ \beta = 0$)				($\alpha = 2,\ \beta = 2$)			
		Mag. (pu)		θ (deg.)		Mag. (pu)		θ (deg.)	
		MNR	RBFNN	MNR	RBFNN	MNR	RBFNN	MNR	RBFNN
	1	0.9527	0.9526	0.0000	0.0000	0.9702	0.9703	0.0000	0.0000
	2	0.9727	0.9725	−0.0320	−0.0318	0.9780	0.9782	−0.1682	−0.1683
	3	0.9437	0.9437	0.4744	0.4741	0.9656	0.9655	−2.4125	−2.4129
	4	0.9786	0.9786	−0.5100	−0.5102	1.0020	1.0020	−0.2974	−0.2970
	5	0.9860	0.9861	−0.4579	−0.4577	0.9938	0.9937	0.0136	0.0135
	6	0.9522	0.9520	0.4666	0.4671	0.9708	0.9707	−2.5864	−2.5873
		f (pu)							
		($\alpha = 0,\ \beta = 0$)				($\alpha = 2,\ \beta = 2$)			
		MNR		RBFNN		MNR		RBFNN	
		1.00065		1.00065		1.00059		1.00059	
	Max. Error of V (%)	Mag.		θ		Mag.		θ	
	Absolute	0.0002		0.0005		0.0002		0.0004	
	Relative	0.0210		0.6250		0.0209		0.9456	
		Iterations number, used CPU time, and time used for each iteration							
		MNR		RBFNN		MNR		RBFNN	
	Iterations no.	4		3		4		3	
	Used CPU time (s)	0.9122		0.4430		0.9143		0.4571	
	Used time for each iteration (s)	0.2281		0.1477		0.2286		0.1524	

Case 3 (P−V−ω and Q−V−ω droop functions)	Bus	V							
		Case 1 ($\alpha = 0$, $\beta = 0$)				Case 1 ($\alpha = 2$, $\beta = 2$)			
		Mag. (pu)		θ (deg)		Mag. (pu)		θ (deg)	
		MNR	RBFNN	MNR	RBFNN	MNR	RBFNN	MNR	RBFNN
	1	0.9300	0.9301	0.0000	0.0000	0.9387	0.9386	0.0000	0.0000
	2	0.9470	0.9471	−0.3410	−0.3414	0.9535	0.9534	−0.2961	−0.2959
	3	0.9296	0.9295	−1.5781	−1.5780	0.9383	0.9382	−1.3713	−1.3715
	4	0.9588	0.9586	−0.2896	−0.2893	0.9638	0.9637	−0.2507	−0.2510
	5	0.9641	0.9641	−0.4510	−0.4513	0.9684	0.9685	0.3919	0.3921
	6	0.9367	0.9366	−1.7070	−1.7072	0.9444	0.9445	−0.1483	−0.1483
		f (pu)							
		Case 1 ($\alpha = 0$, $\beta = 0$)				Case 1 ($\alpha = 2$, $\beta = 2$)			
		MNR		RBFNN		MNR		RBFNN	
	0.999698			0.999698		0.999735		0.999735	
	Max. Error of V (%)		Mag.	θ		Mag.		θ	
	Absolute		0.0002	0.0004		0.0001		0.0003	
	Relative		0.0209	0.1173		0.0107		0.1197	

Continued

	Iterations number, used CPU time, and time used for each iteration			
	MNR	RBFNN	MNR	RBFNN
Iterations no.	4	3	4	3
Used CPU time (s)	0.9150	0.446	0.9146	0.4574
Used time for each iteration (s)	0.2288	0.1487	0.2287	0.1525

MG, microgrid; *MNR*, modified Newton–Raphson; *RBFNN*, radial basis function neural network.

Table 3.4 Results of 38-bus MG [16] for RBFNN-based algorithm and MNR method.

	V mag. (pu)		Θ (deg.)		P_L		Q_L	
Bus. No.	MNR	RBFNN	MNR	RBFNN	MNR	RBFNN	MNR	RBFNN
1	0.9802	0.9802	0.0000	0.0000	—	—	—	—
2	0.9802	0.9802	0.0000	0.0000	−0.0980	−0.0982	−0.0554	−0.0553
3	0.9790	0.9790	−0.0275	−0.0277	−0.0895	−0.0897	−0.0353	−0.0356
4	0.9787	0.9786	−0.0585	−0.0574	−0.1159	−0.1161	−0.0745	−0.0748
5	0.9787	0.9787	−0.0930	−0.0932	−0.0587	−0.0591	−0.0275	−0.0272
6	0.9796	0.9796	−0.1699	−0.1697	−0.0597	−0.0593	−0.0177	−0.0173
7	0.9825	0.9825	−0.2940	−0.2938	−0.1944	−0.1946	−0.0943	−0.0946
8	0.9834	0.9834	−0.5453	−0.5452	−0.1946	−0.1946	−0.0946	−0.0943
9	0.9834	0.9834	−0.7487	−0.7484	−0.0597	−0.0594	−0.0181	−0.0184
10	0.9838	0.9837	−0.9447	−0.9447	−0.0584	−0.0587	−0.0190	−0.0187
11	0.9838	0.9838	−0.9762	−0.9759	−0.0438	−0.0436	−0.0284	−0.0286
12	0.9839	0.9839	−1.0380	−1.0378	−0.0590	−0.0592	−0.0328	−0.0326
13	0.9784	0.9784	−1.1254	−1.1252	−0.0579	−0.0581	−0.0326	−0.0328
14	0.9764	0.9764	−1.1961	−1.1963	−0.1172	−0.1174	−0.0728	−0.0725
15	0.9752	0.9751	−1.2307	−1.2305	−0.0577	−0.0578	−0.0092	−0.0090
16	0.9739	0.9738	−1.2534	−1.2536	−0.0596	−0.0595	−0.0171	−0.0174
17	0.9722	0.9722	−1.3230	−1.3228	−0.0574	−0.0578	−0.0182	−0.0179
18	0.9716	0.9715	−1.3330	−1.3331	−0.0894	−0.0892	−0.0337	−0.0332
19	0.9807	0.9806	0.0212	0.0214	−0.0882	−0.0880	−0.0370	−0.0375
20	0.9872	0.9872	0.2245	0.2239	−0.0881	−0.0883	−0.0383	−0.0386
21	0.9893	0.9893	0.3112	0.3115	−0.0897	−0.0895	−0.0376	−0.0373
22	0.9939	0.9940	0.5066	0.5063	−0.0893	−0.0891	−0.0391	−0.0394

Continued

Table 3.4 Results of 38-bus MG [16] for RBFNN-based algorithm and MNR method.—cont'd

Bus. No.	V mag. (pu)		Θ (deg.)		P_L		Q_L	
	MNR	RBFNN	MNR	RBFNN	MNR	RBFNN	MNR	RBFNN
23	0.9785	0.9785	−0.0149	−0.0151	−0.0869	−0.0867	−0.0465	−0.0462
24	0.9783	0.9783	0.0149	0.0153	−0.4055	−0.4058	−0.1859	−0.1856
25	0.9811	0.9812	0.0880	0.0881	−0.4073	−0.4076	−0.1878	−0.1882
26	0.9796	0.9796	−0.1136	−0.1134	−0.0581	−0.0583	−0.0234	−0.0236
27	0.9798	0.9798	−0.0342	−0.0344	−0.0597	−0.0599	−0.0222	−0.0222
28	0.9796	0.9795	0.3296	0.3295	−0.0581	−0.0584	−0.0187	−0.0180
29	0.9799	0.9799	0.6138	0.6142	−0.1162	−0.1166	−0.0654	−0.0657
30	0.9767	0.9767	0.6969	0.6966	−0.1926	−0.1925	−0.5549	−0.5549
31	0.9730	0.9730	0.6191	0.6192	−0.1460	−0.1463	−0.0628	−0.0629
32	0.9722	0.9721	0.5982	0.5983	−0.2042	−0.2046	−0.0894	−0.0892
33	0.9720	0.9719	0.5915	0.5918	−0.0574	−0.0574	−0.0364	−0.0367
					Generation (P_G)		Generation (Q_G)	
34	0.9664	0.9665	−0.7724	−0.7725	0.3670	0.3650	0.6777	0.6771
35	0.9993	0.9992	1.2862	1.2865	1.2465	1.2470	0.3203	0.3194
36	0.9971	0.9971	−1.2054	−1.2058	0.4155	0.4158	0.6446	0.6441

37	0.9973	0.9972	0.6195	0.6192	0.8310	0.8322	0.2554	0.2539
38	0.9847	0.9847	0.1856	0.1853	0.8310	0.8322	0.3000	0.3000
		P_G	**Q_G**	**P_L**	**Q_L**	**P_{loss}**	**Q_{loss}**	
Total		3.6922	2.1945	3.6213	2.1262	0.0709	0.0683	
Max. V error	Mag.				θ			
Absolute	0.0001				0.0011			
Relative	0.0103				2.6846			
Iterations number, used CPU time, and time used for each iteration								
	MNR				RBFNN			
Iterations no.	5				3			
Used CPU time (s)	1.1562				0.5624			
Used time for each iteration (s)	0.2312				0.1875			

MG, microgrid; *MNR*, modified Newton–Raphson; *RBFNN*, radial basis function neural network.

Table 3.5 Results of 25-bus MG [16] for proposed algorithm and NTR method.

Bus	NTR method [26]												Proposed RBFNN-based algorithm											
	Phase A				Phase B				Phase C				Phase A				Phase B				Phase C			
	V_{an} (pu, degree)				V_{bn} (pu, degree)				V_{cn} (pu, degree)				V_{an} (pu, degree)				V_{bn} (pu, degree)				V_{cn} (pu, degree)			
	ag.	ng.	L	L	ag.	ng.	L	L	ag.	ng.	L	L	ag.	ng.	L	L	ag.	ng.	L	L	ag.	ng.	L	L
1	.97 91	.000 0	.000 0	.000 0	.973 3	119. 933 6	.000 0	.000 0	.970 7	19.90 81	.000 0	.000 0	.979 1	.000 0	.000 0	.000 0	.970 7	119. 908 1	.000 0	.000 0	.970 7	19.90 81	.000 0	.000 0
2	.97 91	.000 0	.000 0	.000 0	.973 3	119. 933 6	.000 0	.000 0	.097 1	19.90 81	.000 0	.000 0	.979 1	.000 0	.000 0	.000 0	.970 7	119. 908 3	.000 0	.000 0	.097 1	19.90 83	.000 0	.000 0
3	.98 00	.056 8	.008 0	.006 0	.974 6	119. 887 9	.012 0	.009 0	.972 0	19.99 02	.016 0	.012 0	.980 1	.056 8	.008 0	.006 0	.972 1	119. 990 5	.012 1	.009 1	.972 0	19.99 02	.016 1	.012 1
4	.98 21	.143 2	.012 0	.010 0	.976 2	119. 816 5	.018 0	.013 0	.973 7	20.09 24	.018 0	.013 0	.982 3	.143 4	.012 0	.010 1	.973 7	120. 092 5	.018 1	.013 0	.973 7	20.09 23	.018 0	.013 0
5	.98 04	.145 0	.010 0	.007 0	.975 2	119. 810 0	.012 0	.009 0	.972 5	20.08 89	.014 0	.011 0	.980 4	.145 0	.010 1	.007 0	.972 7	120. 088 9	.012 2	.009 0	.972 5	20.08 89	.014 0	.011 0
6	.97 70	0.06 64	.010 0	.007 0	.970 4	119. 982 8	.013 0	.010 0	.967 4	19.82 77	.013 0	.009 0	.977 1	0.06 64	.010 1	.007 0	.967 5	119. 827 7	.013 0	.010 1	.967 2	19.82 77	.013 1	.009 1
7	.97	0.13	.000	.000	.969	120.	.000	.000	.966	19.75	.000	.000	.976	0.13	.000	.000	.966	119.	.000	.000	.966	19.75	.000	.000

	65	79	0	0	8	041 1	0	0	2	47	0	0	5	77	0	0	2	754 7	0	0	2	47	0	0
8	.97 54	0.06 27	.010 0	.007 0	.968 3	119. 969 7	.012 0	.009 0	.965 0	19.82 06	.014 0	.011 0	.975 4	0.06 27	.010 0	.007 1	.965 2	119. 820 6	.012 1	.009 1	.965 0	19.82 06	.013 9	.011 0
9	.98 12	0.21 82	.014 0	.011 0	.976 1	120. 148 5	.015 0	.012 0	.973 2	19.70 98	.018 0	.013 0	.981 2	0.21 82	.014 0	.011 2	.973 2	119. 709 5	.015 2	.012 0	.973 2	19.70 98	.018 0	.013 2
10	.98 71	0.29 96	.008 0	.006 0	.983 8	120. 264 0	.012 0	.009 0	.981 6	19.67 22	.016 0	.012 0	.987 2	0.29 98	.008 1	.006 0	.981 6	119. 672 2	.012 0	.008 9	.981 3	19.67 22	.016 2	.012 1
11	.99 10	0.34 88	.011 0	.008 0	.989 0	120. 337 0	.010 0	.007 0	.987 4	19.65 40	.014 0	.011 0	.991 1	0.34 88	.011 2	.008 0	.987 4	119. 654 0	.010 1	.006 9	.987 4	19.65 10	.014 0	.011 0
12	.99 03	0.34 75	.012 0	.008 0	.987 9	120. 333 4	.018 0	.013 0	.986 3	19.65 82	.018 0	.014 0	.990 3	0.34 75	.012 1	.008 1	.986 3	119. 658 2	.018 0	.013 0	.986 2	19.65 82	.018 0	.014 0
13	.99 74	0.41 56	.008 0	.007 0	.997 4	120. 415 6	.013 0	.010 0	.997 4	19.58 44	.014 0	.011 0	.997 4	0.41 58	.008 1	.007 0	.997 2	119. 584 8	.013 2	.010 1	.997 4	19.58 43	.014 2	.010 9
14	.97 22	0.13 06	.012 0	.008 0	.963 9	119. 997 5	.015 0	.012 0	.959 1	19.73 02	.022 0	.016 0	.972 2	0.13 06	.012 0	.008 0	.959 1	119. 730 2	.015 1	.012 0	.959 1	19.73 02	.022 0	.016 0
15	.97	0.12	.032	.024	.961	119.	.040	.030	.956	19.72	.048	.036	.970	0.12	.032	.024	.956	119.	.040	.030	.956	19.72	.048	.035

Continued

Table 3.5 Results of 25-bus MG [16] for proposed algorithm and NTR method.—cont'd

	06	68	0	0	7	984 1	0	0	8	30	0	0	5	68	0	0	7	723 0	0	0	4	30	0	8
16	.97 57	0.13 61	.010 0	.007 0	.968 7	120. 034 5	.012 0	.009 0	.965 0	19.75 12	.014 0	.012 0	.975 7	0.13 64	.010 0	.007 0	.965 0	119. 751 2	.012 1	.009 0	.965 0	19.75 12	.014 1	.012 0
17	.97 13	0.12 82	.010 0	.007 0	.962 8	119. 994 4	.010 0	.007 0	.957 5	19.73 07	.016 0	.012 0	.971 3	0.12 82	.010 0	.007 1	.957 5	119. 730 7	.010 0	.007 0	.957 5	19.73 09	.016 0	.011 9
18	.97 97	.018 2	.010 0	.007 0	.975 0	119. 934 4	.012 0	.009 0	.972 6	19.97 11	.014 0	.011 0	.979 7	.018 2	.010 0	.007 0	.972 6	119. 971 1	.012 2	.009 0	.972 6	19.97 11	.014 1	.011 0
19	.98 64	0.08 99	.014 0	.011 0	.986 4	120. 089 9	.015 0	.010 0	.986 4	19.91 01	.018 0	.014 0	.986 5	0.08 97	.014 2	.011 0	.986 4	119. 910 4	.015 0	.010 0	.986 3	19.91 01	.018 0	.014 2
20	.98 19	0.02 50	.008 0	.006 0	.979 0	120. 001 7	.012 0	.009 0	.977 1	19.96 37	.016 0	.012 0	.981 9	0.02 50	.008 0	.006 1	.977 1	119. 963 7	.012 0	.009 1	.977 1	19.96 37	.016 0	.012 2
21	.97 70	.024 0	.010 0	.007 0	.971 3	119. 922 9	.010 0	.007 0	.968 1	19.98 09	.016 0	.012 0	.977 0	.024 1	.010 0	.007 1	.968 2	119. 980 9	.010 1	.007 0	.968 1	19.98 09	.015 9	.012 0
22	.97 55	.026 7	.012 0	.008 0	.968 9	119. 915 4	.018 0	.013 0	.965 7	19.98 92	.018 0	.014 0	.975 5	.026 7	.012 1	.008 0	.965 7	119. 989 2	.018 1	.013 0	.965 6	19.98 95	.018 0	.014 0
23	.98	.218	.014	.011	.981	119.	.015	.012	.978	20.18	.018	.013	.985	.218	.014	.011	.978	120.	.015	.012	.978	20.18	.017	.013

	51	4	0	0	2	757 1	0	0	9	50	0	0	1	3	0	0	9	185 3	0	1	7	50	8	1
24	.99 00	.281 4	.008 0	.006 0	.987 3	119. 706 0	.013 0	.010 0	.985 3	20.28 17	.014 0	.011 0	.990 1	.281 4	.008 0	.006 0	.985 4	120. 281 7	.013 1	.010 1	.985 2	20.28 19	.014 2	.011 1
25	.00 03	.413 8	.014 0	.011 0	.000 3	119. 586 2	.015 0	.009 0	.000 3	20.41 38	.018 0	.013 0	.000 2	.413 6	.014 1	.010 9	.000 2	120. 413 7	.015 2	.009 0	.000 1	20.41 38	.018 1	.013 0

	P_{L-A}	Q_{L-A}	P_{L-B}	Q_{L-B}	P_{L-C}	Q_{L-C}	$P_{L-total}$	$Q_{L-total}$	P_{L-loss}	QL-loss		
	0.2570	0.1900	0.3220	0.2380	0.3870	0.2930	0.9660	0.7210	0.2570	0.19 00		

Phase	A				B				C			
Max. V Error (%)	Mag.	Ang.	P_L	Q_L	Mag.	Ang.	P_L	Q_L	Mag.	Ang.	P_L	Q_L
Absolute	0.0002	0.0003	0.0002	0.0002	0.0053	0.8308	0.0002	0.0001	0.0004	0.0030	0.00 02	0.00 02
Relative	0.0204	0.4167	1.8182	1.8182	0.5505	0.6920	1.6667	1.4286	0.0418	0.0025	1.42 86	1.66 67

Iterations number, used CPU time, and time used for each iteration

	NTR method [24]	**Proposed RBFNN-based algorithm**
Iterations No.	6	4
Used CPU time (s)	1.6187	*0. 6835*
Used time for each iteration (s)	0.2698	0.17088

MG, microgrid; *NTR*, Newton trust region; *RBFNN*, radial basis function neural network.

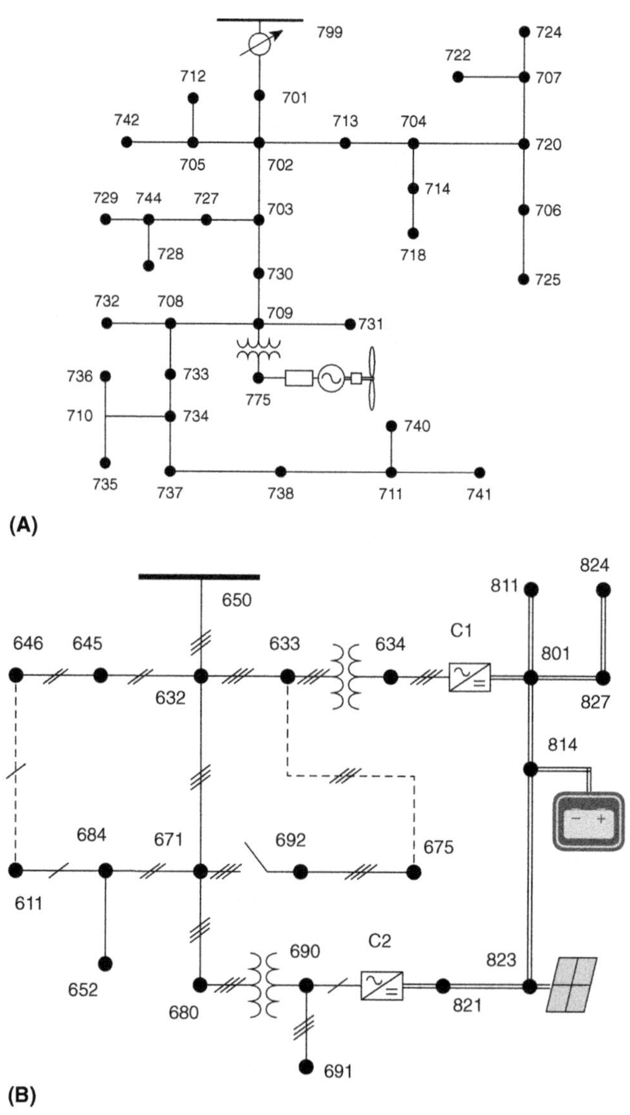

Figure 3.6 (A) IEEE 37-bus three-phase network [27] and (B) modified IEEE 13-bus network including AC subnetwork and DC subnetwork [29].

The proposed algorithm has been developed in a MATLAB environment. It is obvious that its performance can be improved after compiling developed program in C, C++, Pascal, or FORTRAN, which is able to solve the PFP with less computational effort.

Table 3.6 Results of modified 37-bus system [27] for RBFNN-based algorithm and MLIPFA.

Bus no.	V_{ab} (pu)		V_{bc} (deg.)		V_{ca} (deg.)	
	MLIPFA [27,28]	RBFNN	MLIPFA [27,28]	RBFNN	MLIPFA [27,28]	RBFNN
799	1.0000	1.0000	1.0000	1.0000	1.0000	1.0000
RG7	1.0437	1.0437	1.0250	1.0250	1.0345	1.0345
701	1.0141	1.0141	0.9971	0.9971	0.9981	0.9981
702	0.9937	0.9938	0.9797	0.9798	0.9757	0.9756
703	0.9684	0.9684	0.9599	0.9599	0.9498	0.9498
727	0.9673	0.9673	0.9592	0.9592	0.9488	0.9488
744	0.9666	0.9665	0.9588	0.9588	0.9484	0.9484
728	0.9662	0.9662	0.9584	0.9584	0.9480	0.9482
729	0.9662	0.9662	0.9588	0.9586	0.9483	0.9483
730	0.9427	0.9427	0.9390	0.9390	0.9254	0.9253
709	0.9381	0.9383	0.9356	0.9356	0.9214	0.9214
708	0.9356	0.9356	0.9346	0.9344	0.9191	0.9191
732	0.9355	0.9354	0.9345	0.9345	0.9186	0.9184
733	0.9315	0.9315	0.9328	0.9329	0.9152	0.9152
734	0.9280	0.9280	0.9313	0.9313	0.9118	0.9119
710	0.9275	0.9275	0.9303	0.9302	0.9102	0.9103
735	0.9274	0.9274	0.9301	0.9301	0.9096	0.9096
736	0.9270	0.9269	0.9288	0.9288	0.9099	0.9099
737	0.9246	0.9244	0.9303	0.9303	0.9096	0.9096
738	0.9234	0.9234	0.9299	0.9299	0.9085	0.9085
711	0.9231	0.9231	0.9297	0.9297	0.9075	0.9075
740	0.9230	0.9231	0.9295	0.9295	0.9069	0.9069
741	0.9230	0.9232	0.9296	0.9296	0.9072	0.9073
731	0.9262	0.9264	0.9242	0.9242	0.9100	0.9100
XF7	0.9381	0.9381	0.9356	0.9356	0.9214	0.9214
DFIG	0.9354	0.9354	0.9333	0.9333	0.9193	0.9193
705	0.9931	0.9931	0.9784	0.9784	0.9744	0.9744
712	0.9930	0.9930	0.9782	0.9781	0.9737	0.9737
742	0.9928	0.9928	0.9775	0.9775	0.9742	0.9740
713	0.9924	0.9924	0.9778	0.9778	0.9738	0.9738
704	0.9907	0.9908	0.9752	0.9752	0.9720	0.9720
714	0.9904	0.9904	0.9751	0.9752	0.9719	0.9721
718	0.9891	0.9891	0.9749	0.9749	0.9715	0.9715
720	0.9894	0.9894	0.9719	0.9721	0.9696	0.9696
706	0.9893	0.9893	0.9714	0.9714	0.9694	0.9694
725	0.9892	0.9895	0.9711	0.9711	0.9694	0.9694
707	0.9876	0.9875	0.9667	0.9667	0.9680	0.9681

Continued

Table 3.6 Results of modified 37-bus system [27] for RBFNN-based algorithm and MLIPFA.—cont'd

	V_{ab} (pu)		V_{bc} (deg.)		V_{ca} (deg.)	
Bus no.	**MLIPFA [27,28]**	**RBFNN**	**MLIPFA [27,28]**	**RBFNN**	**MLIPFA [27,28]**	**RBFNN**
722	0.9875	0.9875	0.9662	0.9664	0.9679	0.9681
724	0.9873	0.9873	0.9658	0.9658	0.9679	0.9679
Max. V error (%)	V_{ab} (pu)		V_{bc} (deg.)		V_{ca} (deg.)	
Absolute	0.0003		0.0002		0.0002	
Relative	0.0303		0.0214		0.0218	
	MLIPFA [35,36]			RBFNN		
Iterations no.	7			4		
Used CPU time (s)	1.6187			0.7335		
Used time for each iteration (s)	0.231242857			0.183375		

DFIG, doubly fed induction generator; *MLIPFA*, modified ladder iterative power flow algorithm; *RBFNN*, radial basis function neural network.

Based on the obtained outcomes, it is obvious that the suggested RBFNN-based PFA is able to solve the ill-conditioned systems like radial networks with high R/X, interconnected distribution networks, and MGs, which are loaded heavily. It can simultaneously solve the combination of equations developed for droop-controlled ECDERs and other components, also in case of $n \neq m$. It has less CT and more efficient in comparison with different NR and GS methods. Also, it is reliable and is simple in terms of implementation.

4. Probabilistic power flow problem

Noticing the intermittent and uncertain nature of wind and solar irradiation, the WG units and PV systems can create crucial challenges in planning and operation phases of power systems. Also, plug-in hybrid electric vehicles (PHEVs), WG units, and PV panels can affect the reliability and security of MGs [2,43,45]. In MGs, the uncertainty in consumption, i.e., load side, cannot be neglected as well. Thus, the MG steady-state condition

Table 3.7 Results of modified IEEE 13-bus systems for RBFNN-based algorithm and NR method of Ref. [35].

	NR method [35]			RBFNN-based algorithm		
Phase	**a**	**b**	**c**	**a**	**b**	**c**
Operation of converter	Balanced					
$[V_{abc}]_{634}$	0.9259< −1.2856°	0.9353< −121.6087°	0.9296 < 118.557°	0.926< −1.2855°	0.9354< −121.6085°	0.9296 < 118.556°
$[V_{abc}]_{690}$	0.9203< −1.1295°	0.9724< −123.6124°	0.9521 < 116.7561°	0.9202< −1.1294°	0.9723< −123.6125°	0.9522 < 116.7561°
Operation of converter	Unbalanced					
$[V_{abc}]_{634}$	0.9292< −1.9347°	0.9538< −121.6106°	0.9324 < 118.8129°	0.9291< −1.9347°	0.9537< −121.6105°	0.9324 < 118.813°
$[V_{abc}]_{690}$	0.9246< −1.8103°	0.9803< −123.4495°	0.9504 < 116.9151°	0.9246< −1.8102°	0.9803< −123.4494°	0.9503 < 116.9150°
Phase	a		b		c	
	Mag.	Θ	Mag.	Θ	Mag.	Θ
Max. V error (%)	0.0001	0.0001	0.0001	0.0002	0.0001	0.0010
Absolute	0.0109	0.0089	0.0107	0.0002	0.0105	0.0008
Relative	0.0001	0.0001	0.0001	0.0002	0.0001	0.0010
	NR method [35]			RBFNN-based algorithm		
Iterations no.	7			4		
Used CPU time (s)	1.2139			0.6704		
Used time for each iteration (s)	0.1734			0.1676		

NR, Newton–Raphson; *RBFNN*, radial basis function neural network.

Table 3.8 Iterations No. to convergence for various algorithms considering the impact of *R/X* and load multiplier.

	Case study 1 [1]		Case study 2 [1]		Case study 3 6-bus system [16] [19]		Case study3 IEEE 38-bus network [19]		Case study 3 – Unbalanced PFA 25-bus network [19]		Case study 4 IEEE 37-bus network including WG [26]		Case study 5 IEEE 13-bus network including AC and DC subsystems [35]	
Scenarios	RBFNN	Two-step NR method [1]	RBFNN	2-step NR method [1]	RBFNN	MNR [19]	RBFNN	MNR [19]	RBFNN	MNR [19]	RBFNN	MLIPFA [27]-[28]	RBFNN	NR [35]
R/X							LM = 1							
1	3	5	3	6	3	4	3	5	4	6	4	7	4	7
1.15	3	5	3	6	3	4	4	7	4	8	4	9	4	8
1.3	4	6	4	7	3	5	7	9	5	9	5	9	6	9
1.45	4	7	4	7	4	5	9	11	7	9	7	10	8	9
1.6	4	7	5	8	4	6	10	15	9	11	9	12	9	10
1.75	6	8	5	8	5	8	11	17	9	14	11	13	9	12
1.9	7	8	6	8	5	9	12	23	11	14	12	14	11	14
2.05	7	9	6	9	6	11	13	–	13	16	16	18	14	17
2.2	8	9	7	9	7	13	15	–	14	–	16	24	14	21
2.35	8	10	8	10	9	–	15	–	14	–	19	26	16	25
2.5	10	–	9	12	9	–	19	–	19	–	21	–	17	–
2.65	14	–	9	–	14	–	–	–	22	–	23	–	22	–
2.8	19	–	14	–	19	–	–	–	–	–	–	–	–	–

LM	$R/X = 1$													
1	3	5	3	6	3	4	3	5	4	6	4	7	4	7
1.15	4	6	4	7	4	5	4	6	5	7	5	8	5	8
1.3	4	6	4	7	4	5	5	8	5	10	5	11	5	10
1.45	5	7	5	8	4	6	8	10	6	11	6	11	7	11
1.6	5	8	5	8	5	6	11	12	8	11	8	12	10	11
1.75	5	9	6	10	5	7	12	15	11	13	11	14	11	12
1.9	7	10	6	10	6	10	13	19	11	17	13	16	11	14
2.05	8	10	7	10	6	11	14	27	13	17	14	17	13	17
2.2	8	11	8	11	7	13	15	—	16	19	19	19	17	20
2.35	10	13	8	11	8	16	18	—	17	—	22	22	17	25
2.5	12	16	10	12	10	—	19	—	17	—	25	25	19	30
2.65	13	—	12	14	11	—	23	—	23	—	28	—	20	—
2.8	17	—	15	—	14	—	—	—	26	—	—	—	24	—

LM, load multiplier; *MLIPFA*, modified ladder iterative power flow algorithm; *MNR*, modified Newton–Raphson; *NR*, Newton–Raphson; *PFA*, power flow analysis; *R/X*, resistance to reactance; *RBFNN*, radial basis function neural network; *WG*, wind generation.

cannot accurately be determined using deterministic solutions (deterministic power flow algorithms) such as NR-based methods or forward–backward sweep method. Also, it is impossible to consider all possible states of an MG, including different combinations of loads, generations, and topology. Also, MGs have a high LM and R/X ratio, which must be handled by new robust algorithms [2]. To overcome this obstacle, probabilistic power flow (PPF) algorithms should be used with good accuracy and low computational effort [45–49]. They lead to more realistic results, also for electricity market participants [50].

To handle the power system uncertainties, various methods have been presented [48–75]. These methods can be divided into three groups: simulation-based methods [52,53], approximate solutions [55–59], and analytical approaches [60–62].

Monte Carlo simulation (MCS) is able to obtain acceptable solutions for random output variables of the PPF problem by means of a probability density function (PDF) or cumulative distribution function (CDF). It is straightforward, but it cannot be applied to MGs due to their high correlations.

To obtain a solution for the PPF, different methods such as point, two-point, 2n+1, and fast point estimate methods (PEM [55,56], 2PEM [57], 2n+1 PEM [58], and FPEM [59], respectively) have been used. They are efficient and simple, but they cannot obtain PDF for results, and increasing the uncertain variables leads to an increase in CT. Therefore, they are impractical for the problems, which have many components like MGs.

The analytical approaches are using some simplifications and also assumptions, which decrease the accuracy of PPF results. Also, they should calculate JM and its inverse. Also, integration of RERs in these methods and handling correlated variables are difficult. To overcome these issues, and analyze correlation among loads and resources, the well-known Nataf transformation has been proposed in Refs. [48,63–66], mean and covariance nonlinear transformation has been suggested, and in Ref. [67], a modified unscented transformation (UT) has been applied, which considers loads and WG system uncertainties. In Ref. [69], a nonparametric density estimation method, which is based on linear diffusion method [68], has been presented, and in Refs. [70,71], it has been used to model correlations among WG systems and specify the PDF and WG uncertainties. Also, other approaches can be seen in the literature, which have been used to solve the PPF [72–77]. In Refs. [78–80], conventional algorithms and forward–backward sweep method and, in Refs. [81,82], heuristic and multiobjective optimization algorithms have been used. A review on uncertainty modeling and PPF methods has been presented in Ref. [83], which can be used for further readings.

In the case of unavailability of sufficient data for variables, the PDF cannot be determined, and it can be represented possibilistically. In a system with probabilistic and possibilistic variables, the system cannot be studied by a pure probabilistic or pure possibilistic method, and a combination must be considered. A novel possibilistic PPF (PPPF) will be suggested in this section, which uses fuzzy inference system and possibility theory, and it does not need calculation of JM and its inverse. Thus, it can be used for PPF problem of heavily loaded MGs with high LM and R/X ratio. Here, Weibull and normal PDFs are, respectively, employed for wind speed and irradiation, and the loads are modeled by their PDFs. Then, the proposed PPPF algorithm is developed using an RBFNN and compared in various MGs with other PPF methods such as analytical, approximate, and heuristic methods and MCS.

4.1 Uncertain elements models in microgrid

4.1.1 Load probabilistic model

To model the load uncertainty, Weibull, beta, and normal PDFs have been suggested [84,85]. In this section, the normal distribution is used to study the load behavior, as follows [85]:

$$f(P_L) = \frac{1}{\sigma\sqrt{2\pi}} \exp\left(-\frac{(P_L - \mu)^2}{2\sigma^2} \right) \tag{3.61}$$

4.1.2 Distributed energy resource probabilistic model

The probabilistic models of WG and PV systems will be presented in the following subsections.

4.1.2.1 Wind generation probabilistic model

The Weibull PDF represents wind speed variations, as follows [81]:

$$f_v(v) = \begin{cases} \frac{\rho}{\alpha} \times \left(\frac{v}{\rho}\right)^{\rho-1} \times \exp\left(\left(\frac{v}{\alpha}\right) \times \rho\right) & v \geq 0 \\ 0 & \text{otherwise} \end{cases} \tag{3.62}$$

Using Eq. (3.62), the real power of the WG turbine is specified as follows:

$$P_{WTv}(v) = \begin{cases} N_{WT}P_{r,WT} \times \left(\dfrac{v - v_{cut-in}}{v_r - v_{cut-in}}\right) & v_{cut-in} \leq v \leq v_r \\ N_{WT}P_{r,WT} & v_r \leq v \leq v_{cut-out} \\ 0 & \text{otherwise} \end{cases} \tag{3.63}$$

4.1.2.2 Probabilistic model of photovoltaic

The beta PDF is used for the solar irradiation distribution function [85].

$$f(r: \alpha_\beta, \beta_\beta) = \frac{\Gamma(\alpha_\beta + \beta_\beta)}{\Gamma(\alpha_\beta)\Gamma(\beta_\beta)} r^{\alpha_\beta - 1}(1 - r)\beta_\beta \tag{3.64}$$

Then, using this equation, the PV output power is specified as follows [86]:

$$P_{PV}(r) = \begin{cases} N_{PV}P_{rs} \times \left(\dfrac{r^2}{R_{STD}R_C}\right) & 0 \leq r \leq R_C \\ N_{PV}P_{rs}\dfrac{r}{R_{STD}} & R_C \leq r \leq R_{STD} \\ N_{PV}P_{rs} & r \geq R_{STD} \end{cases} \tag{3.65}$$

To model the irradiation and air temperature, the normal PDF is employed. The module power can be calculated as follows [87]:

$$P_{PV} = N_{PV}P_{STC} \times \frac{G_{ING}}{G_{STC}} \times (1 + k(T_c - T_r)) \tag{3.66}$$

where in the standard condition, we have $G_{STC} = 1000\ \text{W/m}^2$ and $T_r = 25°\text{C}$.

4.1.3 Probabilistic model of plug-in hybrid electric vehicles

The charging and discharging of PHEVs are considered as stochastic variables. To model these vehicles at charging stations, queuing theory [87] and the model presented in Ref. [72] are used. Also, they are modeled by binomial PDF, which indicates that PHEVs are in charging and discharging modes, 90% and 10% of the times, respectively.

4.2 Probabilistic power flow problem

4.2.1 Deterministic power flow problem

The PFP is based on Eqs. (3.51) and (3.52), including the following inequality constraints of bus voltages at PQ buses and generators reactive power at PV buses [2].

$$V_i^{\min-PQ} \leq V_i^{PQ} \leq V_i^{\max-PQ} \tag{3.67}$$

$$Q_i^{\min-PV} \leq Q_i^{PV} \leq Q_i^{\max-PV} \tag{3.68}$$

In AC/DC MGs, the following power balance equation must be considered [2].

$$P_{DC,i} = [PG_{DC}]_i - [PD_{DC}]_i = V_{DC,i} \sum_{j=1}^{NDC} \left(YB_{ij} \cdot V_{DC,j} \right) \tag{3.69}$$

The integrated AC and DC equations are written as follows [2]:

$$\begin{bmatrix} (PG-PD)_{AC} \\ (QG-QD)_{AC} \\ \cdots \\ (PG-PD)_{DC} \end{bmatrix} = \begin{bmatrix} V_{AC} \\ V_{AC} \\ \cdots \\ V_{DC} \end{bmatrix} \left(\begin{bmatrix} YBUS_{AC} & \vdots & \\ \cdots & \cdots & \cdots \\ & \vdots & YBUS_{DC} \end{bmatrix} \begin{bmatrix} V_{AC} \\ V_{AC} \\ \cdots \\ V_{DC} \end{bmatrix} \right)^* \tag{3.70}$$

4.2.2 Probabilistic power flow

The results of the probabilistic power flow must be input variables function as given by Eq. (3.60). The input vector X and output vector Y are given by the following equations:

$$X = \begin{bmatrix} P_D & Q_D & P_{DER} & Q_{DER} & \ldots \end{bmatrix}^T \tag{3.71}$$

$$Y = \begin{bmatrix} V & \delta & P_{slack} & Q_{slack} & \ldots \end{bmatrix}^T \tag{3.72}$$

4.2.3 Correlation between two uncertain variables

When in a system, there is a dependency between x and y variables, a change in one leads to a change in the other one. The covariance matrix or correlation coefficient matrix represents this concept as follows:

$$\rho_{x,y} = \frac{\text{cov}(x,y)}{\sigma_x \sigma_y} = \frac{E\left[(x-\mu_x)(y-\mu_y)\right]}{\sigma_x \sigma_y}, x, y = 1, 2, \ldots, n \quad (3.73)$$

The correlation coefficient can be between −1 and −1, and it is equal to +1 and −1 for a full positive and negative linear relation, respectively [88]. For independent or uncorrelated variables, the correlation coefficient is equal to 0.

4.3 Suggested radial basis function neural network–based probabilistic power flow

4.3.1 Radial basis function neural network

The algorithms, which are based on RBFNN, will be modified in this section to use for solving the PPPF problem in MGs, including the models of uncertain elements.

4.3.2 Enhanced unscented transformation

As stated before, the MCS is a time-consuming algorithm and cannot be suggested for online applications. The other option, i.e., the 2PEM method decomposes Eq. (3.60) into subproblems and selects two (deterministic) values for each stochastic variable, which one of them is more and the other one is less than this variable mean value [57]. For these two values, the deterministic power flow is carried out two times. In this method, it is assumed that we have independent input variables; therefore, this method cannot be used for variables, which are correlated [57]. These correlated variables may be handled by the UT.

The UT and its variants, like fuzzy UT (FUT), must select appropriate samples of input variables. These samples should obtain sufficient information concerning the input variables.

Assume that $\boldsymbol{X}$ is an n-dimensional random vector and its mean value and covariance are $\boldsymbol{m}$ and $\boldsymbol{P_x}$, respectively. Also, assume that the random vector $\boldsymbol{Y}$ is related to $\boldsymbol{X}$ by f, which can be defined by Eq. (3.60). To get the

$\boldsymbol{Y}$ vector statistics, a matrix $\boldsymbol{\chi}$ of $2L+1$ sigma vectors (χ_i) considering their weights (W_i) must be written by the following equations:

$$\begin{aligned}\chi_0 &= m \\ \chi_i &= m + \left(\sqrt{L+\lambda}\,P_x\right)_i i = 1, \ldots, L \\ \chi_i &= m + \left(\sqrt{L+\lambda}\,P_x\right)_{i-L} i = L+1, \ldots, 2L\end{aligned} \tag{3.74}$$

$$\begin{aligned}W_0^{(m)} &= \lambda/(L+\lambda) \\ W_0^{(c)} &= \lambda/(L+\lambda) + \left(1-\alpha^2+\beta\right) \\ W_0^{(m)} &= W_0^{(c)} = 1/(2(L+\lambda)) i = 1, \ldots, 2L\end{aligned} \tag{3.75}$$

where β is applied to combine distribution prior knowledge and $\lambda = \alpha^2(L+\kappa)\text{-}L$ is used as a parameter for scaling parameter, which depends on α *and* κ. α denotes sigma points deployment around $\boldsymbol{m}$, which is generally adjusted to an amount (which must be positive and small, e.g., 10^{-3}), and κ represents secondary parameter for scaling generally adjusted to 0. Note that $\sum_{i=0}^{2L} W_i^{(c)} = 1$. $\left(\sqrt{(L+\lambda)P_x}\right)_i$ presents the matrix square root ith row. The square root of a positive definite matrix like $\boldsymbol{P}$ is defined by $A = \sqrt{P}$ such that $\mathrm{P} = \mathrm{AA}^{\mathrm{T}}$. The sigma vectors, as shown in Fig. 3.7, are propagated through the nonlinear function (3.60), as follows [90]:

$$y_i = f(\chi_i) i = 0, \ldots, 2L \tag{3.76}$$

The posterior sigma points that mean and covariance are used as an approximation for mean and covariance of $\boldsymbol{y}$, as follows:

$$\bar{y} \approx \sum_{i=0}^{2L} W_i^{(m)} y_i \tag{3.77}$$

$$P_y \approx \sum_{i-0}^{2L} W_i^{(c)} \{y_i - \bar{y}\} \cdot \{y_i - \bar{y}\}^T \tag{3.78}$$

4.3.3 Theory of possibility

The application of the probability theory can face the following challenges [89]:

- Any external variable may have an effect on other variables.
- Choosing a suitable PDF for uncertain variables based on inadequate or imprecise data is difficult.

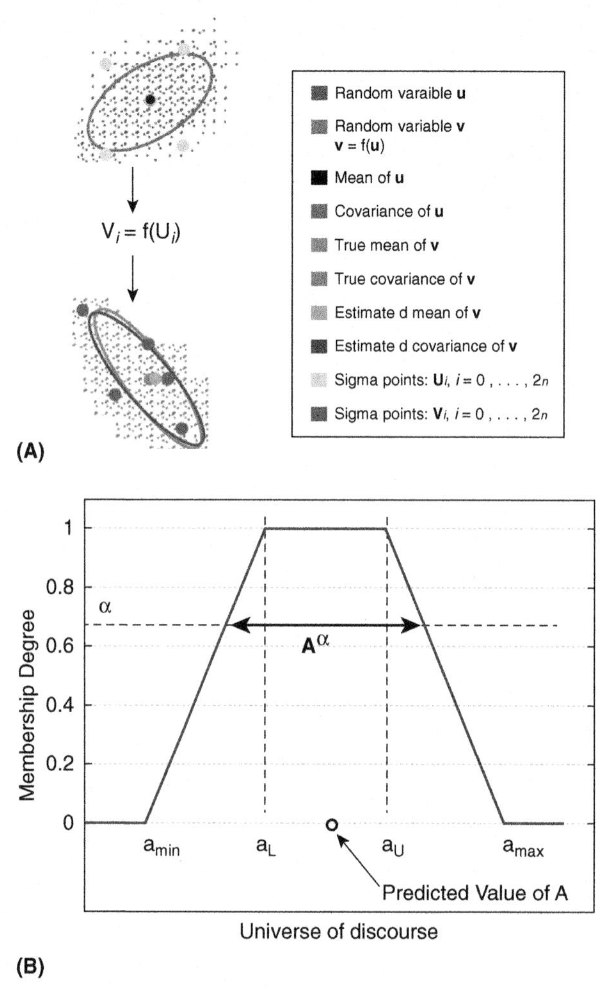

Figure 3.7 (A) Unscented transformation [90] and (B) a fuzzy trapezoidal membership function [89].

In this case, the possibility theory can be suggested as an effective option. In this theory, to model the epistemic uncertainty, the possibility of distribution $\pi_{\widetilde{X}}(x)$ is used for each uncertain value of $\widetilde{X}$. In this possibility distribution, $\pi_{\widetilde{X}}(x) = 0$ and $\pi_{\widetilde{X}}(x) = 1$, known as membership degree of 0 and 1, signify that x is an impossible and possible event, respectively. Also, the possible function is used for the possibility and necessity measures (possibility bounds). This measure for event A is denoted by $Pos(A)$ and given by the following equation [89,90]:

$$Pos(A) = \sup_{[x\in A]} \pi_{\widetilde{X}}(x) \tag{3.79}$$

where *inf(.)* and *sup(.)* are understudy gap lower and upper limits. As a result, the event necessity measure, denoted by *Nec(A)*, is determined by the following equation:

$$Nec(A) = 1 - Pos(A) = 1 - \sup_{[x\in A]} \pi_{\widetilde{X}}(x) \tag{3.80}$$

Different kinds of membership functions have been presented, to characterize the membership degrees of possibilistic uncertain variables. As shown in Fig. 3.7B, fuzzy trapezoidal numbers are used in this chapter with the notation A = $\{a_{min}, a_l, a_u, a_{max}\}$.

4.3.4 Alpha-cut method

In case of having a known possibility distribution for an uncertain input variable, i.e., $\widetilde{X}$, the possibility distribution of $\widetilde{Y}$ is determined using the α-cut method [89,90]. In this method, the α-cut of $\widetilde{X}$ is defined by the following equation [89,90]:

$$A^{\alpha} = \left\{x\in U \middle| \pi_{\widetilde{X}}(x) \geq \alpha, 0 \leq \alpha \leq 1\right\} \tag{3.81}$$

$$A^{\alpha} = \left\lfloor A^{\alpha}_{-}, A^{-\alpha}\right\rfloor \tag{3.82}$$

where U is the range of possible values of $\widetilde{X}$, also known as the universe of discourse. Lower and upper bands of the A^{α} are A^{α}_{-} and $A^{-\alpha}$, respectively. Fig. 3.7B shows an alpha-cut of a trapezoidal membership function. The α-cut of $\widetilde{Y}$ is specified by α-cut of $\widetilde{X}$, as follows [89,90]:

$$Y^{\alpha} = \left[Y^{\alpha}_{-}, Y^{-\alpha}\right] \tag{3.83}$$

$$Y^{\alpha}_{-} = \inf\left\lfloor h\left(F^{\alpha}_{\widetilde{X}_1}, F^{\alpha}_{\widetilde{X}_2}, \ldots, F^{\alpha}_{\widetilde{X}_N}\right)\right\rfloor \tag{3.84}$$

$$Y^{-\alpha} = \inf\left\lfloor h\left(F^{\alpha}_{\widetilde{X}_1}, F^{\alpha}_{\widetilde{X}_2}, \ldots, F^{\alpha}_{\widetilde{X}_N}\right)\right\rfloor \tag{3.85}$$

where $F^{\alpha}_{\widetilde{X}_1}$ is i-th possibilistic input variable alpha-cut.

4.3.5 Defuzzification

The centroid method will be employed here for conversion and defuzzification of a fuzzy number into a crisp one. The fuzzy quantity, i.e. $\widetilde{X}$, is converted to the defuzzied one by the following equation:

$$X^* = \frac{\int \pi_{\widetilde{X}}(x) x \mathrm{d}x}{\int \pi_{\widetilde{X}}(x) \mathrm{d}x} \tag{3.86}$$

4.3.6 Proposed algorithm

As mentioned before, the suggested method can be applied to NLES with $n \neq m$, and this approach does not need to calculate JM and its inverse. The proposed algorithm training algorithm aims at finding X^* such that $Y^* = f(X = X^*)$. The suggested solution is carried out in four steps, as follows:

Step 1: Generate $2L+1$ points using Eq. (3.74).

Step 2: Compute weights for each $\mathbf{X}$ using Eq. (3.75).

Step 3: Apply generated points one-by-one to (3.60) as follows:

Step 3.1: For $\alpha = 0, \Delta\alpha, 2\Delta\alpha, \ldots, 1$ ($\Delta\alpha$ is the step size of the α-cut method).

Step 3.1.1: The corresponding alpha-cuts of possibility distributions $\left(\pi_{\widetilde{X}k+1}(x), \ldots, \pi_{\widetilde{X}N}(x)\right)$, as the intervals of possibilistic variables $\left(\widetilde{X}_{k+1}, \ldots, \widetilde{X}_N\right)$, are calculated.

Step 3.1.2: The output maximum and minimum of the model $h\left(\widetilde{X}_1, \widetilde{X}_2, \ldots \widetilde{X}_k, \widetilde{X}_{k+1}, \widetilde{X}_N\right)$, denoted by h_{j-}^{α} and $h_j^{-\alpha}$ are, respectively, determined. Here, the probabilistic variables are set to the values generated in step 1$\left(x_1^j, x_2^j, \ldots, x_k^j\right)$. In contrast, the possibilistic variables can have all the values of the α-cut defined in Step 3.1.1. It should be noted that h is the PFP presented in Eq. (3.60). Each sample point is applied to this equation in the following substeps.

Step 3.1.2.1: The random vectors X_i are generated in the feasible space of the problem. As mentioned before, these vectors contain the MG and DERs variables.

Step 3.1.2.2: X_i vectors generate vectors Y_i using the mapping of f.

Step 3.1.2.3: Vectors Y_i and X_i are applied to RBFNN for its training.

Step 3.1.2.4: The vector X^0 as the output is provided using the RBFNN and Y^* as the input. The kernel generates the least response at the output and is selected as the input vector farthest center. This center (input, output) sets may be employed as the replaced set in step 3.1.2.4.

Step 3.1.2.5: After mapping X^0 by f, we have $Y^0 = f(X^0)$, and $Y^0 - Y^*\|^2$ can be determined. If this value is less than a prespecified small value (ε), the procedure will stop. In this case, X^0 should be the problem solution. Otherwise, the procedure should pass to step 3.1.2.6.

Step 3.1.2.6: When Y^0 is closer to Y^* than the other Y_is, one of the (input, output) pairs (Y_i, X_i) must be replaced. Now, the algorithm should pass to step 3.1.2.3. It must be noted that in case of violation of maximum or minimum values of DER operational parameters, their amounts must be fixed to the maximum or minimum values, and then the procedure must go to step 3.1.2.5.

Step 3.1.3: The values of h_{j-}^{α} and $h_j^{-\alpha}$, i.e., the lower and upper bounds of the alpha-cuts of the output variables $h(x_1^j, \dots x_k^j, \widetilde{X}_{k+1}, \widetilde{X}_N)$ must be preserved.

Step 3.2: The higher and lower bounds of various alpha-cuts of $h(x_1^j, \dots x_k^j, \widetilde{X}_{k+1}, \widetilde{X}_N)$ are employed for approximation of the possibility distribution of the output, i.e., π_j^Y. This denotes that the output variable possibility distribution can be determined by the following:

- Setting the probabilistic variables to j-th sampling vector
- Applying the fuzzy alpha-cuts on possibilistic variables
- Using transformation function for the sample points
- Saving h_{j-}^{α} and $h_j^{-\alpha}$

Step 4: Use Eqs. (3.77) and (3.78), to compute $\boldsymbol{Y}$ mean and covariance.

The same as a conventional power flow, due to reactive power limitations, a PV bus can be changed to a PQ bus. In this case, the algorithm must rearrange the PV and PQ buses and then moves to step 2.

In the case of combining the UT and the RBFNN, steps 3.1.1, 3.1.2, 3.1.3, and 3.2 must be removed, and the program will be carried out without using any fuzzification and can be considered as PPF.

4.4 Simulation results

In this subsection, based on the suggested PPF/PPPF algorithm, the program, developed in MATLAB environment, will be used to study various test networks. In all the simulations, the tolerance, ε, is 10^{-6}, and the maximum number of iterations is 300 for each run.

4.5 Six-Bus system with wind generation system and photovoltaic panel

In this subsection, the six-bus grid, with the parameters given in Ref. [91], is tested using the suggested algorithm, which is based on UT/RBFNN. The loads have normal PDF. Their mean value equal to those of Ref. [91], and their standard deviation (STD) is 5% of their mean values. In this case study, we have four uncertain variables, and two cases will be studied.

In the first case, the network does not have any WG system, and in the second one, bus 4 has a WG system, modeled using the data given in Ref. [67].

The output variables mean value and STD are listed in Table 3.9 for two cases. It can be seen that the application of the WG system reduces the losses but increases the STD. To investigate correlated variables effect on the results, the same as Ref. [67], it is assumed that the loads are correlated with a coefficient of +0.3, and the WG is correlated to the loads at bus 4 with coefficient of −0.25. It can be seen that for loads, which are positively correlated, the random variables have more variations. In the case of negatively correlated coefficients, the STD of $|V4|$ is fewer than the uncorrelated ones.

Table 3.10 summarizes the comparison results of the suggested PPF algorithm, which is based on UT/RBFNN with PEM [81], and MCS methods. The WG and PV systems data can be found in Ref. [81]. In this comparison, the first case is the one with uncorrelated variables. In the correlated case, the coefficient matrix of buses 4, 5, and 6 loads, WG system, and PV unit connected to bus 6 is as follows [81]:

$$\rho = \begin{bmatrix} 1 & 0.5 & 0.3333 & 0.25 & 0.2 \\ 0.5 & 1 & 0 & 0 & 0 \\ 0.3333 & 0 & 1 & 0.0833 & 0.0667 \\ 0.25 & 0 & 0.0833 & 1 & 0.2 \\ 0.2 & 0 & 0.0667 & 0.2 & 1 \end{bmatrix} \tag{3.87}$$

Table 3.9 Comparison of results of six-bus power system between suggested PPF and UT algorithms [67].

	Uncorrelated variables								Correlated variables			
	WG-less				Including WG							
	UT [67]		Suggested PPF method		UT [67]		Suggested PPF method		UT [67]		Suggested PPF method	
Parameter	μ	Σ	μ	σ	μ	σ	μ	σ	μ	σ	μ	σ
Losses(MW)	7.885	0.362	7.883	0.359	7.572	0.6854	7.57	0.6854	7.48	0.754	7.46	0.753
$P_{1\text{-}2}$(MW)	28.69	1.28	28.68	1.29	27.589	2.38	27.589	2.38	27.82	4.95	27.81	4.94
$P_{1\text{-}4}$(MW)	43.58	1.29	43.59	1.3	42.03	3.11	42.03	3.11	42.35	8.14	42.36	8.13
$P_{1\text{-}5}$(MW)	33.6	1.09	33.61	1.09	34.94	1.62	34.94	1.62	35.04	2.56	35.04	2.57
$P_{2\text{-}4}$(MW)	33.09	0.952	33.08	0.954	31.81	2.5	31.81	2.5	32.02	5.49	32.04	5.48
$P_{4\text{-}5}$(MW)	4.08	0.44	4.09	0.46	4.46	0.8137	4.46	0.8137	4.38	0.892	4.381	0.893
$\|V_1\|$(p.u.)	1.05	0	1.05	0	1.05	0	1.05	0	1.05	0	1.05	0
$\|V_2\|$(p.u.)	1.05	0	1.06	0	1.05	0	1.05	0	1.05	0	1.05	0
$\|V_3\|$(p.u.)	1.07	0	1.07	0	1.07	0	1.07	0	1.07	0	1.07	0
$\|V_4\|$(p.u.)	0.9894	0.0017	0.9893	0.0019	0.992	0.0051	0.992	0.0051	0.9918	0.0032	0.9917	0.003
$\|V_5\|$(p.u.)	0.9854	0.0019	0.9854	0.0018	0.985	0.002	0.985	0.002	0.9854	0.0021	0.9859	0.002
$\|V_6\|$(p.u.)	1.004	0.0015	1.004	0.0017	1.004	0.0015	1.004	0.0015	1.0045	0.0016	1.0046	0.0018

PPF, probabilistic power flow; *UT*, unscented transformation; *WG*, wind generation.

Table 3.10 Suggested, MCS, and 2PEM methods simulation results for six-bus power system [67].

Cases	Uncorrelated						Correlated					
Methods	**MCS method**		**2PEM [67]**		**Suggested**		**MCS method**		**2PEM [67]**		**Suggested**	
Parameter	μ	σ	μ	σ	μ	σ	μ	σ	μ	σ	μ	σ
Losses(MW)	6.6912	0.7738	6.6759	0.7232	6.6838	0.7765	6.6985	0.8359	6.6758	0.7939	6.7092	0.857
P_{1-2}(MW)	23.8718	3.0334	23.8266	2.8511	23.85	3.0430	23.8992	3.2338	23.8300	3.0935	23.9216	3.3172
P_{1-4}(MW)	37.8101	3.6396	37.7445	3.3418	37.7793	3.6578	37.8378	3.7069	37.7471	3.4778	37.8835	3.8012
P_{1-5}(MW)	32.4588	2.3399	32.4424	2.2521	32.4583	2.3459	32.4915	2.5650	32.4430	2.4932	32.5158	2.6016
P_{2-3}(MW)	2.3411	0.8654	2.3359	0.8613	2.3386	0.8656	2.3362	0.8302	2.3360	0.8265	2.3363	0.832
P_{2-4}(MW)	30.3178	2.8199	30.2541	2.5778	30.2865	2.8325	30.3209	2.2577	30.2554	2.0572	30.3304	2.3616
P_{2-5}(MW)	15.8146	0.7210	15.7982	0.7170	15.8087	0.7212	15.7995	0.6406	15.7984	0.6395	15.8000	0.6411
P_{2-6}(MW)	24.8029	1.6172	24.8025	1.6110	24.8027	1.6172	24.8044	1.5896	24.8026	1.5794	24.8048	1.5942
P_{3-5}(MW)	19.8166	0.7327	19.8016	0.7304	19.8094	0.7328	19.8060	0.7160	19.8020	0.7111	19.8080	0.7182
P_{3-6}(MW)	42.5058	1.2043	42.4959	1.2029	42.5010	1.2044	42.4998	1.2091	42.4957	1.1986	42.5019	1.2144
P_{4-5}(MW)	4.9486	1.0515	4.9216	0.9981	4.9430	1.0542	4.9334	0.7834	4.9200	0.7427	4.9402	0.8023
P_{5-6}(MW)	0.4540	0.9891	0.4384	0.9885	0.4463	0.9891	0.4407	0.9917	0.4381	0.9843	0.4414	0.9924
Pg1(MW)	94.1407	8.8283	94.0135	8.2621	94.0791	8.8584	94.2284	9.4015	94.0201	8.9653	94.2306	9.6263
Max. Error in comparison with MCS (%)			3.4451	8.5858	1.7091	0.4465	—	—	0.5841	8.8807	0.2930	4.6794

2PEM, two-point estimate method; *MCS*, Monte Carlo simulation.

Considering the simulation results for the correlated case, an increase in system losses and the mean (μ) and STD (σ) of the power generated at slack bus can be seen due to the correlation between DERs.

The error index is defined by Eq. (3.88) to study the accuracy of the suggested algorithm. This error is compared concerning the MCS method as the base index.

$$\varepsilon_l^j = \left| \frac{100\left(\ell_{MCS}^j - \ell_{Method}^j\right)}{\ell_{MCS}^j} \right| (\%) \tag{3.88}$$

In comparison with the MCS, it can be seen in Table 3.10 that the PPF problem has been solved by the suggested algorithm with higher accuracy.

4.5.1 14-Bus grid containing wind generation systems and plug-in hybrid electric vehicle charging station

As a test system, the modified IEEE 14-bus grid is used here. This modified grid has two WG systems modeled at buses 3 and 4, using the data given in Ref. [49]. The generations and loads have the normal PDF with σ-s equal to 5% of μ-s given in Ref. [91]. Bus 4 load sustainability is considered as 70% of the times determined by a binomial PDF. The WG system active power is determined by Eq. (3.63), and it is assumed that it can inject reactive power, uniformly distributed between 0.8 and 0.85. Also, the wind speed has a Weibull distribution. It is assumed that we have a charging station for PHEVs modeled at bus 4 of the test system using the date of Ref. [87]. Its probabilistic charging model can be found in Ref. [72]. The PHEVs have binomial PDF, i.e., 90% of the time; they consume power; and 10% of the time, they inject power to the system.

The test system is studied using the suggested PPF method, which is based on UT/RBFNN, and other methods, including the PPF method presented in Ref. [49], which is based on Parzen's window density estimator. In the test system, we have 30 uncertain variables. Their features are listed in Table 3.11 for uncorrelated and correlated variables with the coefficient matrix presented by Eq. (3.89) [49]. The correlated variables are bus 3 WG and injected active and reactive powers (i.e., ($P_{wf\text{-}3}$, $Q_{wf\text{-}3}$) and

Table 3.11 Comparison of results of 14-bus grid between suggested PPF and other methods.

Variables types		Uncorrelated						Correlated					
Methods	Parameters	$\lvert V_4 \rvert$ (p.u.)	$\lvert V_5 \rvert$ (p.u.)	P_{1-2} (p.u.)	P_{g-1} (p.u.)	Q_{4-5} (p.u.)	δ_2 (°)	$\lvert V_4 \rvert$ (p.u.)	$\lvert V_5 \rvert$ (p.u.)	P_{1-2} (p.u.)	P_{g-1} (p.u.)	Q_{4-5} (p.u.)	δ_2 (°)
MCS	Mean	1.0240	1.0240	1.2830	1.9150	0.1500	−4.0300	1.0180	1.0200	0.9500	1.4300	0.1800	−2.9100
	STD	0.0040	0.0030	0.1750	0.2600	0.0180	0.5800	0.0100	0.0100	0.1800	0.2470	0.0200	0.5600
2n+1 PEM [58]	Mean	1.0240	1.0240	1.2820	1.9140	0.1500	−4.0300	1.0180	1.0300	1.1000	1.4100	0.1800	−2.9000
	STD	0.0030	0.0020	0.1750	0.2600	0.0210	0.5800	0.0030	0.0050	0.2000	0.3000	0.0300	0.6700
2n PEM [57]	Mean	1.0240	1.0240	1.2800	1.9200	0.1500	−4.0400	—	—	—	—	—	—
	STD	0.0030	0.0020	0.1760	0.2600	0.0210	0.5800	—	—	—	—	—	—
Diffusion [69]	Mean	1.0240	1.0240	1.2850	1.9300	0.1600	−4.2300	1.0170	1.0200	0.9500	1.4300	0.1850	−2.9500
	STD	0.0040	0.0020	0.1810	0.2900	0.0200	0.5900	0.0020	0.0020	0.1710	0.3120	0.0300	0.5400
PW [49]	Mean	1.0240	1.0240	1.2700	1.9200	0.1500	−4.0200	1.0170	1.0200	0.9500	1.4260	0.1700	−2.9000
	STD	0.0040	0.0030	0.1850	0.2700	0.0170	0.5950	0.0050	0.0030	0.1800	0.2230	0.0200	0.5500
Suggested UT/ RBFNN PPF	Mean	1.0240	1.0240	1.2810	1.9150	0.1500	−4.0200	1.0180	1.0200	0.9500	1.4300	0.1800	−2.9000
	STD	0.0040	0.0030	0.1750	0.2620	0.0181	0.5800	0.0100	0.0090	0.1800	0.2410	0.0300	0.5400
Suggested FUT/ RBFNN PPF	Mean	1.0240	1.0240	1.2820	1.9150	0.1501	−4.0200	1.0180	1.0200	0.9502	1.4300	0.1800	−2.9000
	STD	0.0040	0.0030	0.1750	0.2610	0.0181	0.5802	0.0100	0.0090	0.1802	0.2430	0.0250	0.5400

FUT, fuzzy unscented transformation; *MCS*, Monte Carlo simulation; *PEM*, point estimate method; *PPF*, probabilistic power flow; *PW*, Parzen's window; *RBFNN*, radial basis function neural network; *STD*, standard deviation; *UT*, unscented transformation.

(P_3, Q_3), respectively) and bus 4 WG and charging station active and reactive powers (i.e., ($P_{wf\text{-}4}$, $Q_{wf\text{-}4}$) and ($P_{PHEV\text{-}4}$, $Q_{PHEV\text{-}4}$), respectively).

$$\rho = \begin{bmatrix} 1 & 0.8 & -0.5 & -0.5 & 0.5 & 0.5 & -0.2 & -0.2 \\ 0.8 & 1 & -0.5 & -0.5 & 0.5 & 0.5 & -0.2 & -0.2 \\ -0.5 & -0.5 & 1 & 0.5 & -0.2 & -0.2 & 0.1 & 0.1 \\ -0.5 & -0.5 & 0.5 & 1 & -0.2 & -0.2 & 0.1 & 0.1 \\ 0.5 & 0.5 & -0.2 & -0.2 & 1 & 0.8 & -0.5 & -0.5 \\ 0.5 & 0.5 & -0.2 & -0.2 & 0.8 & 1 & -0.5 & -0.5 \\ -0.2 & -0.2 & 0.1 & 0.1 & -0.5 & -0.5 & 1 & 0.5 \\ -0.2 & -0.2 & 0.1 & 0.1 & -0.5 & -0.5 & 0.5 & 1 \end{bmatrix} \tag{3.89}$$

Using Eq. (3.88), Table 3.12 lists various solutions average error of μ and σ for magnitude and phase of voltage (i.e., V and δ, respectively), lines powers (P_{br}, Q_{br}), and generation powers (P_g, Q_g). It is obvious that the suggested algorithm leads to higher accuracy in solving the PPF problem.

4.5.2 118-Bus grid including wind generation systems

The IEEE 118-bus grid has been changed for simulation studies [87]. Four WG systems of Ref. [49] have been added to this system at buses 28, 29, 78, and 79. The characteristics of the generation and loads are the same as in the previous section. The slack bus is bus 69 with the base of $S_{base} = 100\ MVA$. The test system has random variables. It is assumed that we have a binomial PDF for the load modeled at the bus 47, and the sustainability is equal to 95. Comparison results of the PFA of the modified grid between the suggested PPF method, which is based on UT/RBFNN, and other methods are presented in Table 3.13. The various solutions average error is listed in Table 3.14. It must be noted that in this case, the coefficient matrix is the same as the previous section. Considering the results presented in these two tables, it can be said that the suggested solution has better performance in comparison with other methods.

4.5.3 Comparison of computation time in different methods

Table 3.15 summarizes the CT of the suggested solution with other PPF algorithms. The developed program has used MATLAB 7.11.0 version and

Table 3.12 IEEE 14-bus power system average errors of suggested and other algorithms.

Variables types		Uncorrelated						Correlated					
Methods	**Parameters**	**Voltage mag.**	**Voltage phase.**	**P_{br}**	**Q_{br}**	**P_g**	**Q_g**	**Voltage mag.**	**Voltage phase.**	**P_{br}**	**Q_{br}**	**P_g**	**Q_g**
2n+1 PEM [58]	$\overline{\varepsilon_\mu}$	0.0019	0.2279	0.373	0.8335	0.337	1.47	0.002	0.226	0.37	0.853	1.07	1.46
	$\overline{\varepsilon_\sigma}$	0.2665	0.11	0.141	0.3431	0.041	0.252	0.268	0.107	0.12	0.333	0.818	
2n PEM [57]	$\overline{\varepsilon_\mu}$	0.002	0.235	0.253	0.3	0.276	0.7	—	—	—	—	—	—
	$\overline{\varepsilon_\sigma}$	145	40	148	15	29	14.8	—	—	—	—	—	—
Diffusion [69]	$\overline{\varepsilon_\mu}$	0.002	0.2	0.25	0.71	0.24	0.78	0.002	0.154	0.227	0.751	0.841	1.02
	$\overline{\varepsilon_\sigma}$	0.157	0.14	0.112	0.41	0.05	0.194	0.154	0.102	0.114	0.123	0.712	0.214
PW [49]	$\overline{\varepsilon_\mu}$	0.002	0.0042	0.008	0.0024	0.006	0.033	0.002	0.009	0.009	0.021	0.016	0.013
	$\overline{\varepsilon_\sigma}$	0.025	0.0348	0.031	0.042	0.083	0.086	0.031	0.047	0.035	0.063	0.005	0.025
Suggested UT/RBFNN PPF	$\overline{\varepsilon_\mu}$	0.002	0.031	0.005	0.003	0.52	0.061	0.002	0.007	0.008	0.02	0.019	0.01
	$\overline{\varepsilon_\sigma}$	0.019	0.031	0.029	0.038	0.069	0.071	0.023	0.037	0.031	0.055	0.003	0.03
Suggested FUT/RBFNN PPF	$\overline{\varepsilon_\mu}$	0.002	0.030	0.005	0.003	0.52	0.060	0.002	0.007	0.008	0.02	0.018	0.01
	$\overline{\varepsilon_\sigma}$	0.019	0.030	0.028	0.037	0.067	0.071	0.022	0.037	0.030	0.054	0.003	0.03

FUT, fuzzy unscented transformation; *PEM*, point estimate method; *PPF*, probabilistic power flow; *PW*, Parzen's window; *RBFNN*, radial basis function neural network; *UT*, unscented transformation.

Table 3.13 Comparison of results of modified 118-bus grid between suggested and other algorithms.

Type of variables		Uncorrelated						Correlated					
Methods	**Parameters**	$\|V_{47}\|$ (p.u.)	$\|V_{78}\|$ (p.u.)	$P_{47\text{-}49}$ (p.u.)	$P_{g\text{-}69}$ (p.u.)	$Q_{78\text{-}79}$ (p.u.)	δ_{89} (°)	$\|V_{47}\|$ (p.u.)	$\|V_{78}\|$ (p.u.)	$P_{47\text{-}49}$ (p.u.)	$P_{g\text{-}69}$ (p.u.)	$Q_{78\text{-}79}$ (p.u.)	δ_{89} (°)
MCS	Mean	1.017	1.005	−0.1	4.6	−0.21	−10.38	1.017	1.003	−0.12	4	−0.12	−11.4
	STD	0.001	0.0007	0.006	0.685	0.02	2.69	0.001	0.0004	0.06	0.71	0.015	2.7
2n+1 PEM [58]	Mean	1.017	1.005	−0.08	4.92	−0.22	−10.2	1.017	1.007	−0.09	4.1	−0.12	−10.9
	STD	0.001	0.0003	0.005	0.72	0.013	2.69	0.001	0.001	0.06	0.8	0.07	2.7
2n PEM [57]	Mean	1.017	1.005	−0.08	4.9	−0.22	−10.1	—	—	—	—	—	—
	STD	0.001	0.0003	0.004	0.75	0.015	2.59	—	—	—	—	—	—
Diffusion [69]	Mean	1.01	1.005	−0.08	4.81	−0.22	−10.21	1.017	1.004	−0.1	4.05	−0.11	−11.24
	STD	0.002	0.0005	0.05	0.71	0.025	2.71	0.001	0.002	0.057	0.77	0.04	2.61
PW [49]	Mean	1.017	1.005	−0.1	4.7	−0.21	−10.07	1.017	1.003	−0.11	4	−0.12	−11.53
	STD	0.001	0.0007	−0.06	0.693	0.024	2.6	0.001	0.004	0.055	0.75	0.014	2.69
Suggested UT/ RBFNN PPF	Mean	1.017	1.005	−0.1	4.57	−0.2	−10.38	1.017	1.003	−0.115	4	−0.12	−11.38
	STD	0.001	0.0007	0.005	0.684	0.01	2.62	0.001	0.0004	0.058	4	0.013	2.69
Suggested FUT/ RBFNN PPF	Mean	1.016	1.004	−0.1	4.56	−0.2	−10.36	1.016	1.003	−0.114	4	−0.12	−11.36
	STD	0.001	0.0007	0.005	0.682	0.01	2.62	0.001	0.0004	0.056	4	0.013	2.68

FUT, fuzzy unscented transformation; *MCS*, Monte Carlo simulation; *PEM*, point estimate method; *PPF*, probabilistic power flow; *PW*, Parzen's window; *RBFNN*, radial basis function neural network; *STD*, standard deviation; *UT*, unscented transformation.

Table 3.14 Comparison of average errors of 118-bus grid between proposed and other algorithms.

Type of variables		Uncorrelated						Correlated					
Methods	**Parameters**	**Voltage mag.**	**Voltage phase.**	P_{br}	Q_{br}	P_g	Q_g	**Voltage mag.**	**Voltage phase.**	P_{br}	Q_{br}	P_g	Q_g
2n+1 PEM [58]	$\overline{\varepsilon_\mu}$	0.0004	0.4	0.125	0.24	0.03	0.05	0.0005	0.6	0.146	0.3	0.03	0.07
	$\overline{\varepsilon_\sigma}$	0.147	0.214	0.227	0.231	0.01	0.14	0.155	0.264	0.278	0.281	0.011	0.195
2n PEM [57]	$\overline{\varepsilon_\mu}$	0.0007	0.6	0.245	0.581	0.8	0.12	—	—	—	—	—	—
	$\overline{\varepsilon_\sigma}$	3.241	1.478	2.671	2.7	1.24	1.345	—	—	—	—	—	—
Diffusion [69]	$\overline{\varepsilon_\mu}$	0.0002	0.3	0.11	0.3	0.04	0.15	0.0004	0.51	0.16	0.3	0.04	0.05
	$\overline{\varepsilon_\sigma}$	0.12	0.15	0.21	0.17	0.02	0.07	0.125	0.185	0.31	0.254	0.12	0.341
PW [49]	$\overline{\varepsilon_\mu}$	0.0005	0.013	0.09	0.1	0.02	0.34	0.0003	0.025	0.16	0.1	0.04	0.43
	$\overline{\varepsilon_\sigma}$	0.041	0.03	0.035	0.045	0.038	0.048	0.053	0.077	0.051	0.054	0.046	0.06
Suggested UT/RBFNN PPF	$\overline{\varepsilon_\mu}$	0.0003	0.011	0.05	0.091	0.02	0.31	0.0002	0.02	0.15	0.09	0.03	0.39
	$\overline{\varepsilon_\sigma}$	0.032	0.02	0.03	0.027	0.034	0.039	0.048	0.072	0.04	0.044	0.041	0.05
Suggested FUT/ RBFNN PPF	$\overline{\varepsilon_\mu}$	0.0003	0.011	0.05	0.091	0.02	0.31	0.0002	0.02	0.15	0.09	0.03	0.39
	$\overline{\varepsilon_\sigma}$	0.032	0.02	0.03	0.027	0.034	0.039	0.048	0.072	0.04	0.044	0.041	0.05

FUT, fuzzy unscented transformation; *PEM*, point estimate method; *PPF*, probabilistic power flow; *PW*, Parzen's window; *RBFNN*, radial basis function neural network; *UT*, unscented transformation.

Table 3.15 Comparison of CT of different test systems for suggested and other algorithms.

Methods	Studied grids				
	6-Bus grid	14-Bus grid	118-Bus grid	33-Bus grid	574-Node grid
MCS	32.25	1129.60	1243.83	3762.022	211.8525
2n+1 PEM [20]	–	1.7092	10.7348	–	–
2n+1 PEM [58]	0.6089	1.7309	11.4917	–	14.4103
PW [49]	–	1.9013	3.0971	–	–
Gram–Charlie [61]	–	–	0.1315	–	–
Diffusion [69]	–	–	8.8155	–	–
UT [67]	0.6179	–	–	–	14.75420
ICA method [17]	–	–	–	385.9237	
Suggested UT/RBFNN PPF	0.0806	0.1023	0.1652	0.2345	1.269
Suggested FUT/ RBFNN PPF	0.081	0.1024	0.1654	0.2345	1.272

FUT, fuzzy unscented transformation; *MCS*, Monte Carlo simulation; *PEM*, point estimate method; *PPF*, probabilistic power flow; *PW*, Parzen's window; *RBFNN*, radial basis function neural network; *UT*, unscented transformation.

has been carried out on Pentium IV (CPU 2.2 GHz-RAM 1.96 GB). It can be seen that the suggested algorithm has almost less CT in comparison with other methods.

4.6 Main result

To solve the PPF problem, a novel algorithm was suggested and applied to different grids. Comparison of simulation results shows that the proposed solution, which is based on FUT/RBFNN, had higher accuracy and less CT. This algorithm does not need calculation of the JM and its inverse, and therefore, it has less CT with good accuracy and robustness for MGs with high LM and R/X ratio. It must be mentioned that the proposed nonlinear approach may be applied to other PFA algorithms, which will be discussed in the next subsections.

5. Extension of harmonic power flow

5.1 Proposed radial basis function neural network–based harmonic power flow

In this subsection, the solution method, which is based on RBFNN, is enhanced to solve the PFP known as problem of harmonic power flow (HPF).

5.1.1 Harmonic power flow problem

The conventional PFP must be enhanced to include the equations of nonlinear devices (NLDs) in the HPF problem. Using this formulation, the power consumption must be calculated considering the fundamental and harmonics voltages.

$$S_i = \sum_{k=1,3,...} \overline{V}_i^k \left(\overline{I}_i^k\right)^* (i = g + 1, ..., n) \tag{3.90}$$

To solve the HPF, two procedures have been suggested [92]. The first one is based on complete harmonic power flow ($CHPF_F$), which considers the solution at the fundamental frequency. The other one used in this subsection is the fundamental frequency, and harmonic frequencies (known as $CHPF_H$) are considered. The inputs and unknowns can be seen in Table 3.16. The following equations form the HPF [92]:

- As PV-bus constraint, we have

Table 3.16 Summary of $CHPF_F$ and $CHPF_H$ [92] (slack bus, $i = 1$; PV buses, $i = 2, ..., g$; PQ buses, $i = g+1, ..., c$, buses with NLD, $i = c+1, ..., N$ and h and k, odd harmonic orders.).

	$CHPF_F$	$CHPF_H$		
Type of buses	**Input**	**Unknowns**	**Input**	**Unknowns**
Slack	$\overline{V}_1^1$	$\overline{V}_1^h$	$\overline{V}_1^1$	$\overline{V}_1^h$
PV	P_i, U_i	$\overline{V}_i^k$	P_i, U_i	$\overline{V}_i^k$
PQ	P_i, Q_i	$\overline{V}_i^k$	P_i, Q_i	$\overline{V}_i^k, Y_i^1$
NLD	$B_i^1, ..., B_i^r$	$\overline{V}_i^k, \beta_i^1, ..., \beta_i^r$	$B_i^1, ..., B_i^r$	$\overline{V}_i^k, \beta_i^1, ..., \beta_i^r$

NLD, nonlinear device; *PV*, photovoltaic.

$$P_i = \mathrm{Re}\left\{ \overline{V}_i^1 \left(\sum_{j=1}^{n} \overline{Y}_{ij}^1 \overline{V}_j^1 \right)^* \right\}$$
$$U_i = \sqrt{\sum_{k=1,3,\ldots} \left(V_i^k\right)^2 (i = 2, \ldots, g)} \tag{3.91}$$

- As PQ-bus constraint, we have

$$\overline{S}_i = \overline{V}_i^1 \left(\sum_{j=1}^{n} \overline{Y}_{ij}^1 \overline{V}_j^1 \right)^* (i = g+1, \ldots, c) \tag{3.92}$$

- The harmonic currents balance of the buses with conventional loads is written as follows:

$$\overline{Y}_i^k . \overline{V}_i^k + \sum_{j=1}^{n} \overline{Y}_{ij}^k \cdot \overline{V}_j^k = 0(i = 1, \ldots, c; k = 1, 3, 5, \ldots) \tag{3.93}$$

The load admittance is determined by the following equation [92]:

$$\overline{Y}_i^1 = \frac{\overline{S}_i^*}{\left(V_i^1\right)^2}; \overline{Y}_i^h = f_i\left(Y_i^1, h\right)(i = g+1, \ldots, c; h = 3, 5, \ldots) \tag{3.94}$$

- The harmonic currents balance at the buses with the NLDs is calculated by the following equation:

$$\sum_{j=1}^{n} \overline{Y}_{ij}^k . \overline{V}_j^k = \overline{I}_j^k (i = c+1, \ldots, n; k = 1, 3, 5, \ldots) \tag{3.95}$$

where $\overline{I}_j^k$ is the current injected by the NLD, which can be determined as follows:

$$B_i^j = nl^j \left(\overline{V}_i^1, \overline{V}_i^h, \beta_i^1, \ldots, \beta_i^r \right) (i = c+1, \ldots, n; j = 1, \ldots, r; h = 3, 5, \ldots) \tag{3.96}$$

$$\overline{I}_i^k = \overline{d}^k \left(\overline{V}_i^1, \overline{V}_i^h, \beta_i^1, \ldots, \beta_i^r \right) (i = c+1, \ldots, n; h = 3, 5, \ldots) \tag{3.97}$$

Using Eq. (3.94), the fundamental admittances of PQ buses can be added as an unknown since these admittances cannot be computed.

For $CHPF_H$, the HPF NLES can be written as follows:

- As PV-bus constraint, we have

$$P_i = \mathrm{Re}\left\{ \sum_{k=1,3,\ldots} \overline{V}_i^k \left(\sum_{j=1}^{n} \overline{Y}_{ij}^k \overline{V}_j^k \right)^* \right\}$$
$$U_i = \sqrt{\sum_{k=1,3,\ldots} \left(V_i^k\right)^2 (i = 2, \ldots, g)} \tag{3.98}$$

- As PQ-bus constraint, we have

$$\overline{S}_i = \sum_{k=1,3,\ldots} \overline{V}_i^k \left(\sum_{j=1}^{n} \overline{Y}_{ij}^k \overline{V}_j^k \right)^* (i = g+1, \ldots, c) \tag{3.99}$$

- The harmonics currents balance of conventional loads and NLDs are given by Eqs. (3.94) and (3.95).
- The NLD behavior is represented through Eq. (3.96).
- The admittances of PQ-bus loads can be determined by the following equations:

$$\overline{S}_i = \sum_{k=1,3,\ldots} \left(V_i^k\right)^2 \left(\overline{Y}_i^k\right)^* (i = g+1, \ldots, c) \tag{3.100}$$

$$\overline{Y}_i^k = f_i\left(\overline{Y}_i^1, k\right) k = 1, 3, 5, \ldots \tag{3.101}$$

References

[1] H. Nikkhajoei, R. Iravani, Steady-state model and power-flow analysis of electronically-coupled distributed resource units, IEEE Trans. Power Deliv. 22 (1) (2007) 721–728.

[2] H.R. Baghaee, M. Mirsalim, G.B. Gharehpetian, H.A. Talebi, Three phase AC/DC power-flow for balanced/unbalanced microgrids including wind/solar, droop-controlled and electronically-coupled distributed energy resources using RBF neural networks, IET Power Electron. 10 (3) (March 2017) 313–328, https://doi.org/10.1049/iet-pel.2016.0010.

[3] H.R. Baghaee, M. Mirsalim, G.B. Gharehpetian, H. Talebi, Generalized three phase robust load-flow for radial and meshed power systems with and without uncertainty in energy resources using dynamic radial basis functions neural networks, J. Clean. Prod. 174 (part C) (February 2018) 96–113, https://doi.org/10.1016/j.jclepro.2017.10.316.

[4] T. Ding, R. Bo, F. Li, Q. Guo, H. Sun, W. Gu, G. Zhou, Interval power-flow analysis using linear relaxation and optimality-based bounds tightening (OBBT) methods, IEEE Trans. Power Syst. 30 (1) (2015) 177–188.
[5] H. Sun, Q. Guo, B. Zhang, Y. Guo, Z. Li, J. Wang, Master–slave-splitting based distributed global power-flow method for integrated transmission and distribution analysis, IEEE Trans. Smart Grid 6 (3) (2015) 1484–1492.
[6] H.H. Müller, M.J. Rider, C.A. Castro, Artificial neural networks for power-flow and external equivalents studies, Elec. Power Syst. Res. 80 (1) (2010) 1033–1041.
[7] C. Carmona-Delgado, E. Romero-Ramos, J. Riquelme-Santos, Probabilistic power-flow with versatile non-Gaussian power injections, Elec. Power Syst. Res. 119 (5) (2015) 266–277.
[8] W.C. Briceno Vicente, R. Caire, N. Hadjsaid, Probabilistic power-flow for voltage assessment in radial systems with wind power, Int. J. Electr. Power Energy Syst. 41 (1) (2012) 27–33.
[9] F.J. Ruiz-Rodriguez, J.C. Hernández, F. Jurado, 'Probabilistic power-flow for photovoltaic distributed generation using the Cornish–Fisher expansion', Elec. Power Syst. Res. 89 (1) (2012) 129–138.
[10] A.C. Lisboa, L.S.M. Guedes, D.A.G. Vieira, R.R. Saldanha, A fast power-flow method for radial networks with linear storage and no matrix inversions, Int. J. Electr. Power Energy Syst. 63 (1) (2014) 901–907.
[11] H. Li, Y. Jin, A. Zhang, X. Shen, C. Li, B. Kong, An improved hybrid power-flow calculation algorithm for weakly-meshed power distribution system, Int. J. Electr. Power Energy Syst. 74 (1) (2016) 437–445.
[12] L. Ramos de Araujo, D.R.R. Penido, S.C. Júnior, J.L.R. Pereira, P. Augusto, Comparisons between the three-phase current injection method and the forward/backward sweep method, Int. J. Electr. Power Energy Syst. 32 (7) (2010) 825–833.
[13] A. Karami, M.S. Mohammadi, Radial basis function neural network for power system power-flow, Int. J. Electr. Power Energy Syst. 30 (1) (2008) 60–66.
[14] V.L. Paucar, M.J. Rider, Artificial neural networks for solving the power-flow problem in electric power systems, Elec. Power Syst. Res. 62 (28) (2002) 139–144.
[15] M.Z. Kamh, R. Iravani, A unified three-phase power-flow analysis model for electronically coupled distributed energy resources, IEEE Trans. Power Deliv. 26 (2) (2011) 899–909.
[16] M.M.A. Abdelaziz, H.E. Farag, E.F. El-Saadany, Y.A.R.I. Mohamed, A novel and generalized three-phase power-flow algorithm for islanded microgrids using a Newton Trust region method, IEEE Trans. Power Syst. 28 (1) (2013) 190–201.
[17] N. Nikmehr, S.N. Ravadanegh, Heuristic probabilistic power-flow algorithm for microgrids operation and planning, IET Gener. Transm. Distrib. 9 (11) (2015) 985–995.
[18] A.C.Z.S.M. Santos, M. Castilla, J. Miret, Voltage security in AC microgrids: a power-flow-based approach considering droop-controlled inverters, IET Renewable Power Gen. 9 (8) (2015) 954–960.
[19] F. Mumtaz, M.H. Syed, M.A. Hosani, H.H. Zeineldin, A novel approach to solve power-flow for islanded microgrids using modified Newton-raphson with droop control of DG, IEEE Trans. Sustain. Energy 7 (2) (2016) 493–503.
[20] P.C. Krause, Analysis of Electric Machinery and Drive Systems, third ed., Wiley-IEEE Press, 1986, 2013.
[21] R. Payasi, A. Singh, D. Singh, Effect of voltage step constraint and load models in optimal location and size of distributed generation, in: Proc. Int. Conf. Power Energy Control (ICPEC), Dindigul, Tamil Nadu, India, February 2013, pp. 710–716.
[22] W.W. Price, et al., Load representation for dynamic performance analysis, IEEE Trans. Power Syst. 8 (2) (1993) 472–482.

[23] P. Kundur, N. Balu, M. Lauby, Power System Stability and Control, McGraw-Hill, New York, NY, USA, 1994.
[24] J. Guerrero, J. Matas, L.G. de-Vicuna, M. Castilla, J. Miret, Decentralized control for parallel operation of distributed generation inverters using resistive output impedance, IEEE Trans. Ind. Electron. 54 (2) (2007) 994–1004.
[25] IEEE Standard 1547, IEEE Standard for Interconnecting Distributed Resources with Electric Power Systems, 2003.
[26] W. Yao, M. Chen, J. Matas, J. Guerrero, Z.M. Qian, Design and analysis of the droop control method for parallel inverters considering the impact of the complex impedance on the power sharing, IEEE Trans. Ind. Electron. 58 (2) (2011) 576–588.
[27] C. Opathella, B.N. Singh, D. Cheng, B. Venkatesh, Intelligent wind generator models for power-flow studies in PSS®E and PSS®SINCAL, IEEE Trans. Power Syst. 28 (2) (2013) 1149–1159.
[28] A. Dadhania, Modeling of Doubly Fed Induction Generators for Distribution System Power-Flow Analysis (M.A.S. thesis), Dept. Elect. Comput. Eng., Univ. Ryerson, Toronto, ON, Canada, 2010.
[29] M.J.E. Alam, K.M. Muttaqi, D. Sutanto, A three-phase power-flow approach for integrated 3-wire MV and 4-wire multigrounded LV networks with roof top solar PV, IEEE Trans. Power Electr. 28 (2) (2013) 1728–1737.
[30] T. Alquthami, H. Ravindra, M.O. Faruque, M. Steurer, T. Baldwin, Study of photovoltaic integration impact on system stability using custom model of PV arrays integrated with PSS/E, in: Proc. 2010 North American Power Symp, NAPS, Arlington, TX, USA, 2010, pp. 1–8.
[31] T. Xu, B. Venkatesh, C. Opathella, B.N. Singh, Artificial neural network model of photovoltaic generator for power-flow analysis in PSS®SINCA, IET Gen. Trans. Dist. 8 (7) (2014) 1346–1353.
[32] G.M. Masters, Renewable and Efficient Electric Power Systems, second ed., Wiley-IEEE Press, New York, NY, USA, 2004, 2013.
[33] M.G. Villalva, J.R. Gazoli, E.R. Filho, Comprehensive approach to modeling and simulation of photovoltaic arrays, IEEE Trans. Power Electron. 24 (5) (2009) 1198–1208.
[34] A.J. Wood, B.F. Wollenberg, G.B. Sheblé, Power Generation, Operation and Control, third ed., Wiley-Interscience, NY, USA, 1984, 2013.
[35] C. Opathella, B. Venkatesh, Three-phase unbalanced power-flow using π-model of controllable AC-DC converters, IEEE Trans. Power Syst. 99 (2016) 1–11.
[36] H. Demuth, M. Beale, Neural Network Toolbox for Use with MATLAB, fourth ed., The MathWorks Massachuset, MA, USA, 1993, 2015.
[37] S. Chen, C.F.N. Cowan, P.M. Grant, Orthogonal least squares learning algorithm for radial basis function networks, IEEE Trans. Neural Networks 2 (2) (1991) 302–309.
[38] J.A. Leonard, M.A. Kramer, Radial basis function networks for classifying process faults, IEEE Trans. Control Syst. 11 (3) (1991) 31–38.
[39] I. Tarassenko, S. Roberts, Supervised and unsupervised learning in radial basis function classifiers, IET Proc. Vision Image Signal Process. 141 (4) (1994) 210–216.
[40] D. Singh, R.K. Misra, D. Singh, Effect of load models in distributed generation planning, IEEE Trans. Power Syst. 22 (4) (2007) 2204–2212.
[41] G.K.V. Raju, P.R. Bijwe, Efficient reconfiguration of balanced and unbalanced distribution systems for loss minimisation, IET Gen. Transm. Distrib. 2 (1) (2008) 7–12.
[42] H.R. Baghaee, M. Mirsalim, G.B. Gharehpetian, Multi-objective optimal power management and sizing of a reliable wind/PV microgrid with hydrogen energy storage using MOPSO, J. Intell. Fuzzy Syst. 31 (2) (2016) 1–15, https://doi.org/10.3233/JIFS-152372.

[43] H.R. Baghaee, M. Mirsalim, G.B. Gharehpetian, H.A. Talebi, Reliability/cost based multi-objective pareto optimal design of stand-alone wind/PV/FC generation microgrid system, Energy 115 (1) (2016) 1022–1041, https://doi.org/10.1016/j.energy.2016.09.007.
[44] N.W.A. Lidula, A.D. Rajapakse, Microgrids research: a review of experimental microgrids and testsystems', Renew. Sustain. Energy Rev. 15 (1) (2011) 186–202.
[45] IEEE Power Engineering Society, "Distribution test feeders—13 bus feeder." Available from: http://ewh.ieee.org/soc/pes/dsacom/testfeeders/index.html.
[46] H.R. Baghaee, M. Mirsalim, G.B. Gharehpetian, H.A. Talebi, Application of RBF neural networks and unscented transformation in probabilistic power-flow of micro-grids including correlated wind/PV units and plug-in hybrid electric vehicles, Simulat. Model. Pract. Theory 72 (C) (2017) 51–68, https://doi.org/10.1016/j.simpat.2016.12.006.
[47] H.R. Baghaee, M. Mirsalim, G.B. Gharehpetian, H.A. Talebi, Fuzzy unscented transform for uncertainty quantification of correlated wind/PV microgrids: possibilistic-probabilistic power flow based on RBFNNs, IET Renew. Power Gener. 11 (6) (May 2017) 867–877, https://doi.org/10.1049/iet-rpg.2016.0669.
[48] Y. Chen, J. Wen, S. Cheng, Probabilistic load flow method based on Nataf transformation and Latin hypercube sampling, IEEE Trans. Sustain. Energy 4 (2) (2013) 294–301.
[49] M. Rouhani, M. Mohammadi, A. Kargarian, Parzen window density estimator-based probabilistic power flow with correlated uncertainties, IEEE Trans. Sustain. Energy 99 (2016) 1–12.
[50] M. Aien, M. Fotuhi-Firuzabad, M. Rashidinejad, Probabilistic optimal power flow in correlated hybrid wind–photovoltaic power systems, IEEE Trans. Smart Grid 5 (1) (2014) 130–138.
[51] B. Borkowska, Probabilistic load flow, IEEE Trans. Power App. Syst. 93 (3) (1974) 752–759.
[52] P. Jorgensen, J. Christensen, J. Tande, Probabilistic load flow calculation using Monte Carlo techniques for distribution network with wind turbines, in: Proc. 8th Conf. Harm. Qual. of Power, 1998, pp. 1146–1151.
[53] R.Y. Rubinstein, D.P. Kroese, Simulation and the Monte Carlo Method, first ed., Wiley, Hoboken, NJ, USA), 2011.
[54] M. Hajian, W.D. Rosehart, H. Zareipour, Probabilistic power flow by Monte Carlo simulation with Latin supercube sampling, IEEE Trans. Power Syst. 28 (2) (2013) 1550–1559.
[55] C.S. Saunders, Point estimate method addressing correlated wind power for probabilistic optimal power flow, IEEE Trans. Power Syst. 29 (3) (2014) 1045–1054.
[56] X. Ai, J. Wen, T. Wu, W.J. Lee, A discrete point estimate method for probabilistic load flow based on the measured data of wind power, IEEE Trans. Ind. Appl. 49 (5) (2013) 2244–2252.
[57] G. Verbic, C.A. Canizares, Probabilistic optimal power flow in electricity markets based on a two-point estimate method, IEEE Trans. Power Syst. 21 (4) (2006) 1883–1893.
[58] J.M. Morales, J.P. Ruiz, Point estimate schemes to solve the probabilistic power flow, IEEE Trans. Power Syst. 22 (4) (2007) 1594–1601.
[59] M. Mohammadi, Probabilistic harmonic load flow using fast point estimate method, IET Gen. Trans. Dist. 9 (13) (2015) 1790–1799.
[60] Y. Yuan, J. Zhou, P. Ju, J. Feuchtwang, Probabilistic load flow computation of a power system containing wind farms using the method of combined cumulants and Gram-Charlier expansion, IET Ren. Power Gen. 5 (6) (2013) 448–454.

[61] T. Williams, C. Crawford, Probabilistic load flow modeling comparing maximum entropy and gram-Charlier probability density function reconstructions, IEEE Trans. Power Syst. 28 (1) (2013) 272–280.
[62] W. Wu1, K. Wang, G. Li, X. Jiang, Z. Wang, Probabilistic load flow calculation using cumulants and multiple integrals, IET Gen. Trans. Dist. 10 (7) (2016) 1703–1709.
[63] C. Chen, W. Wu, B. Zhang, H. Sun, Correlated probabilistic load flow using a point estimate method with Nataf transformation, Elect. Power Energy Syst. 65 (1) (2015) 325–333.
[64] S. Fang, H. Cheng, G. Xu, Q. Zhou, H. He, P. Zeng, A probabilistic load flow method based on modified Nataf transformation and quasi Monte Carlo simulation, in: Proc. IEEE PES Asia-Pacific Power and Energy Engineering Conference (APPEEC), Brisbane, Queensland, AUS, November 2015, pp. 1–5.
[65] S. Julier, J. Uhlmann, F.D. Whyte, 'A new method for the nonlinear transformation of means and covariances in filters and estimators, IEEE Trans. Auto. Cont. 45 (3) (2000) 477–482.
[66] S. Julier, J. Uhlmann, Unscented filtering and nonlinear estimation, IEEE Proc. 92 (3) (2004) 401–422.
[67] M. Aien, M. Fotuhi-Firuzabad, F. Aminifar, Probabilistic load flow in correlated uncertain environment using unscented transformation, IEEE Trans. Power Syst. 27 (4) (2012) 2233–2241.
[68] Z.I. Botev, J.F. Grotowski, D.P. Kroese, Kernel density estimation via diffusion, Ann. Statist. 38 (5) (2010) 2916–2957.
[69] N. Soleimanpour, M. Mohammadi, Probabilistic load flow by using nonparametric density estimators, IEEE Trans. Power Syst. 28 (4) (2013) 3747–3755.
[70] A. Khosravi, S. Nahavandi, Combined nonparametric prediction intervals for wind power generation, IEEE Trans. Sustain. Energy 4 (4) (2013) 849–856.
[71] C. Wan, Z. Xu, J. Ostergaard, Z.Y. Dong, K.P. Wong, Discussion of combined nonparametric prediction intervals for wind power generation, IEEE Trans. Sustain. Energy 5 (3) (2014) 1021.
[72] G. Li, X. Zhang, Modeling of plug-in hybrid electric vehicle charging demand in probabilistic power flow calculations, IEEE Trans. Smart Grid 3 (1) (2012) 492–499.
[73] H. Wu, Y. Zhou, S. Dong, Y. Song, Probabilistic load flow based on generalized polynomial chaos, IEEE Trans. Power Syst. 99 (2016) 1–2.
[74] D. Cai, D. Shi, J. Chen, Probabilistic load flow computation using Copula and Latin hypercube sampling, IET Gen. Trans. Dist. 8 (9) (2014) 1539–1549.
[75] Z. Ren, W. Li, R. Billinton, W. Yan, Probabilistic power flow analysis based on the stochastic response surface method, IEEE Trans. Power Syst. 31 (3) (2016) 2307–2315.
[76] Y.Y. Hong, F.J. Lin, T.H. Yu, Taguchi method-based probabilistic load flow studies considering uncertain renewables and loads, IET Gen. Trans. Dist. 10 (2) (2016) 221–227.
[77] J. Tang, F. Ni, F. Ponci, A. Monti, Dimension-adaptive sparse grid interpolation for uncertainty quantification in modern power systems: probabilistic power flow, IEEE Trans. Power Syst. 31 (2) (2016) 907–919.
[78] A.C. Melhorn, A. Dimitrovski, Three-phase probabilistic load flow in radial and meshed distribution networks, IET Gen. Trans. Dist. 9 (16) (2015) 2743–2750.
[79] E. Janecek, D. Georgiev, Probabilistic extension of the backward/forward load flow analysis method, IEEE Trans. Power Syst. 27 (2) (2016) 695–704.
[80] N. Gupta, 'Probabilistic load flow with de tailed wind generator models considering correlated wind generation and correlated loads, Ren. Energy 94 (1) (2016) 96–105.
[81] M. Aien, M. Fotuhi-Firuzabad, M. Rashidinejad, Probabilistic power flow of correlated hybrid wind-photovoltaic power systems, IET Ren. Power Gen. 8 (6) (August 2014) 649–658.

[82] H.R. Baghaee, M. Mirsalim, G.B. Gharehpetian, A. Kashefi Kaviani, Security/cost-based optimal allocation of multi-type FACTS devices using multi-objective particle swarm optimization, Simulation: Int. Trans. Soc. Model. Simulation 88 (8) (2012) 999–1010, https://doi.org/10.1177/0037549712438715.
[83] M. Aein, A. Hajebrahimi, M. Fotuhi-Firuzabad, A comprehensive review on uncertainty modeling techniques in power system studies, Renew. Sustain. Energy Rev. 57 (1) (2016) 1077–1089.
[84] S.W. Heunis, R. Herman, A probabilistic model for residential consumer loads, IEEE Trans. Power Syst. 17 (3) (2002) 621–625.
[85] A. Seppala, Load Research and Load Estimation in Electricity Distribution (Ph.D. thesis), Helsinki University of Technology, 1996.
[86] T. Ishii, K. Otani, T. Takashima, S. Kawai, Estimation of the maximum power temperature coefficients of PV modules at different time scales, Sol. Energy Mater. Sol. Cells 95 (1) (2011) 386–389.
[87] J.G. Vlachogiannis, Probabilistic constrained load flow considering integration of wind power generation and electric vehicles, IEEE Trans. Power Syst. 24 (4) (Nov. 2009) 1808–1817.
[88] R.L. Mason, R.F. Gunst, J.L. Hess, Statistical Design and Analysis of Experiments: With Applications to Engineering and Science, first ed., Wiley-Inter-Science, NY, USA, 2003.
[89] M. Aien, M. Rashidinejad, M. Fotuhi-Firuzabad, On possibilistic and probabilistic uncertainty assessment of power flow problem: a review and a new approach, Renew. Sustain. Energy Rev. 37 (1) (2014) 883–895.
[90] A. Soroudi, M. Ehsan, 'A possibilistic–probabilistic tool for evaluating the impact of stochastic renewable and controllable power generation on energy losses in distribution networks—a case study', Renewable Sustain. Energy Rev. 15 (1) (2011) 794–800.
[91] R.D. Zimmerman, C.E. Murillo-Sanchez, R.J. Thomas, Matpower's extensible optimal power flow architecture, in: Proc. IEEE Power and Energy Society General Meeting, 2009, pp. 1–7.
[92] S. Herraiz, L. Sainz, J. Clua, Review of harmonic load-flow formulations, IEEE Trans. Power Del. 18 (3) (July 2003) 1079–1087.

Further reading

[1] H.R. Baghaee, A.K. Kaviani, M. Mirsalim, G.B. Gharehpetian, Harmonic optimization in single DC source multi-level inverters using RBF neural networks, in: Proc. 3rd Power Electronics and Drive Systems Technology, PEDSTC, Tehran, Iran, 2012, pp. 403–409.
[2] R.P. Mukund, Wind & Solar Power Systems, CRC, Boca-Rat., FL, USA, 1999.
[3] J. Park, W. Liang, J. Choi, A. El-Keib, M. Shahidehpour, R. Billinton, A probabilistic reliability evaluation of a power system including solar/photovoltaic cell generator, in: Proc. IEEE PES Power & Energy Society General Meeting (PES'09), Calgary, AB, CA, July 2009, pp. 1–6.
[4] E.A. Wan, R. Van-der-Merwe, The Unscented Kalman Filter for Nonlinear Estimation', Proc. IEEE Adaptive Systems for Signal Processing, Communications, and Control Symposium (SPCC 2000), Lake Louise, Alberta, CA, October 2000, pp. 153–158.
[5] H.R. Baghaee, M. Mirsalim, G.B. Gharehpetian, H.A. Talebi, Nonlinear load sharing and voltage compensation of microgrids based on harmonic power-flow calculations using radial basis function neural networks, IEEE Syst. J. 99 (2017) 1–11, https://doi.org/10.1109/JSYST.2016.2645165.

CHAPTER 4

Fault analysis

Nomenclature

C_f	LCL filter capacitance
E	Amplitude of output voltage
E^*	Reference of E
f	Nominal frequency of network
$G(s)$	Inverter TF gain
G_{IC}	Current controller closed-loop transfer function
G_{LC}	LC filter transfer function
$G_P(s)$, $G_Q(s)$	Droop control TFs
$G_{st}(s)$	Novel loop TF
$Gv(s)$	V loop TF
$G_v(s)$, $G_i(s)$	V and I controllers TF
h	Order of harmonics
i_L^{ref}, i'^{ref}_L	Actual and limited inductance current reference
i_{th}	Maximum permissible I or limiting strategy threshold
K	Constant parameter
$k\text{-}pst$, k_{ist}	PI coefficients of $G_{st}(s)$
$k\text{-}pV$, k_{pI}	Proportional coefficients of V and I control blocks
k_{iP}, k_{pQ}	Droop coefficients
k_{pF}, k_{iF}, k_{pE}, k_{iE}	PI controller coefficients
k_{pP}	Virtual inertia/transient droop
k_{rhV}, $k\text{-}rhI$	Resonant coefficients of V and I controllers
$L\text{-}vi$, R_{vi}	L and R of virtual impedance, respectively
P, Q	Powers
p, q	Instantaneous powers
P^*, Q^*	Power references
P^*_G, Q^*_G	Required P_G and Q_G
$P_{DER,i}$, $Q_{DER,i}$	DER P and Q set points
P_G, Q_G	PCC real and reactive powers
TF	Transfer function
$V^1_{vi\alpha\beta}$, $I^1_{o\alpha\beta}$	Fundamental components of voltage and current, respectively
v_c, v_c^*	LCL filter capacitor voltage and v_c reference
$v_{c\alpha\beta}$, $i_{o\alpha\beta}$	Capacitor voltage and current (transferred to $\alpha\beta$-coordinates)
$VI(s)$	Virtual impedance transfer function
V_o^*	Phase voltage nominal value
v_{ref}	Reference voltage
ω	Output voltage angular frequency
ω^*	Reference of ω

Microgrids and Methods of Analysis
ISBN 978-0-12-816172-2
https://doi.org/10.1016/B978-0-12-816172-2.00004-3

ω_c	Cutoff frequency of low-pass filter
$Z_o(s)$	Output impedance of inverter
ω_{rest}, E_{rest}	ω and V deviation, respectively

1. Introduction

Optimal integration of renewable energy systems [1] in microgrids (MGs) should locally overcome energy problems and enhance flexibility in both operation modes [2,3]. But the integration of renewable energy resources and other distributed generations (DGs) in MGs increases the fault level, and therefore, the contribution of distributed energy resources (DERs) and the main grid must be considered for the determination of MG ratings [4,5]. For short-circuit (SC) calculations, inverter-interfaced distributed energy resources (IIDERs) along with a control system must be modeled in the time domain. The IIDERs have lower contributions in SC current [5,6]. However, using this model, it is possible to correct or update the settings of the protection system and design fault current-limiting (FCL) strategies and methods.

Various models have been suggested to represent the DERs during faults. In Ref. [5], an SC calculation method has been presented, which can consider DGs in distribution networks based on IEC 60909 standard [7–11]. The DG-type effect on the SC level has been studied in Ref. [6]. Considering inverter response in load flow analysis and simulations conducted by Electromagnetic Transients Program (EMTP) in Ref. [12], an analytical solution based on an approximated model has been used for fault analysis of distribution networks, which have DERs. The presented solution did not consider the control system effect, the same as the case investigated in Ref. [13]. This subject has been studied in Refs. [14,15] considering inverter control loops in on-grid and off-grid operation modes.

The FCL control approaches have been suggested by different researchers [16–18]. The effect of FCL strategies on IIDERs, which have droop method–based controller, and also, the capability of MG fault ride-through (FRT) have been studied in Refs. [19–21].

As mentioned in previous chapters, the well-known droop-based controllers [22,23] do not need a communication network with a high bandwidth, unlike the systems with centralized controllers [24–26]. But their performance during transients is weak. They have problems with dynamic loads and black starts [23]. To enhance small signal stability, large signal stability and sharing of power among DERs, in Ref. [3], the conventional control system has used a supplementary control loop. The radial basis function neural network (RBFNN) has been implemented in Ref. [28] to adjust active and reactive powers. The hierarchical control application can improve stability margins and performance of overload or overcurrent protection [3,28]. But, in directly voltage-controlled (DVC) DERs, the IIDER does not have the current control loop [29], and as a result, it has a faster transient response [26,30–34]. But this may not necessarily lead to less steady-state error and improved robustness. Also, it cannot restrict the SC and motor starting currents and VSC output power transients [29].

For SC studies, a generalized fault model of IIDERs and FCL strategy, including controllers, is presented in this chapter. The controllers can be a droop method–based controller or DVC such as a decentralized servo-mechanism controller [34] or robust H^{∞} control [36–40].

2. Droop method–based control, hierarchical organization

To enhance performance of decentralized droop-based hierarchical control systems of MGs, a new algorithm using RBFNN power flow has been suggested in Ref. [28] (see Fig. 4.1A). This RBFNN has two complementary functions. The function number 1 is solution of power flow problem [1,2,28]. Failure of MG communication system needs application of the second function. It will determine the harmonics approximately and estimate reference values for suggested controller. The decentralized feature of the suggested control scheme can be realized by this function.

2.1 Droop control method

The primary control level of the droop controller adjusts the f and v_{ref} amplitude, as follows [3]:

$$\omega = \omega^{*} - G_P(s)\cdot(P - P^{*}) \tag{4.1}$$

$$E = E^{*} - G_Q(s)\cdot(Q - Q^{*}) \tag{4.2}$$

where we have

$$\begin{aligned} G_P(s) &= \left(k_{pP} + \frac{k_{iP}}{s}\right) \\ G_Q(s) &= \left(k_{pQ}\right) \end{aligned} \tag{4.3}$$

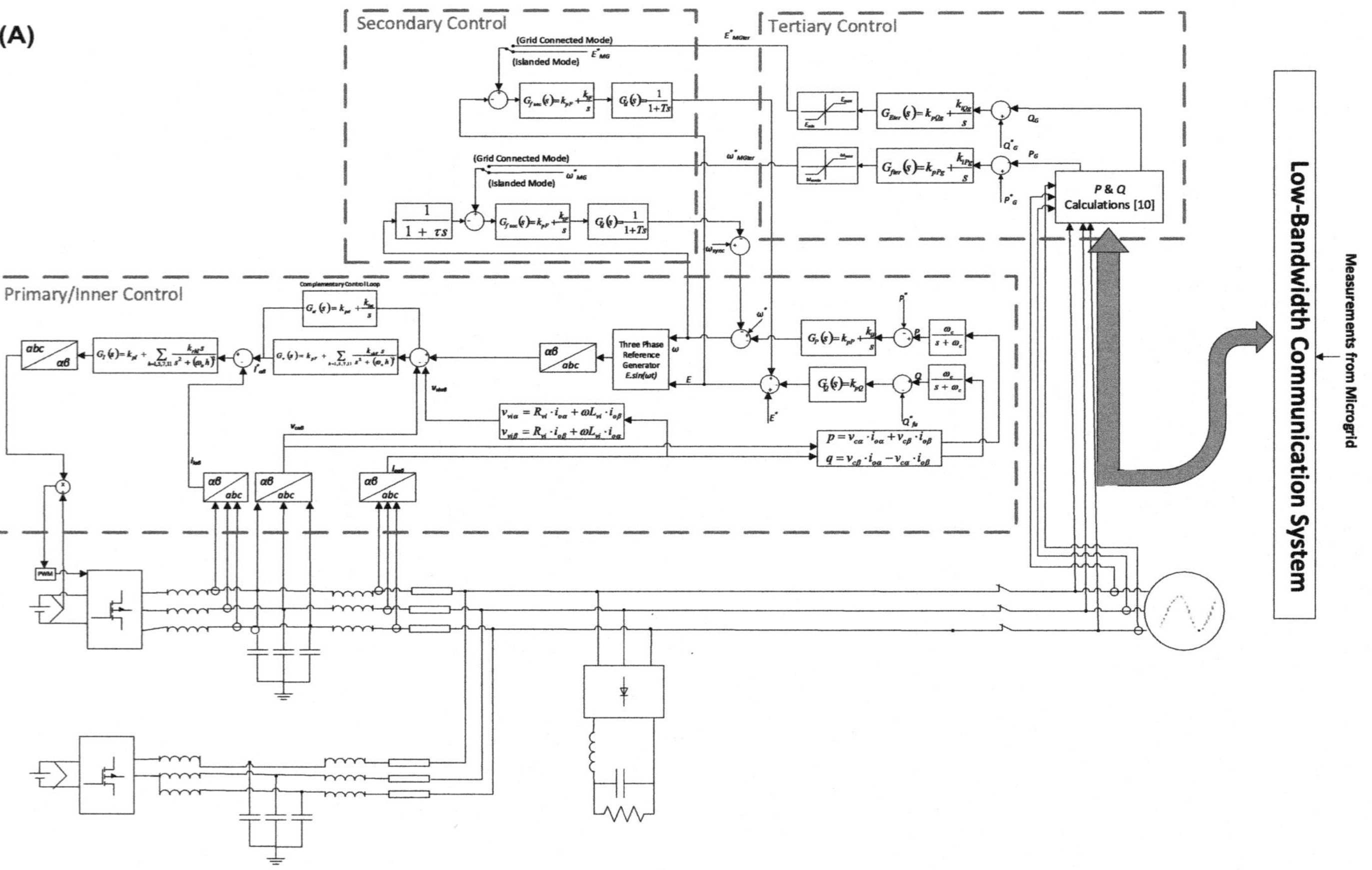

Figure 4.1 (A) Hierarchical organization of an MG consisting of two DERs and a common load [3,28], (B) Robust controller [27,29], and (C) DVC DER and its controller [29]. *DERs*, distributed energy resources; *DVC*, directly voltage-controlled; *MG*, microgrid.

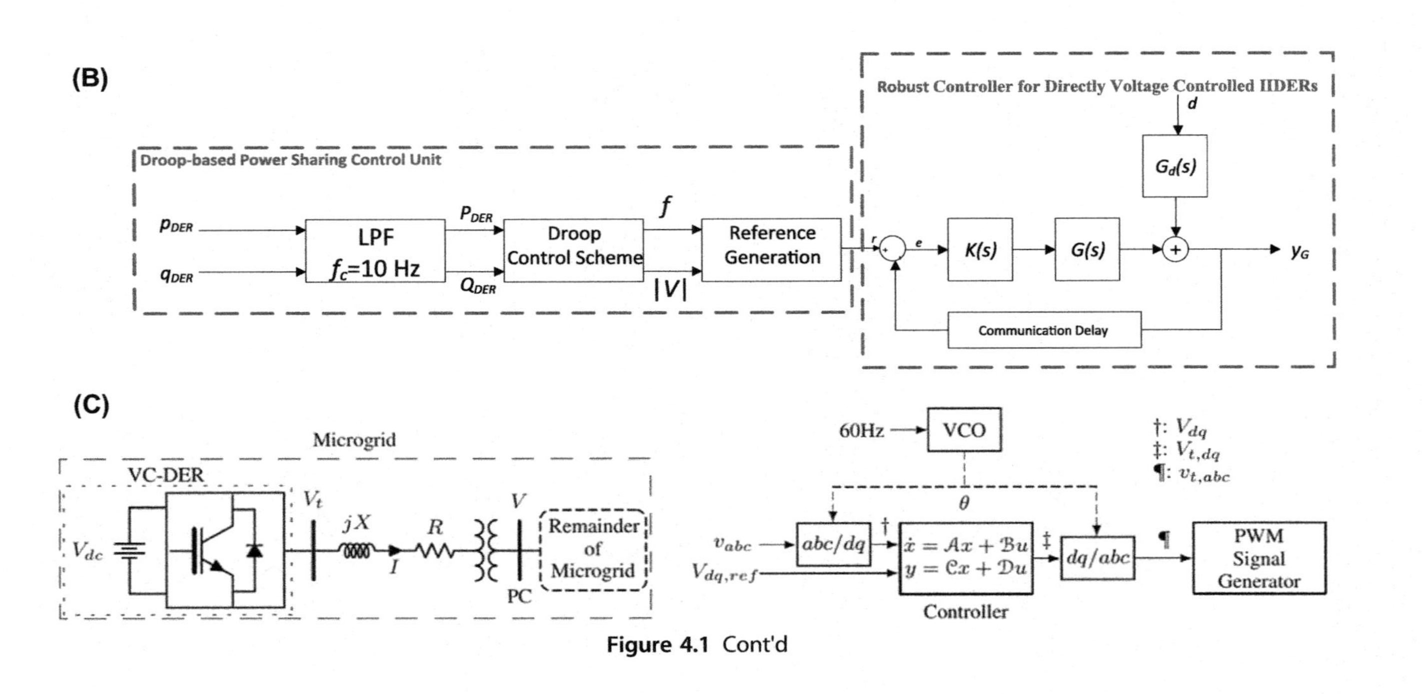

Figure 4.1 Cont'd

Based on Clark's transformation, the instantaneous powers can be written using Eq. (4.4) in $\alpha\beta$ coordinate.

$$\begin{aligned} p &= v_{c\alpha} \cdot i_{o\alpha} + v_{c\beta} \cdot i_{o\beta} \\ q &= v_{c\beta} \cdot i_{o\alpha} - v_{c\alpha} \cdot i_{o\beta} \end{aligned} \tag{4.4}$$

The outputs of low-pass filters are the quasi—steady-state powers, as follows:

$$(P, Q) = \frac{\omega_c}{s + \omega_c} \cdot (p, q) \tag{4.5}$$

2.2 Virtual impedance

To ensure the inductive behavior of the system, a virtual impedance modeling method has been proposed in Ref. [3]. Here, the same has been used, as follows:

$$\begin{aligned} v_{vi\alpha} &= R_{vi} \cdot i_{o\alpha} + \omega L_{vi} \cdot i_{o\beta} \\ v_{vi\beta} &= R_{vi} \cdot i_{o\beta} + \omega L_{vi} \cdot i_{o\alpha} \end{aligned} \tag{4.6}$$

2.3 Internal control loops

In Ref. [3], the design of inner controllers has been presented. A proportional integral (PI) controller has obtained the output voltage and current, as follows:

$$\frac{d\varphi_\alpha}{dt} = v_\alpha^* - v_{c\alpha}, \quad \frac{d\varphi_\beta}{dt} = v_\beta^* - v_{c\beta} \tag{4.7}$$

$$\frac{d\gamma_\alpha}{dt} = i_\alpha^* - i_{l\alpha}, \quad \frac{d\gamma_\beta}{dt} = i_\beta^* - i_{l\beta} \tag{4.8}$$

To adjust the fundamental and the harmonics of order of $h = 5{,}7$, and 11, proportional resonant (PR) terms have been implemented in V and I loops using the following transfer functions (TFs) [3]:

$$G_v(s) = k_{pV} + \sum_{h=1,5,7,11} \frac{k_{rhV} s}{s^2 + (\omega_o h)^2} \tag{4.9}$$

$$G_I(s) = k_{pI} + \sum_{h=1,5,7,11} \frac{k_{rhI}s}{s^2 + (\omega_o h)^2} \tag{4.10}$$

To reduce the resonance effects of the voltage controller, the following supplementary PI controller has been used.

$$G_{st}(s) = k_{pst} + \frac{k_{ist}}{s} \tag{4.11}$$

2.4 Secondary control

The following PI controllers have been applied to frequency and voltage, to compensate for their deviations [3]:

$$\omega_{rest} = \left(k_{pF} + \frac{k_{iF}}{s}\right) \cdot \left(\omega_{MG}^{*} - \omega_{MG}\right) \tag{4.12}$$

$$E_{rest} = \left(k_{pE} + \frac{k_{iE}}{s}\right) \cdot \left(E_{MG}^{*} - E_{MG}\right) \tag{4.13}$$

2.5 Tertiary control

As shown in Fig. 4.1A, the P_G and Q_G are measured and compared with P^*_G and Q^*_G, i.e., the required powers, to adjust power flow through the PCC. Therefore, we have [3]

$$\omega_{MGter}^{*} = \left(k_{pPg} + \frac{k_{iPg}}{s}\right) \cdot \left(P_G^{*} - P_G\right) \tag{4.14}$$

$$E_{MGter}^{*} = \left(k_{pQg} + \frac{k_{iQg}}{s}\right) \cdot \left(Q_G^{*} - Q_G\right) \tag{4.15}$$

3. Direct voltage control

3.1 Power management of microgrid and its synchronization

For a radial MG, a design has been developed in Ref. [27] as controller. It has local controllers (LCs), a frequency control system, and a power management system. The MG frequency has been adjusted by an open-loop LC independent oscillator. A GPS-based common time reference has been used to synchronize all the LCs. In this study, using a low-bandwidth communication network, DERs and loads' instantaneous active and reactive powers have been transmitted to the power management system and LCs, to obtain DERs' required power sharing and control the voltage of DER_i, which is connected to PCC_i. It must be mentioned that considering load sharing strategy and active and reactive powers demand of MG, $P_{DER,i}$ and $Q_{DER,i}$ and also, the voltage reference values of the DERs buses are determined by the following well-known load flow equations [1,2,28,35]:

$$P_{DER,i} = \sum_{j=1}^{N} |V_i||V_j||Y_{ij}|\cos(\delta_i - \delta_k - \theta_{ij}) \tag{4.16}$$

$$Q_{DER,i} = \sum_{j=1}^{N} |V_i||V_j||Y_{ij}|\sin(\delta_i - \delta_k - \theta_{ij}) \tag{4.17}$$

3.2 Local controllers of inverter-based distributed energy resources

A control scheme with two degrees of freedom has been designed in Ref. [31] for two plants of a system. In this section, one controller, shown in Fig. 4.1B, is used, which results in robust stabilization and disturbance attenuation. Using the theory of H^{∞} robust control, this case can be modeled by a set of linear matrix inequalities. The design procedure of this controller has been presented in Refs. [30,31]. As seen in Fig. 4.1A and B, this control system is used in the hierarchical scheme primary level [30].

This DVC IIDER control system, which should adjust the PCC voltage, is transferred to *dq*-reference frame and indicated in Fig. 4.1C. The parameters of the MG, robust controllers of LCs, and DVC IIDER have been given in Refs. [27,29].

4. Fault model: general form

4.1 Droop-based control of inverter-interfaced distributed energy resources

In this section, the proposed control system impact on an IIDER is investigated during fault period. The following equation presents output current ($\boldsymbol{i_o}$) and capacitor voltage ($\boldsymbol{v_c}$) connection [3,15,28]:

$$V_c(s) = G(s)V_c^*(s) - Z_o(s)I_o(s) \tag{4.18}$$

where

$$G(s) = \frac{G_V(s)G_I(s)}{sC_f(1 + G_{st}(s)G_V(s)) - G_v(s)G_I(s)} \tag{4.19}$$

$$Z_o(s) = -\frac{(VI(s)G_I(s) - 1)(1 + G_{st}(s)G_V(s))}{sC_f(1 + G_{st}(s)G_V(s)) + G_v(s)G_I(s)} \tag{4.20}$$

As indicated in Fig. 4.2, the inverter, using this equation, can be represented by a two-port network for the hierarchical droop-based control presented in Refs. [3,28]. In applying the current-limiting strategy, the model shown on the right side of the same figure can be used.

4.2 Inverter current limiting

The current of an IIDER can be limited using a virtual impedance [18]. However, the current limiting strategies can be divided into two groups: latched limit (LL) strategy and instantaneous saturation limit (ISL) strategy, respectively studied in Refs. [15,19]. The current-limiting strategy and its control systems have the possibility to be realized in the following frames:

- NARF: natural reference frame (*abc* coordinates)
- SYRF: synchronous reference frame (*dq0* coordinates)
- STRF: stationary reference frame ($\alpha\beta\gamma$ coordinates) [36].

Using the ISL strategy in NARF and STRF can lead to distortions in inverter output voltage and current for all fault types. However, if the fault occurs symmetrical, the waveforms are sinusoidal in SYRF. If we have unbalanced conditions or the fault is asymmetrical, the sinusoidal waveform ripples will appear at the frequency of 2ω [19,21]. However, the LL strategy

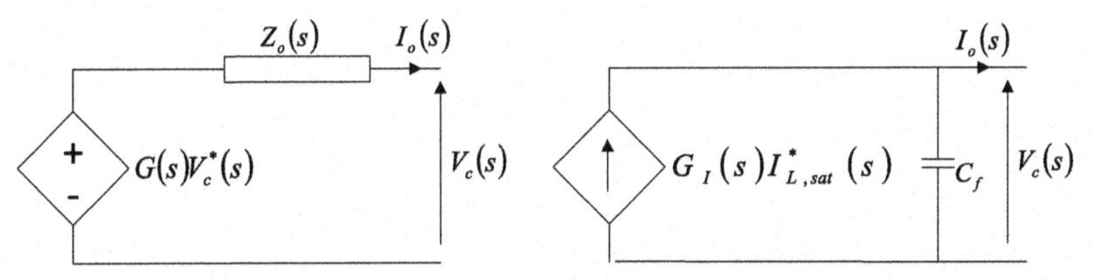

Figure 4.2 Inverter two-port equivalent network in off-grid mode, using (A) hierarchical droop-based control [3,28], and (B) current-limiting strategy.

results in better situations. The inverter current shall be sinusoidal and controlled. Due to an asymmetrical fault, the healthy phases will have overvoltages in SYRF or STRF [15,20]. Thus, it can be said that both strategies have advantages and disadvantages in each reference frame.

A policy for FCL will be suggested here for a hierarchical controller and DVC inverter-based DERs. A supplementary control loop is added in the droop-based hierarchical control of Refs. [3,28] to enhance MG FRT capability and large and small signal stability. Here, the $G_{st}(s)$ affects frequency response of the TFs presented by Eqs. (4.19) and (4.20), resulting in less fault level. The current-limiting function presented by Eq. (4.21) is the suggested strategy for current control in NARF.

$$i_{L,j}^{\prime ref} = = \frac{i_{L,j}^{\prime ref} \bullet i_{ith}}{\sqrt{2}\, I_{L,j}^{ref}} u\left(I_{L,j}^{ref} - \frac{i_{th}}{\sqrt{2}} \right) + u\left(\frac{i_{th}}{\sqrt{2}} - I_{L,j}^{ref} \right); \; j = a, b, c \tag{4.21}$$

In case of SYRF or STRF, we have

$$i_{L,j}^{\prime ref} = = \frac{i_{L,j}^{\prime ref} \bullet i_{ith}}{\sqrt{2}\, I_{L}^{ref}} u\left(I_{L}^{ref} - \frac{i_{th}}{\sqrt{2}} \right) + u\left(\frac{i_{th}}{\sqrt{2}} - I_{L}^{ref} \right); \; j = d(\alpha), q(\beta), 0(\gamma) \tag{4.22}$$

In the aforementioned equations, $u(t)$ denotes the unit step function. i_{th} equals 2 pu (i.e., two times of the inverter rated I), and I_L^{ref} is its maximum value for all three phases. This new approach is combined with the hierarchical control presented in Section 2. This limiting strategy will improve the MG protection system.

4.3 Directly voltage-controlled inverted-interfaced distributed energy resources

In this case, there is no current control loop. Therefore, to develop the model of DVC IIDERs, the supplementary stabilizing controller and virtual impedance terms must be removed, and Eqs. (4.19) and (4.20) are rewritten, as follows:

$$G(s) = \frac{G_V(s)}{sC_f - G_v(s)} \tag{4.23}$$

$$Z_o(s) = \frac{1}{sC_f + G_v(s)} \tag{4.24}$$

The protection of the DVC IIDER is challenging, as there is no limiting policy and the current control loop. They must use overcurrent/overload protection schemes, which increase costs.

4.4 Equivalent phase network

The equivalent phase networks of IIDERs can be developed by the model developed in previous subsections. In off-grid operation mode, the IIDER control system requires three control loops. In STRF and SYRF, after applying Park and Clark transformations [36] and integrating the current limiting function, we have

$$v_{o,d(\alpha)} = \left(G(s) i_{L,d(\alpha)}^{Jref} - I_{o,d(\alpha)} + \omega C_f v_{o,q(\beta)}\right) \frac{1}{sC_f} \tag{4.25}$$

$$v_{o,q(\alpha)} = \left(G(s) i_{L,q(\beta)}^{Jref} - I_{o,q(\beta)} + \omega C_f v_{o,d(\alpha)}\right) \frac{1}{sC_f} \tag{4.26}$$

$$v_{o,0(\gamma)} = \left(G(s) i_{L,0(\gamma)}^{Jref} - I_{o,d(0)}\right) \frac{1}{sC_f} \tag{4.27}$$

The control system is usually designed to have a unity gain for the closed-loop TF at f. After using the inverse Park and Clark transformations for Eqs. (4.25)−(4.27), the fault model of the inverter can be developed as shown in Fig. 4.2.

$$v_{oj} = \left(i_{Lj}^{Jref} - I_{oj}\right) \frac{1}{sC_f};\ j = a, b, c \tag{4.28}$$

The current sources are determined using the symmetrical components, as follows:

$$i_{La}^{Jref} = \sqrt{2/3}\, i_{Ldq(\alpha\beta)}^{Jref} e^{j\phi_{dq(\alpha\beta)}};\ j = \sqrt{-1} \tag{4.29}$$

$$i_{Lb}^{J\ ref} = a^2 i_{La}^{J\ ref} \tag{4.30}$$

$$i_{Lc}^{Jref} = a i_{La}^{Jref} \tag{4.31}$$

$$\begin{aligned} i_{Ldq(\alpha\beta)}^{Jref} &= \sqrt{\left(i_{Ld(\alpha)}^{Jref}\right)^2 + \left(i_{Lq(\beta)}^{Jref}\right)^2} \\ \phi_{dq(\alpha\beta)} &= \tan^{-1}\left(i_{Lq(\beta)}^{Jref} / i_{Ld(\alpha)}^{Jref}\right);\ i_{L0}^{Jref} = 0\ A \end{aligned} \tag{4.32}$$

In the aforementioned equations, we have $a = -0.5 + 0.5\sqrt{3}$. Authors of Ref. [37] have suggested a control system with several loops in NARF. In this frame, the controllers can independently adjust all voltages of three phases one-by-one. They also restrict SC current separately, which is not the case in the SYRF and STRF. The strategy for current limiting generates the reference current (sinus waveform), which equals the inverter maximum current in healthy phases. Therefore, for example, the current in phase *a* will be restricted, in case of single-line/phase to ground (SLG) fault occurrence at phase *a*, and in phases *b* and *c*, we will have the voltage-controlled mode. In this case, Fig. 4.3A, as a Thevenin equivalent, is used to develop the SLG fault equivalent network, indicated as Fig. 4.3B. Thus, the current source is i_{La}^{Jref} multiplied by G_{IC}, which is in parallel with $1/sC_f$, and we have

$$G_{IC} = \frac{G_I G_{LC}}{1 + G_I G_{LC}} \tag{4.33}$$

where G_{LC} is the LC filer TF, which can be calculated as follows [15]:

$$\begin{aligned} G_{LC} &= \frac{I_L(s)}{v_i(s) - v_o(s)} \\ &= \frac{s\left(L_f + 2L_n\right) + R_f + 2R_n}{\left(L_f^2 + 3L_fL_n\right)s^2 + \left(2R_fL_f + 3R_fL_n + 3R_nL_f\right)s + \left(R_f^2 + 3R_fR_n\right)} \end{aligned} \tag{4.34}$$

$$Z_{ob}(s) = Z_{oc}(s) = Z_0(s) = -\frac{VI(s)G_I - 1}{sC_f + G_V G_I} \tag{4.35}$$

4.5 Short-circuit contribution of wind generation and photovoltaic systems

The SC current calculation method used in Ref. [5] can consider the effect of DGs considering the requirements of IEC 60909 standard [7–11].

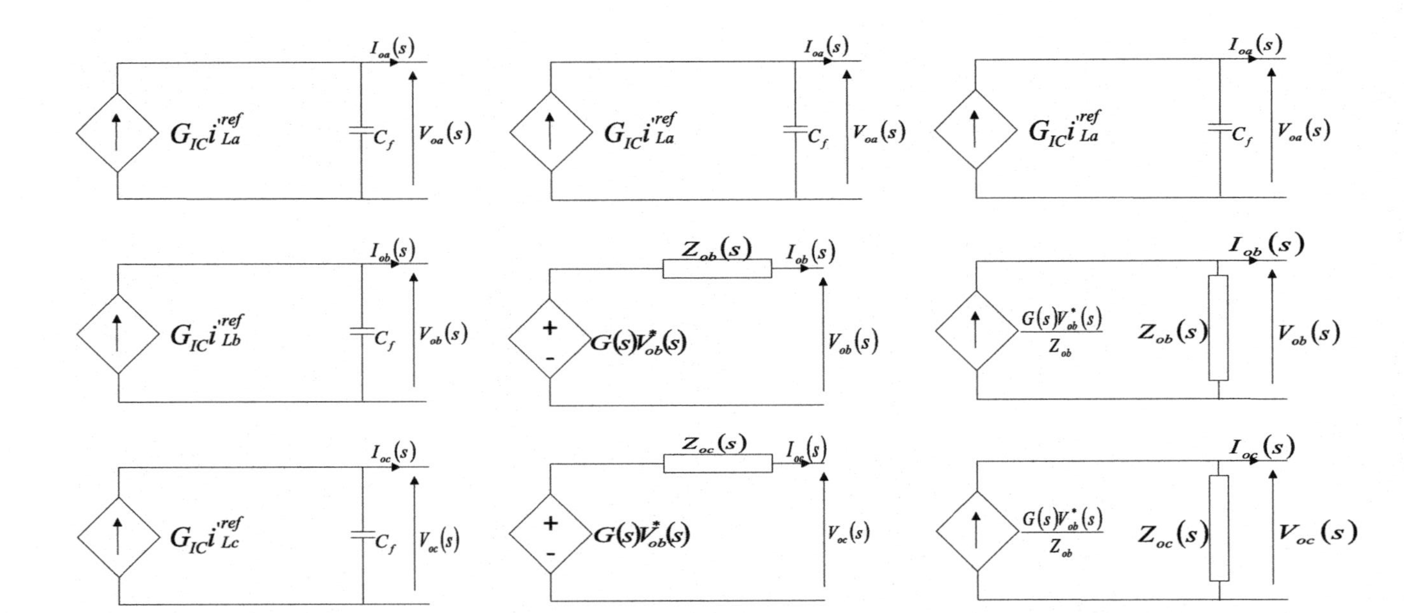

Figure 4.3 IIDER equivalent phase network: (A) All types faults model with controllers in STRF and SYRF, (B) SLG model including controller in NARF, and (C) SLG fault Norton's equivalent circuit including controller in NARF. *IIDER*, inverter-interfaced distributed energy; *NARF*, natural reference frame; *SLG*, single-line/phase to ground; *STRF*, stationary reference frame; *SYRF*, synchronous reference frame.

In Refs. [14–21], the IIDERs model during a fault has been developed by a constant DC voltage source at DC bus of VSC. But DERs dynamic should be considered in practice. For example, wind generation (WG) and photovoltaic (PV) systems are droop-based or DVC DERs, which the proposed method of Ref. [5] must be applied to them to determine their maximum SC current.

The DGs contribution in SC current has been studied in Ref. [22] based on IEC 60909 [7–11]. The DGs can be directly interfaced asynchronous generators, doubly fed inductions generators (DFIGs), converter-interfaced PVs, or microturbines. Among the mentioned DERs, the converter-interfaced DERs are the subject of this research. As discussed in Ref. [5] and IEC 60909, the contribution of DFIGs and other variable speed WG systems in SC current can be represented by an asynchronous generator model. Tests and experiences Refs. [5,38–40] and vendors' data [41,42] show that due to controllers' response, the fault current cannot exceed 200% of the rated current. The same is valid for PV systems [5,43–45].

5. Simulation results

Before presenting the simulation results, it must be mentioned that in previous researches, the following problems and limitations have been seen for current-limiting methods.

In the case of using STRF or SYRF in a four-wire configuration, and also in the case of using all the reference frames in a three-wire configuration, they cannot control the voltage in healthy phases.

In using NARF in a three-wire configuration, the voltage and current power quality are so poor.

In the case of using STRF or SYRF during an SLG fault, overvoltage can occur.

The first and third challenges have been addressed in the control system presented in Refs. [3,28]. But they have not been discussed in Refs. [27,29–31], for DVC-based controllers because there is no study on their robustness versus large-signal disturbances and balanced and unbalanced faults. In this study, there is no concern about power quality because the NARF is not applied to hierarchical droop-controlled and DVC IIDERs. Using the internal loop presented in Section 2.3, the current and voltage controllers adjust the fundamentals and harmonics and can well operate under an unbalanced current [2,3,28]. The DVC DER has not any problem with stability and voltage control. Still, the capability of the DVC method for power quality improvement has not exactly been discussed [27,29–31]. One possible solution for DVC-based methods under unbalanced load conditions is applying a robust controller in positive- and negative-sequence current controllers [46]. Improving DVC DERs under unbalanced and nonlinear loads can be an investigation topic for researchers.

5.1 Performance evaluation of the proposed strategy

To study the suggested strategy applied to islanded MG, the model and equivalent circuits proposed and discussed in Section 4.4 are used in this section. The four-wire and three-wire configurations under various balanced and unbalanced faults will be investigated. The fault is simulated at $t = 0.2$ s and cleared at $t = 0.3$ s. The fault resistance is 1.2 Ω. Table 4.1 summarizes the simulation results, including current and voltage total harmonic distortion (THD) and their peaks. The suggested strategy leads to the best current-limiting effect and better condition for current and voltage THD compared with the current limiting strategies of ISL [19] and LL [15].

5.2 Fault studies on islanded inverter-interfaced distributed energy resources in various reference frames

The studies are conducted in the MATLAB environment for IIDERs in three-wire and four-wire configurations.Fig. 4.4A–C presents the simulation results for the following faults:

- SLG: (a-g) for four-wire configuration including a controller developed in NARF,
- Phase-to-Phase: (a-b) for 3-wire configuration including a controller developed in SYRF and,
- Phase-to-Phase to ground: (a-b-g) for 4-wire configuration including a controller in STRF.

The voltage of the output, inductance current, current of output, and limited inductance current can respectively be seen in each case of these figures. Considering these figures, it can be said that the result of the application of the suggested current limiting strategy is faulted current reduction, voltage (power) quality improvement, and FRT capability enhancement.

Table 4.1

Proposed current-limiting strategy compared with ISL [19] and LL [15] strategies.

Topology	SC fault type	Voltage THD (%)			Current THD (%)			Peak of output current (p.u.)			Peak of output voltage (p.u.)		
		ISL strategy [19]	LL strategy [15]	Proposed strategy	ISL strategy [19]	LL strategy [15]	Proposed strategy	ISL strategy [19]	LL strategy [15]	Proposed strategy	ISL strategy [19]	LL strategy [15]	Proposed strategy
Three-wire	a-g	57.2	26.5	22.6	57.2	26.5	22.6	1.68	1.35	0.6	1.36	1.18	1
	a-b-g	23.4	1.45	9.8	23.3	1.45	9.8	2	2	1.67	1	0.97	1
	a-b	26.4	0.95	1.2	25.2	1.3	1.15	2	2	2	0.98	0.97	0.8
	a-b-c-g	14.1	0.06	0.32	14.1	0.06	0.32	2	2	2	0.16	0.16	0.14
Four-wire	a-g	20.7	0.27	0.91	20.7	0.27	0.91	2	2	2	1	1	1
	a-b-g	21	0.27	1.02	20.9	0.27	1.02	2	2	2	1	1	1
	a-b	34.2	0.37	0.71	19.2	0.37	0.6	2	2	2	1	1	1
	a-b-c-g	20.9	0.13	1.02	20.8	0.13	1.02	2	2	2	0.16	0.16	0.14
Three-wire	a-g	0.54	0.36	0.5	0.54	0.36	0.5	1.65	1.41	1.52	1.59	1.54	1.57
	a-b-g	24.7	0.56	1.42	24.8	0.56	1.42	2.58	2	1.92	0.91	2.32	1
	a-b	17.2	0.58	0.52	25.1	0.63	1.12	2.6	2	1.92	0.64	1.84	0.74
	a-b-c-g	0.42	0.18	0.19	0.42	0.18	0.19	2	2	2	0.16	0.16	0.16
Four-wire	a-g	17.7	0.53	0.42	17.7	0.53	0.42	3.77	2	2	0.98	1.84	0.82
	a-b-g	26.2	0.41	0.4	26.2	0.41	0.4	3.77	2	2	0.82	1.85	0.83
	a-b	25.4	0.42	0.41	33	0.42	1.02	2.76	2	2	0.86	1.84	0.82
	a-b-c-g	0.23	0.26	0.22	0.23	0.26	0.22	2	2	2	0.16	0.16	0.14

ISL, instantaneous saturation limit; *LL*, latched limit; *THD*, total harmonic distortion.

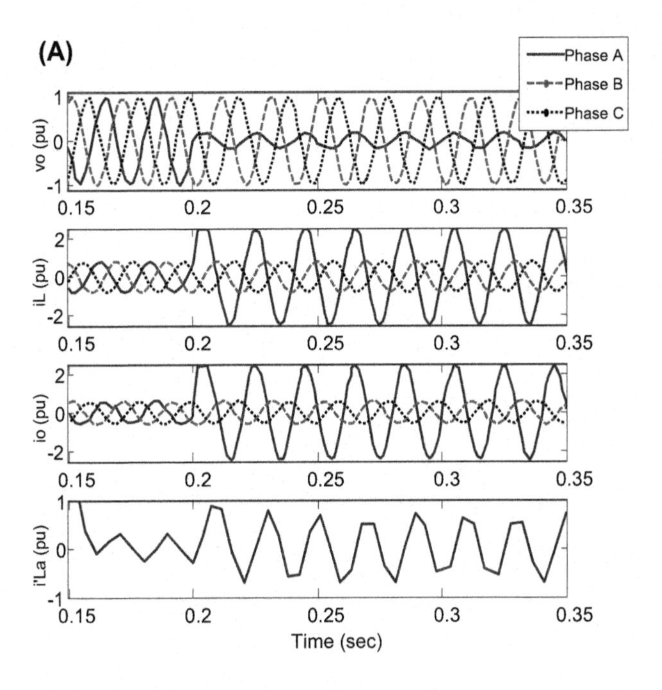

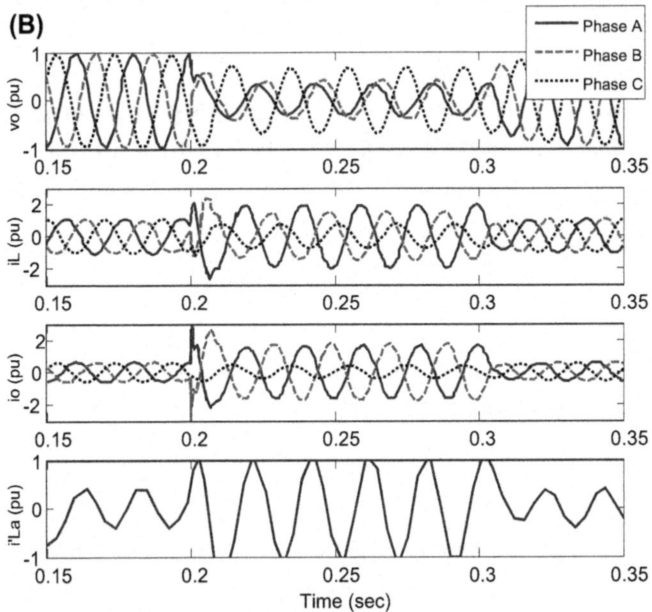

Figure 4.4 The voltage of output, inductance current, current of output, and limited inductance current, respectively, for faults (A) SLG (a–g) for four-wire configuration (controller in NARF), (B) phase-to-phase (a–b) three-wire configuration (controller in SYRF), and (C) phase-to-phase to ground (a-b-g) four-wire configuration (controller in STRF). *NARF*, natural reference frame; *SLG*, single-line/phase to ground; *STRF*, stationary reference frame; *SYRF*, synchronous reference frame.

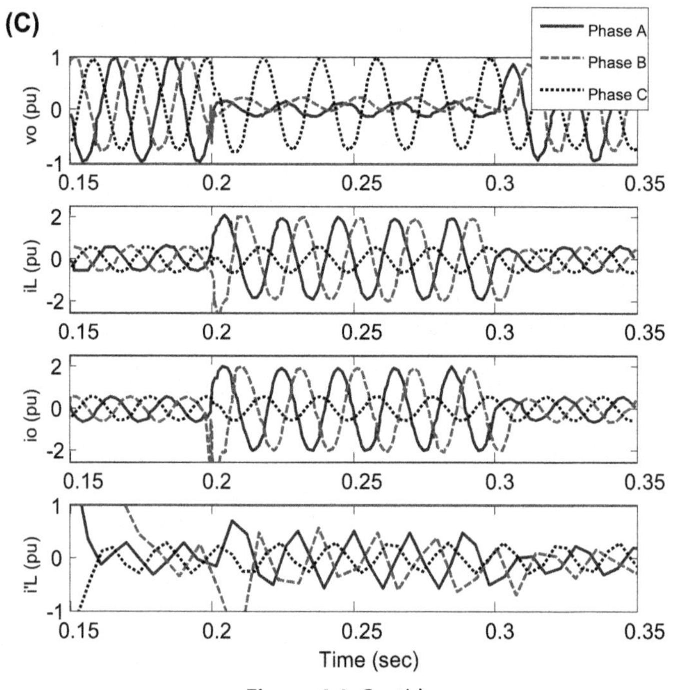

Figure 4.4 Cont'd

5.3 Droop- and directly voltage-controlled–based methods performance comparison versus dynamic loads

A three-phase 132-kW, 4-pole, 400-V, 50-Hz induction motor (IM) is modeled and simulated at bus 2. Its parameters can be found in Ref. [27]. It has a Direct Online (DOL) starting method, and it starts at $t = 0.8$ s under nominal load in the islanded MG. The simulation results, including bus voltages and active and reactive powers, are respectively, presented in Fig. 4.5A–C for the following cases.

- Hierarchical droop-based control [3,28]
- Suggested hierarchical control system with the current limiting strategy
- DVC-based control system [27]

The strategy used for current limiting can reduce IM starting transients. In Fig. 4.5A, it can be observed that after a temporary voltage drop, it is recovered to the rated level. As shown in Fig. 4.5B–C, variations of powers have been reduced. Also, their steady-state error is equal to zero.

It must be said that the hierarchical droop-based control [3,28], which includes the stabilizing supplementary control loop, has a superior

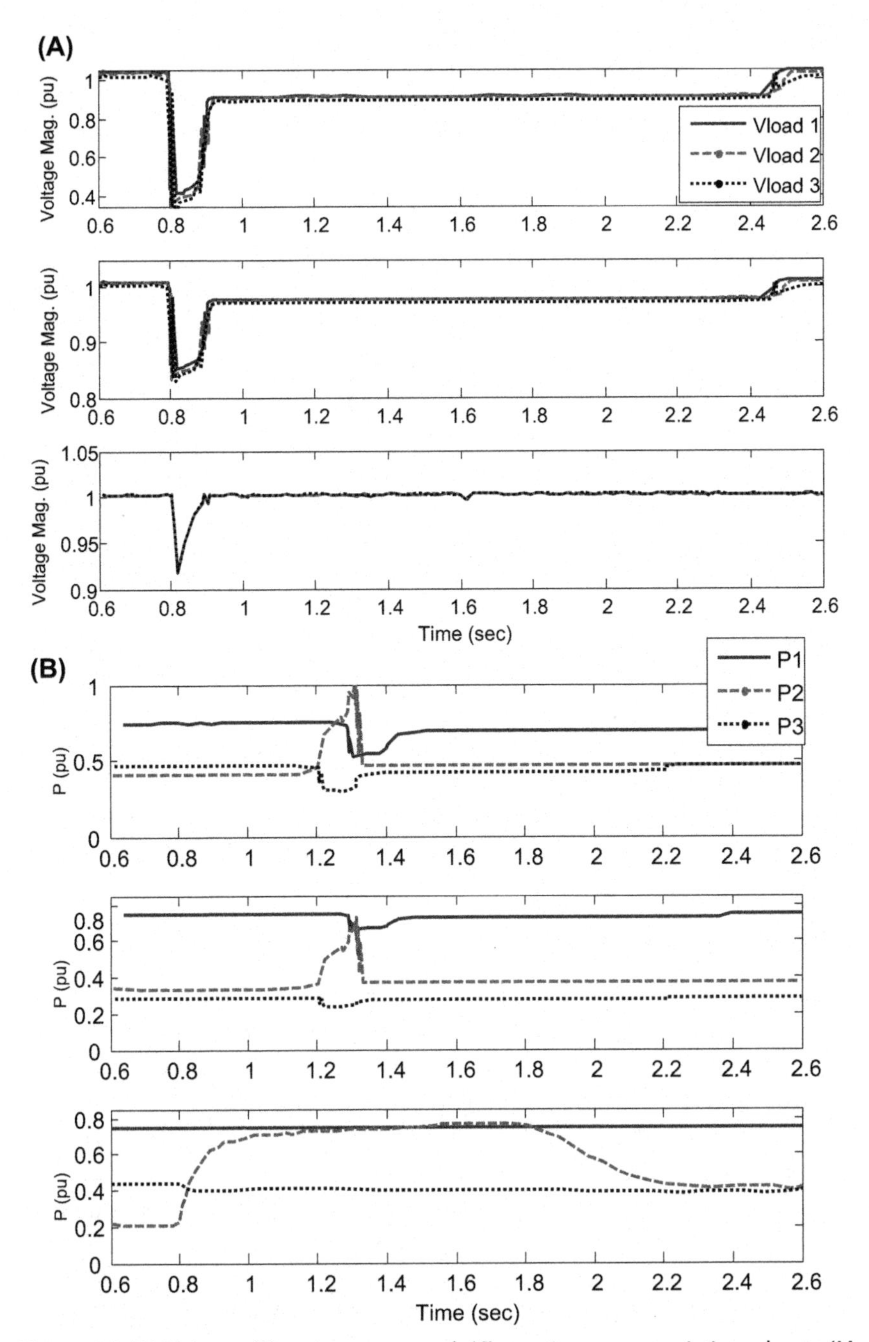

Figure 4.5 (A) Voltage, (B) active power and, (C) reactive power variations due to IM starting, respectively, using droop method controller [3,28], suggested solution including current-limiting and DVC method controller [27]. *DVC*, directly voltage-controlled; *IM*, induction motor.

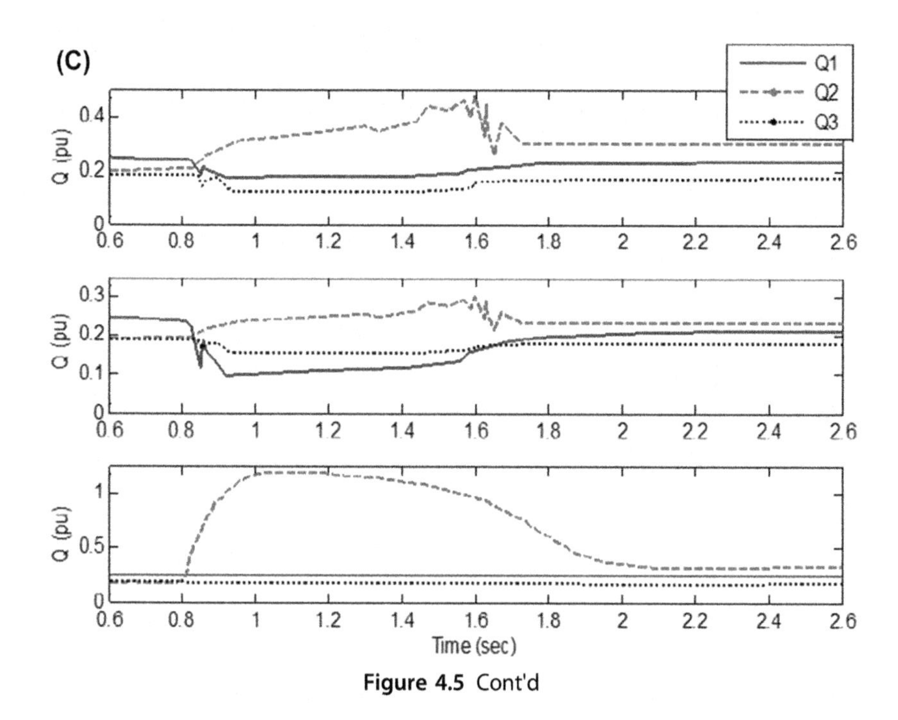

Figure 4.5 Cont'd

accomplishment compared with the conventional droop-based controllers [22,23]. As a result, the conventional droop-based controllers have not been compared with droop and DVC controllers [3,28] discussed in Sections 2 and 4.2.

6. Summary

A model for SC studies was suggested in this chapter for droop-based and DVC inverter-based DERs. The control system effect on IIDERs was studied using an enhanced control system based on the hierarchical droop method. It has a supplementary control loop for stabilizing. The DVC system was reported as well. Also, a novel strategy for current limiting was suggested, which is realizable in various reference frames.

As a case study, WG and PV system's behavior during a fault condition was investigated and discussed. The used model was an analytical one, proposed for three-wire and four-wire configurations under different balanced and unbalanced SCs. This model was verified by simulation results, which show that the suggested loop for current controlling in the hierarchical control system can decrease the MG SC current with and without using the suggested current-limiting strategy.

Future researches should be focused more on the overcurrent protection of MGs for DVC-based and droop-based IIDERs, and also the possible combination of control and protection tasks in converters.

References

[1] H.R. Baghaee, M. Mirsalim, G.B. Gharehpetian, H.A. Talebi, Three phase AC/DC power-flow for balanced/unbalanced microgrids including wind/solar, droop-controlled and electronically-coupled distributed energy resources using RBF neural networks, IET Power Electron. (2016) 1–16, https://doi.org/10.1049/iet-pel.2016.0010 (in press).

[2] H.R. Baghaee, M. Mirsalim, G.B. Gharehpetian, H.A. Talebi, Nonlinear load sharing and voltage compensation of microgrids based on harmonic power-flow calculations using radial basis function neural networks, IEEE Syst. J. (2017) 1–11, https://doi.org/10.1109/JSYST.2016.2645165 (in press).

[3] H.R. Baghaee, M. Mirsalim, G.B. Gharehpetian, Real-time verification of new controller to improve small/large-signal stability and fault ride-through capability of multi-DER microgrids, IET Gener. Transm. Distrib. 10 (12) (2016) 3068–3084, https://doi.org/10.1049/iet-gtd.2016.0315.

[4] IEEE standard for interconnecting distributed resources with electric power systems, IEEE Stand. 1547 (2003).

[5] T.N. Boutsika, S.A. Papathanassiou, Short-circuit calculations in networks with distributed generation, Elec. Power Syst. Res. 78 (1) (2007) 1181–1191.

[6] H.R. Baghaee, M. Mirsalim, G.B. Gharehpetian, M.J. Sanjari, Effect of type and interconnection of DG units in the fault level of distribution networks, in: Proc. 13th Power Elect. & Motion Control Conf., EPE-PEMC, Poznań, Poland, September 2008, pp. 313–319.

[7] Short-circuit Currents in Three-phase A.C. Systems—Part 0: Calculation of Short-Circuit Currents, IEC 60909-0, 2016.

[8] Short-circuit Currents in Three-phase A.C. Systems—Part 1: Factors for the Calculation of Short-Circuit Currents According to IEC 60909-0, IEC 60909-1, 2002.

[9] Short-circuit Currents in Three-phase A.C. Systems - Part 2: Data of Electrical Equipment for Short-Circuit Current Calculations, IEC 60909-2, 2008.

[10] Short-circuit Currents in Three-phase AC Systems - Part 3: Currents during Two Separate Simultaneous Line-To-Earth Short Circuits and Partial Short-Circuit Currents Flowing through Earth, IEC 60909-3, 2013.

[11] Short-circuit Currents in Three-phase A.C. Systems—Part 4: Examples for the Calculation of Short-Circuit Currents, IEC 60909-4, 2000.

[12] M.E. Baran, I. El-Markaby, Fault analysis on distribution feeders with distributed generators, IEEE Trans. Power Syst. 2 (4) (2005) 945–950.

[13] T.S. Ustun, C. Ozansoy, A. Ustun, fault current coefficient and time delay assignment for microgrid protection system with central protection unit, IEEE Trans. Power Syst. 28 (2) (2013) 598–606.

[14] C.A. Plet, M. Graovac, T.C. Green, R. Iravani, Fault response of grid-connected inverter dominated networks, in: Proc. IEEE Power & Energy Society General Meeting, Minneapolis, Minnesota, USA, 2010, pp. 1–8.

[15] C.A. Plet, M. Brucoli, J.D.F. McDonald, T.C. Green, Fault models of inverter-interfaced distributed generators: experimental verification and application to fault analysis, in: Proc. IEEE Power & Energy Society General Meeting, Detroit, Michigan, USA, 2011, pp. 1–8.
[16] H.R. Baghaee, M. Mirsalim, G.B. Gharehpetian, M.J. Sanjari, Fault current reduction in distribution systems with distributed generation units by a new dual functional series compensator, in: Proc. 13th Power Elect. & Motion Control Conf., EPE-PEMC, Poznań, Poland, September 2008, pp. 750–757.
[17] A. Gkountaras, S. Dieckerhoff, T. Sezi, Evaluation of current limiting methods for grid forming inverters in medium voltage microgrids, in: Proc. IEEE Energy Conv. Congress and Exp. (ECCE), Montreal, QC, Canada, September 2015, pp. 1223–1230.
[18] A.D. Paquette, D.M. Divan, Virtual impedance current limiting for inverters in microgrids with synchronous generators, IEEE Trans. Ind. Appl. 50 (2) (2015) 1630–1638.
[19] N. Bottrel, T.C. Green, Comparison of current limiting strategies during fault ride-through of inverters to prevent latch-up and wind-up, IEEE Trans. Power Electron. 29 (7) (2014) 3786–3797.
[20] N. Bottrel, T.C. Green, Investigation into the post-fault recovery time of a droop controlled inverter-interfaced microgrid", in: Proc. IEEE 6th Int. Symp. On Power Elect. for Dist. Gen. Syst. (PEDG), Aachen, Germany, June 2015, pp. 1–7.
[21] I. Sadeghkhani, M.E. Hamedani-Golshan, J.M. Guerrero, A. Mehrizi-Sani, A current limiting strategy to improve fault ride-through of inverter interfaced autonomous microgrids, IEEE Trans. on Smart Grid 5 (4) (2014) 1905–1919.
[22] J.M. Guerrero, M. Chandorkar, T. Lee, P.C. Loh, Advanced control architectures for intelligent microgrids—Part I: decentralized and hierarchical control, IEEE Trans. Ind. Electron. 60 (4) (2012) 1254–1262.
[23] P. Hasanpor-Divshali, A. Alimardani, S.H. Hosseinian, M. Abedi, Decentralized cooperative control strategy of microsources for stabilizing autonomous VSC-based microgrids, IEEE Trans. Power Syst. 27 (4) (2012) 1949–1959.
[24] N. Pogaku, M. Prodanovic, T.C. Green, Modeling, analysis and testing of autonomous operation of an inverter-based microgrid, IEEE Trans. Power Electron. 22 (2) (2007) 613–625.
[25] J.M. Guerrero, L. Hang, J. Uceda, Control of distributed uninterruptible power supply systems, IEEE Trans. Power Electron. 55 (8) (2008) 2845–2859.
[26] J.A.P. Lopes, C.L. Moreira, A.G. Madureira, Defining control strategies for analysing microgrids islanded operation, Proc. IEEE Russia Power Tech. (2005) 1–7.
[27] A.H. Etemadi, E.J. Davison, R. Iravani, A generalized decentralized robust control of islanded microgrids, IEEE Trans. Power Syst. 29 (6) (2014) 3102–3113.
[28] H.R. Baghaee, M. Mirsalim, G.B. Gharehpetian, Power calculation using RBF neural networks to improve power sharing of hierarchical control scheme in multi-DER microgrids, IEEE J. Emerg. & Select. Topic. Power Electron. 4 (4) (2016) 1217–1225, https://doi.org/10.1109/JESTPE.2016.2581762.
[29] A.H. Etemadi, R. Iravani, Overcurrent and overload protection of directly voltage-controlled distributed resources in a microgrid, IEEE Trans. Ind. Electron. 60 (12) (2013) 5629–5638.
[30] M. Hamzeh, S. Emamian, H. Karimi, J. Mahseredjian, Robust control of an islanded microgrid under unbalanced and nonlinear load conditions, IEEE J. Emerg. & Select. Topic. Power Electron. 4 (2) (2016) 512–520.
[31] M. Babazadeh, H. Karimi, A robust two-degree-of-freedom control strategy for an islanded microgrid, IEEE Trans. Power Deliv. 28 (3) (2013) 1339–1347.

[32] H. Karimi, A. Yazdani, R. Iravani, Robust control of an autonomous four-wire electronically-coupled distributed generation unit, IEEE Trans. Power Deliv. 26 (1) (2011) 455–466.
[33] Q. Zhong, T. Hornik, Cascaded current-voltage control to improve the power quality for a grid-connected inverter with a local load, IEEE Trans. Ind. Electron. 60 (4) (2013) 1344–1355.
[34] H. Karimi, H. Nikkhajoei, R. Iravani, Control of an electronically coupled distributed resource unit subsequent to an islanding event, IEEE Trans. Power Deliv. 23 (1) (2008) 493–501.
[35] A.J. Wood, B.F. Wollenberg, G.B. Sheblé, Power Generation, Operation and Control, third ed., Wiley-Interscience, USA, 2013.
[36] P. Krause, O. Wasynczuk, S. Sudhoff, Analysis of Electric Machinery and Drive Systems, second ed., Wiley, Hoboken, NJ, 2002.
[37] N. Abdel-Rahim, J. Quaicoe, Three phase voltage source ups inverter with voltage controlled current regulated feedback control scheme, in: Int. Conf. On Ind. elect., Control & Inst., vol. 1, September 1994, pp. 497–502.
[38] I. Erlich, W. Winter, A. Dittrich, Advanced grid requirements for the integration of wind turbines into the German transmission, in: Proc. IEEE Power Engineering Society (PES) General Meeting, Montreal, Canada, June 2006, pp. 1–7.
[39] A. Morales, X. Robe, M. Sala, P. Prats, C. Aguerri, E. Torres, Advanced grid requirements for integration of wind farms into Spanish transmission system, IET Renew. Power Gener. 2 (1) (2008) 47–59.
[40] E.ON Netz GmbH, Grid Code for High and Extra High Voltage, April 2006.
[41] Vestas Advanced Grid Option 3—V52-850 kW, V66-1.75 MW, V80- 2.0 MW, Technical Description, 2003.
[42] Vestas VCS Frequency Converter, Technical Description, 2001.
[43] A. Datta, A. Ray, D. Mukherjee, H. Saha, Selection of islanding detection methods based on multi-criteria decision analysis for grid-connected photovoltaic system applications, Sustain. Energy Technol. & Assess. 7 (1) (2016) 111–122.
[44] J. Olamaei, S. Ebrahimi, A. Moghassemi, Compensation of voltage sag caused by partial shading in grid-connected PV system through the three-level SVM inverter, Sustain. Energy Technol. & Assess. 18 (1) (2016) 107–118.
[45] P. Satapathy, S. Dhar, P.K. Dash, A mutated hybrid firefly approach to mitigate dynamic oscillations of second order PLL based PV-battery system for microgrid applications, Sustain. Energy Technol. & Assess. 16 (1) (2016) 69–83.
[46] M.M. Rezaei, J. Soltani, A robust control strategy for a grid-connected multi-bus microgrid under unbalanced load conditions, Int. J. Electr. Power Energy Syst. 71 (1) (2015) 68–76.

Further reading

[1] J.M. Guerrero, J. Matas, L.G. de Vicuna, M. Castilla, J. Miret, Decentralized control for parallel operation of distributed generation inverters using resistive output impedance, IEEE Trans. Ind. Electron. 54 (2) (2007) 994–1004.
[2] W. Yao, M. Chen, J. Matas, J.M. Guerrero, Z.M. Qian, Design and analysis of the droop control method for parallel inverters considering the impact of the complex impedance on the power sharing, IEEE Trans. Ind. Electron. 58 (2) (2011) 576–588.

CHAPTER 5
Operation under unbalanced conditions

1. Introduction

In the near future, microgrids are envisaged to prominently alter the shape of energy systems. They offer independent operation, higher efficiency, vast utilization of renewable and distributed energy resources (DERs), environmental benefits, feeding the remote loads, etc. However, crucial challenges arise regarding their power quality and reliability, especially in the off-grid mode of operation. Operation under unbalanced conditions is one of the major issues that relate to both power quality and reliability of the microgrid and the host grid. In microgrids, voltage unbalance can happen due to steady-state asymmetrical loading and configuration as well as temporary unbalanced faults. Uneven distribution of single-phase DG units and unbalanced/nonlinear loads can cause voltage unbalance in the steady-state operation as an important power quality problem. On the other hand, asymmetric faults can impose severe unbalanced conditions, leading to a serious reliability issue.

If not properly managed, unbalanced conditions harm the operation of a microgrid and its components by increasing the loss, degrading the stability, overstressing in dynamic loads, and imposing power oscillations in converter-interfaced distributed generation (DG) units. The IEEE standard [1], International Electrotechnical Commission (IEC), and International Council on Large Electric Systems (CIGRE) [2] all recommend the maximum limit of 2% for voltage unbalance factor in steady-state operation. On the other hand, the new interconnection grid codes necessitate the participation of DGs in supporting power system stability and providing ancillary services, such as low-voltage ride-through (LVRT) capability and unbalance voltage compensation, during temporary gird faults [3]. Therefore, for remaining in the grid-connected mode of operation, a microgrid should ride-through such faults and support the grid voltage. This practice

Microgrids and Methods of Analysis
ISBN 978-0-12-816172-2
https://doi.org/10.1016/B978-0-12-816172-2.00005-5

will have several advantages for microgrids. First, it provides economic benefits since the islanded mode can be avoided and the generated power is not wasted. Second, it prevents harmful impacts of the switching between grid-connected and islanded modes of operation, such as voltage and current deterioration after an islanding, excessive inrush current after a reconnection process, and potential damages to the electrical equipment. In addition to the advantages of microgrids, such support will also benefit power system stability during grid faults [4]. Therefore, it is important to study the operation of a microgrid under unbalanced conditions, caused by either permanent situations or temporary events.

Literature review reveals that numerous methods have been proposed to mitigate the unbalance voltage in microgrids. These methods can be categorized into five main groups:

- ***Load rearrangements:*** Uneven distribution of loads is one of the major causes of steady-state voltage unbalance in microgrids. Therefore, making rearrangements and evenly distributing the loads among phases can reduce voltage unbalance. The latest work in this approach [5] has proposed a dynamic load transfer scheme through static switches and a central controller. By this approach, residential loads transfer from one phase to another phase can reduce voltage unbalance.
- ***Demand-side management:*** Demand response has extensively been studied for various management objectives in microgrids including voltage unbalance mitigation. The most common device to be employed for this purpose is electric vehicles (EVs), as they are becoming more widespread. A recent study [6] has experimentally evaluated EVs' participation in mitigating voltage unbalance by a local charging scheme based on a droop control method.
- ***Generation- and demand-side management:*** The coordination between the generation-side and the demand-side management can also be implemented for enhanced voltage unbalance mitigation. In the latest work of this category [2], a voltage unbalance mitigation scheme has been proposed for an islanded microgrid by coordinating photovoltaic (PV) grid-connected converters and thermostatically controlled loads. The control strategy keeps the voltage unbalance factor within the permissible range by coordinating the PV converters and such residential loads while maintaining the quality of service and customers' thermal comfort.
- ***D-FACTS devices:*** Distributed flexible alternating current transmission system (FACTS) devices have also been deployed to mitigate the voltage unbalance. For instance, the series of active power filters can

compensate the unbalanced voltage by injecting negative sequence voltage in series with the grid. The most recent work in this group [7] has proposed a real-time supervisory control for coordinating active power filters and DG units to optimally collaborate in mitigating the voltage unbalance of a multiarea microgrid.

- ***Converter-interfaced DG units:*** Having several converter-interfaced DG units, microgrids have a great potential to utilize the excess capacity of such units to provide various ancillary services, such as unbalance voltage mitigation and supporting the grid during asymmetrical faults. To accomplish this, numerous studies have thus utilized converter-interfaced DG units for injecting flexible supportive currents with active and reactive components in positive and negative sequences.

Grid-tied microgrids and converter-interfaced DG units are expected to dominate power systems, mainly, thanks to the development of power electronic converters. Therefore, these converters are vital components for the transition of the conventional power systems into future highly integrated energy systems, which will operate in balanced and unbalanced conditions. In this path, several control methodologies and schemas (mainly based on the symmetric sequences) have been proposed to address the concerns and issues related to unbalanced conditions of converter-interfaced units [2–4,8–10]. This chapter will focus on the advanced control techniques to tackle unbalanced conditions in a microgrid by converter-interfaced units.

2. Recent grid code requirements

Conventional power systems rely heavily on their large synchronous generators to support the grid and provide crucial ancillary services, such as delivering large currents during the faults that are very important to support the grid voltage and activation of the protective relays. Unlike these conventional systems, emerging converter-interfaced power units, such as DGs and grid-tied microgrids, have several challenges in supporting grid stability. For instance, these converter-interfaced units, depending on their semiconductor capabilities, can only provide 1–2 pu current during different faults. Besides, the control systems utilized in the converter-interfaced units are sensitive to grid voltage fluctuations. Therefore, increasing requirements, such as LVRT, high-voltage ride-through (HVRT), and reactive current injection (RCI) under short-term grid faults, have been imposed by system operators for the interconnection of new generating units, whether it be a grid-connected microgrid or the DGs inside a microgrid during its grid-connected mode of operation.

The operation of microgrids is prone to even more undesirable functions during unbalanced faults. These vulnerabilities occur during abnormal events such as distortions on the voltage and current of the ac-side, oscillations on the dc-side voltage, and output power oscillations. These undesirable conditions can largely harm the operation of the entire microgrid as well as the host power system. If not managed properly, this can cause cascading failure and serious damages. Hundreds of industrial and academic projects are consequently being conducted across the globe to address the reliability and stability issues in the emerging power systems. "Synchronous Condensers Application in Low Inertia Systems" in Denmark and the "MIGRATE" project under the European Union framework are just two examples of the extensive research and development efforts to tackle the grid stability and reliability challenges in future low-inertia and highly integrated electric grids [11].

2.1 Low-voltage ride-through requirements

Since the past decade, countries with high wind penetration levels such as Germany, Denmark, Spain, and other European countries have started to impose the LVRT requirements by new grid codes. Typically, these codes require DG units to remain connected for a minimum of 150−500 ms and riding through low voltages (sometimes as low as zero voltage in some codes). System characteristics affect the amount of low voltage and its duration. Systems with more vulnerability to instabilities and systems with higher DG penetration usually mandate more stringent requirements to simultaneously preserve the reliability and allow higher integration.

Grid code updates in other parts of the world usually follow Europe based on the integration level of DGs in their distribution systems. Now, different countries aim to regularly and rapidly revise their interconnection codes to increase the utilization of renewable energies, enhance grid performance, and maintain the system stability and reliability at the same time. Followings are some examples of the frequent updates of such codes in different parts of the world. China has been revising its interconnection codes to foster increasing integration of DG units, specifically wind generation, without harming its grid reliability. As a result, the LVRT requirement has been mandated in the Chinese grid code since 2010. In the United States, multiple regulatory institutions (such as the US Federal

Energy Regulatory Commission and North American Electric Reliability Corporation) conduct research and development projects to develop national grid interconnection standards. In compliance with these national regulatory organizations, regional reliability institutions establish their standards and guidelines. The projects accomplished by Independent System Operator – New England in 2009 and the Electric Reliability Council of Texas in 2010 are two well-known examples in this area. These two regions have high wind installation goals for the future. Similarly, more developments and updates are expected for future grid codes in the United States [3]. Table 5.1 provides a summary of time and voltage level in the LVRT curves of 23 countries.

Table 5.1 Time–voltage points in LVRT curves of different grid codes [17,18].

	Time (s)	Voltage (p.u.)
Germany	0.15, 1.5	0, 0.9
Spain	0.5, 1	0.2, 0.8
Ireland/Romania/Cyprus/	0.625, 3	0.15, 0.9
Canada (AESO)	0.6, 3	0.15, 0.8
Poland	0.5, 3	0.15, 0.9
Turkey	0.15, 1.75	0, 0.9
United States (WECC)/	0.15, 1.5	0.15, 0.9
Canada (BC hydro)	0.25, 0.75	0.25, 0.95
Greece	0.1, 3	0.15, 0.9
Denmark/Norway	0.625, 2	0.2, 0.9
India	0.4, 2	0, 0.7
China		
Australia		
Sweden/Finland	0.25, 0.25, 0.75	0, 0.25, 0.9
France	0.15, 0.15, 0.7,	0, 0.5, 0.5, 1
Belgium	1.5	0, 0.5, 0.5, 0.925
Italy	0.2, 0.2, 0.7, 0.7	0.2, 0.75, 0.75, 0.85
New Zealand	0.5, 0.8, 2, 2	0, 0.58, 0.62, 0.9
Brazil	0.2, 0.2, 1, 1	0.2, 0.85, 0.85, 0.9
	0.5, 1, 5, 5	
United Kingdom	0.15, 0.15, 1.25,	0, 0.15, 0.8, 0.8, 0.85
Canada (hydro Quebec)	2.5, 2.5	0, 0.15, 0.75, 0.75, 0.85
	0.15, 1, 1, 2, 2	
United States (IEEE 1547a-2014)	0.16, 0.16, 1, 1, 2, 2	0, 0.45, 0.45, 0.6, 0.6, 0.88
United States (FERC/NERC)	0.15, 0.15, 0.3, 0.3, 2, 2, 3, 3	0, 0.45, 0.45, 0.65, 0.65, 0.75, 0.75, 0.9

LVRT, low-voltage ride-through.

2.2 High-voltage ride-through requirements

Recently, the HVRT requirement is also becoming important in the grid codes. New interconnection standards provide the HVRT guidelines for the regulation of the overvoltages during temporary faults in the system. These temporary overvoltages may happen due to single-phase or double-phase faults and variations in loads as well as in generation units. As a recent example, the latest German interconnection technical standard [15] imposed the HVRT requirement for the connection of new power plants to the German high-voltage grid (60−150 kV) for the first time. According to the HVRT requirements, new power generation units should have a stable operation without disconnecting from the grid as long as the highest line voltage is under the specified curves in the HVRT codes. Table 5.2 shows the time−voltage parameters of the HVRT codes in some countries.

2.3 Reactive current injection requirements

The RCI is another requirement identified by some grid codes for new interconnections to provide the necessary support under the grid faults at the point of common coupling (PCC). Power system operators in countries with large contribution of wind power such as Denmark, Germany, England, Ireland, and Spain now impose RCI requirements for the large DG interconnections in addition to the LVRT, to further support the grid reliability. However, countries with lower wind integration do not stipulate the need for RCI during faults yet. Table 5.3 summarizes the RCI requirements in grid codes of different countries.

The primary RCI codes mainly focused on the required reactive current only under balanced grid faults. However, the most recent grid code,

Table 5.2 Time−voltage points in HVRT curves of different grid codes [17,18].

	Time (s)	Voltage (p.u.)
Canada (AESO)	—	1.1
Italy	0.1, 0.1, 0.5, 0.5	1.25, 1.2, 1.2, 1.15
Canada (hydro Quebec)	0.15, 0.15, 2, 2	1.4, 1.25, 1.25, 1.2
Australia	1, 1, 10, 10	1.3, 1.2, 1.2, 1.1
China	0.2, 0.2, 10, 10	1.2, 1.15, 1.15, 1.1
United States (WECC)/Canada (BC hydro)	1, 1, 2, 2, 3, 3	1.2, 1.18, 1.18, 1.15, 1.15, 1.1
United States (FERC/NERC/ ERCOT)	0.15, 0.15, 0.5, 0.5, 1, 1	1.2, 1.18, 1.18, 1.15, 1.15, 1.1

HVRT, high-voltage ride-through.

Table 5.3 RCI requirements in Europe [11].

Country	Current type	Amount	Rising time (ms)
Denmark	—	≥2% injection for 1% PCC voltage reduction	100
Germany	Positive-sequence additional	≥2% injection for 1% PCC voltage reduction	50 (90%)
Ireland	—	At least proportional to the voltage dip	100 (90%)
Spain	Positive-sequence absolute	3%, 0.75%, or 0.5% injection for 1% PCC voltage reduction (depending on voltage-dip level)	150
United Kingdom	—	Maximum reactive current without exceeding the transient rating limits	—

PCC, point of common coupling; RCI, reactive current injection.

published in 2015 in Germany [15], has even considered the negative-sequence current injection during unbalanced faults. In this regard, recent efforts have been carried out to study the compliance of DG units with the positive- and negative-sequence RCI during unbalanced faults [16,17].

2.4 Frequency control and active power restoration

Under the grid faults, DG units should be able to control their power output according to the defined levels in the interconnection standards. The standards of Germany, Ireland, and Nordic countries require DG units to have the capability of active power curtailment while demanding reactive power injection to support the voltage. Also, most grid codes require large DG units to contribute to supporting the system frequency. The Irish and Danish codes reveal specific response curves that require DGs to alter their output active power for supporting the frequency stability. For instance, DG units must reduce their output active power with a gradient of 40% (of the available power) per each Hz increment in the frequency over 50.2 Hz, according to the German code. The specifications of active power restoration also vary in different systems. According to British and Irish codes, the active power must be rapidly restored to at least 90% of the prefault available value within 1 s after the voltage recovery. According to the German code, active power restoration is required with a rate of (at least) 20% of the nominal output power in a second (reaching 100% in 5 s). Like LVRT curves, the severity of the frequency response requirement and active power restoration corresponds to the grid strength. For example, grid codes in weak systems require faster active power restoration, which is crucial for system stability [3,18].

2.5 Asymmetric ride-through scheme

In addition to the presented requirements so far, a new scheme is provided in this chapter that can be a useful guideline for regulating the operation of microgrids under unbalanced grid faults. This scheme is called asymmetric ride-through (ART) which has been recently proposed [3] and can be adopted by the updated versions of the future interconnection codes. The ART scheme consists of three curves. Fig. 5.1 shows three curves of the ART regulation scheme, i.e., the curve of the lower boundary for the voltage magnitude in the positive sequence (LBVP), the curve of the higher boundary for the voltage magnitude in each phase (HBVP), and the curve of the higher boundary for the voltage magnitude in the negative sequence (HBVN). The general curves of HBVP and LBVP are designed to be able to capture the specifications, respectively, from the existing HVRT and LVRT curves in different grid codes.

By reviewing LVRT codes of 23 countries (reported in Table 5.1), the generic form of the LBVP curve should have at least four "time parameters" (T_i^{lp}) and five voltage parameters (V_i^{lp}) to be able to adapt to the specifications of the LVRT codes in these countries. As Table 5.1 indicates, for

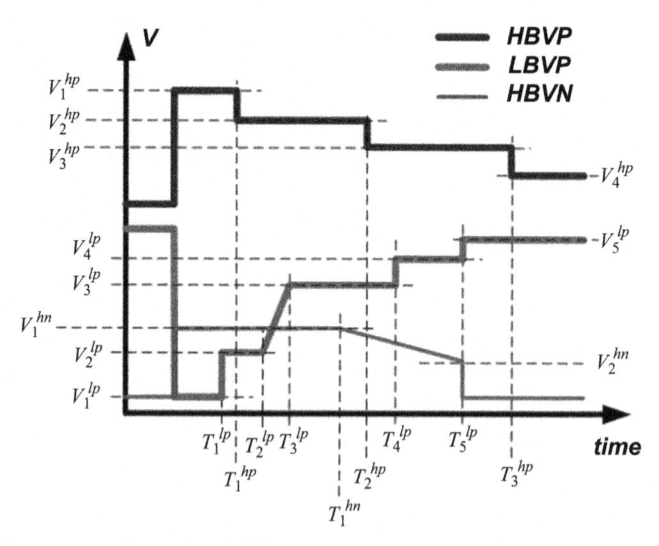

Figure 5.1 General schematic of ART boundary curves for positive sequence, negative sequence, and phase voltage of the PCC. *PCC*, point of common coupling.

instance, only two T^{lp} and two V^{lp} parameters are needed to adopt the LBVP curve of Fig. 5.1 from the LVRT codes of Germany and Ireland. However, adopting from some other codes requires more parameters. For instance, three T^{lp} and three V^{lp} parameters are required to adopt the LBVP curve from France and Italian codes, and five T^{lp} and five V^{lp} parameters are needed for representing British and USA-FERC codes. Furthermore, Table 5.2 indicates the time−voltage parameters of HVRT curves in different grid codes. To adopt the HBVP curve of Fig. 5.1 from the HVRT curves in these codes, three $T^{h\,p}$ parameters and four $V^{h\,p}$ parameters are enough.

To complete the ART scheme, the HBVN curve is also presented to regulate the negative-sequence voltage caused by any unbalanced fault. Three ART curves contain 11 voltage parameters and 9 "time parameters" in total. These curves aim to extend the concept behind the existing LVRT curves to better regulate the operation of the microgrids under asymmetric short-term faults for increasing their reliability in future's highly integrated power systems. Therefore, the followings are three requirements for a successful ART performance, according to Fig. 5.1:

(1) V_p is (or can be regulated) above the LBVP curve,

(2) "V_n" remains (or can be regulated) under the HBVN curve, and

(3) none of the phase voltages exceeds the HBVP curve.

If a microgrid can meet all three requirements, it can still stay connected to the grid to avoid islanding and reconnection procedure. By regulating the operation characteristics of a microgrid under short-term unbalanced grid faults, the control systems are required to regulate the voltage parameters at the PCC to meet the ART guidelines. This can be beneficial for both the microgrid and the grid. Twenty parameters of the ART scheme can be set based on the microgrid specifications and grid characteristics. This will help the microgrid and the host grid to improve their reliability by better regulating the operation under temporary unbalanced faults.

2.5.1 Two-sample asymmetric ride-through schemes

Based on the general curves of Fig. 5.1, two ART curves are presented in this chapter for riding through asymmetrical short-term faults. Fig. 5.2 illustrates the first curve, named ART-1, which comprises two stages, i.e., withstanding (immediately after the fault) and recovery stages. The duration of the first stage in the ART-1 is suggested to be 150 ms after the fault occurrence. This value can vary based on the specification of each grid. However, since most grid codes have the first stage duration of 150 ms in their LVRT curves (as indicated in Table 5.1), it is also suggested here for

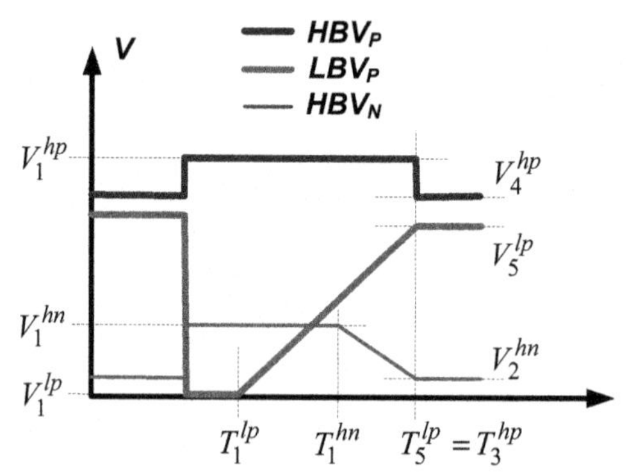

Figure 5.2 ART-1 curve: $V_1^{lp} = 0$, $V_5^{lp} = 0.9pu$, $V_1^{hp} = 1.15pu$, $V_4^{hp} = 1.1pu$, $V_1^{hn} = 0.35pu$, $V_2^{hn} = 0.05pu.T_1^{lp} = 150ms$, $T_5^{lp} = 1.5s$, $150ms \leq T_1^{hn} \leq 1.5s$. *ART*, asymmetric ride-through.

the ART-1 scheme. If the duration of the fault is more than 150 ms and one of the followings happens:

- the positive-sequence voltage magnitude passes LBVP curve, or
- the negative sequence voltage magnitude exceeds the HBVN curve, or
- the phase voltages magnitude surpass the HBVP curve. The microgrid has two options:
  - disconnecting from the grid, or
  - remaining connected to the grid and supporting the PCC voltage (as much as it can) to meet the ART requirements.

Here, the goal is to keep the voltage magnitudes in the allowed boundaries of the ART scheme by boosting the positive-sequence voltage value and reducing the negative-sequence value. The microgrid is required to at least withstand the asymmetric fault up to the time that the connection voltage (without any support) stays within the ART curves. When one of the voltage parameters (without any support) passes the corresponding ART curve, it is no longer necessary to stay connected, but it is still preferred that the DG utilizes its available capacity to further stand connected by regulating its voltage parameters within the allowed ART boundaries. The more one DG can support the PCC to stay at the allowed ART boundaries, the more reliability the host grid can have. For simplicity, 20 ART parameters of Fig. 5.1 are aggregated in six voltage parameters and three "time parameters" (as shown with the suggested values in Fig. 5.2). To consider a single-phase-to-ground fault, V_1^{hn} is suggested 0.35 pu in this chapter. It is selected that the LBVP curve in ART-1 is similar to the LVRT curve of the first group of codes in Table 5.1 (with two tim—-voltage points). This is to show that the general ART scheme in Fig. 5.1 can capture the specifications from the existing LVRT curves.

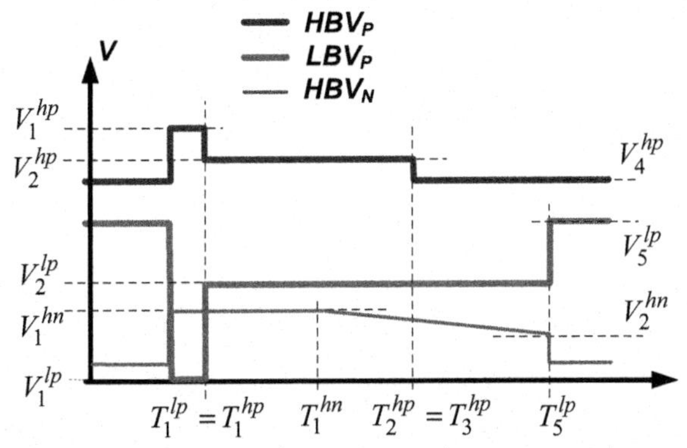

Figure 5.3 ART-2 curve: $V_1^{lp}=0$, $V_2^{lp}=0.5pu$, $V_5^{lp}=0.9pu$, $V_1^{hp}=1.4pu$, $V_2^{hp}=1.2pu$, $V_4^{hp}=1.1pu$, $V_1^{hn}=0.35pu$, $0.05pu \leq V_2^{hn} \leq 0.35pu.T_1^{lp}=150ms$, $T_2^{hp}=1s$, $T_5^{lp}=2s$, $150ms \leq T_1^{hn} \leq 2s$. *ART*, asymmetric ride-through.

Fig. 5.3 also shows another example for riding through asymmetrical faults, named ART-2. According to ART-2 curve, the faults with V_p down to zero can be tolerated up to 150 ms. Also, the voltage sags with a V_p value more than 0.5 pu can be endured up to 2 s. The values of eight voltage parameters and four "time parameters" in ART-2 curves are shown in Fig. 5.3. Again, it is selected here that the LBVP and HBVP curves in ART-2 be, respectively, similar to the LVRT and HVRT curves according to Tables 5.1 and 5.2

Although these curves can be applied to any generic asymmetrical situation, it is beneficial if their application can be further analyzed for common types of unbalanced faults. For better understanding, the ART schemes are studied in the next section under two main unbalanced fault types, i.e., single-phase and double-phase faults. Also, the values of the T_1^{hn}in Fig. 5.2 and V_2^{hn} and T_1^{hn} in Fig. 5.3 can be better understood by the analysis of the next section.

2.5.2 ART-1 and ART-2 schemes with single-phase and double-phase faults

For a typical single-phase or double-phase fault, the following equation simply gives the magnitude of the positive-sequence voltage:

$$V_p = \frac{V_a + V_b + V_c}{3} \tag{5.1}$$

In the case of a single-phase fault, the magnitude of the positive sequence of the voltage is shown in Fig. 5.4A. To interpret this curve for the phase voltage magnitude, the value of $T_{1\phi}$ and $T_{2\phi}$ should be found.

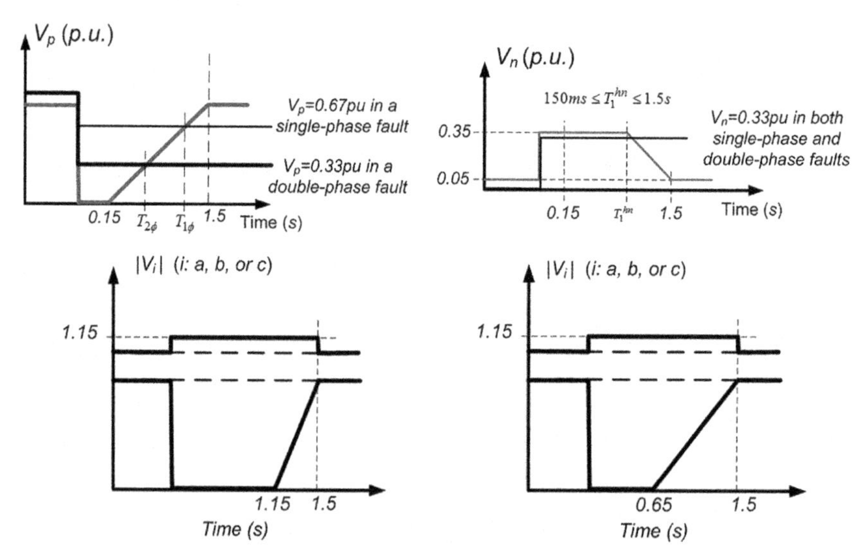

Figure 5.4 ART-1 scheme: (A) LBVP curve and Vp in two fault cases, (B) HBVN curve and Vn in two fault cases, (C) boundaries for the phase voltages under a single-phase fault and (D) boundaries for the phase voltages under a double-phase fault.

From Fig. 5.4A, $T_{1\phi}$ and $T_{2\phi}$ can be easily found as 1.15 and 0.65 s, respectively. If the grid code specifies $T_1^{hn} = T_{1\phi}$, it mandates the DG to withstand the single-phase-to-ground faults up to 1.15 s. However, if the code sets $T_1^{hn} \leq T_{1\phi}$, it allows DG to disconnect under the severe single-phase-to-ground fault before 1.15 s, even if the positive-sequence voltage is still above the LBVP curve. These analyses help to better set the parameters in ART-1 and ART-2 based on the grid requirements.

Since the ART scheme deals with unbalanced faults and boosting all three-phase voltages may cause over-voltage in the un-faulted phases, regulating the phase voltages during a successful ART is also important. Therefore, the ART-1 can be represented for the phase voltage magnitudes in the case of single-phase and double-phase faults as Fig. 5.4B and C, respectively. If the faulted phase voltage magnitude in a single-phase fault stays within the curves indicated in Fig. 5.4C, V_p is above the LBVP curve in ART-1 leading to a successful asymmetrical riding-through period. Similarly, if the phase voltage magnitudes of two faulted phases in a double-phase fault stay within the curves indicated in Fig. 5.4D, the DG will experience a successful ART.

Fig. 5.5 illustrates a similar idea for the ART-2 curve. Fig. 5.5A and C show that even if the faulted phase in a single-phase fault case stays zero up

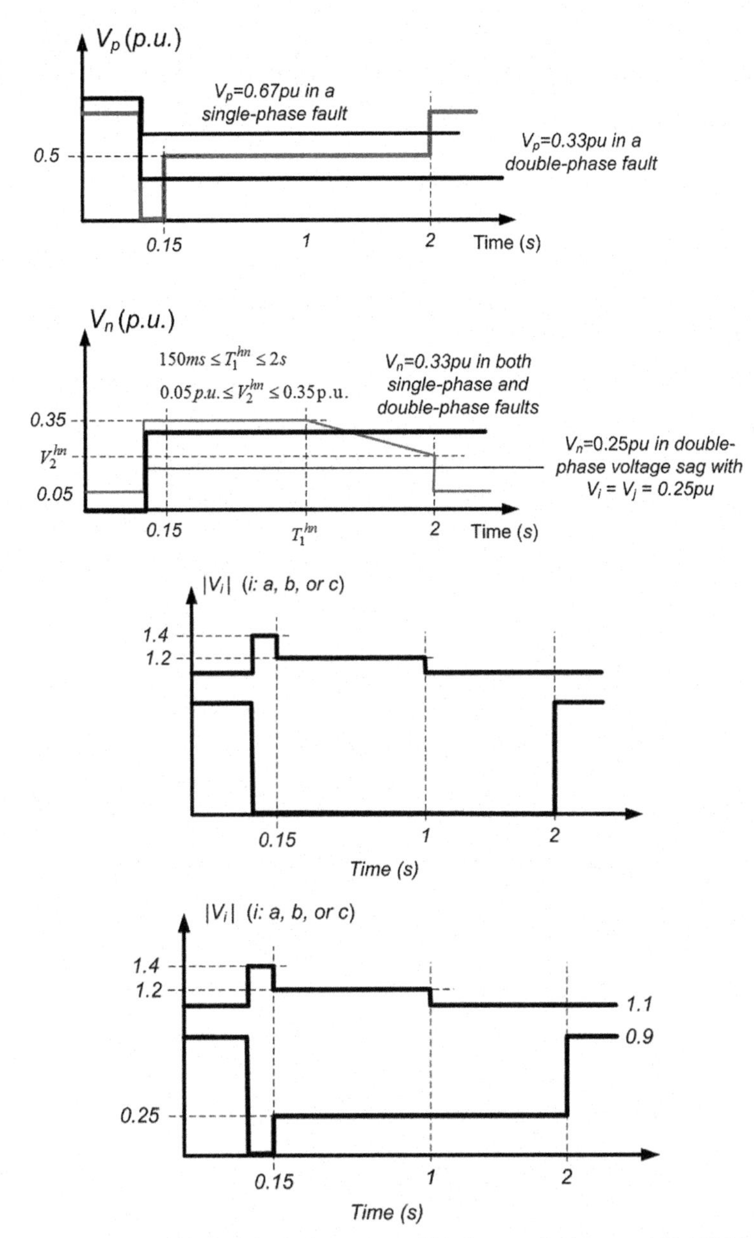

Figure 5.5 ART-2 scheme: (A) LBVP curve and Vp in two fault cases, (B) HBVN curve and Vn in three fault cases, (C) boundaries for the phase voltages under a single-phase fault and (D) boundaries for the phase voltages under a double-phase fault.

to 2 s, the ART process can be successful because V_p is higher than 0.5 pu. However, a grid code can limit the ART requirement by setting T_1^{hn} value to be lower than 2 s, as presented in Fig. 5.5B. On the other hand, the phase voltage magnitudes of two faulted phases should stay within the curves indicated in Fig. 5.5D to satisfy the first requirement in the ART-2 scheme (LBVP curve of Fig. 5.5A), in the case of a double-phase fault.

3. Operation of interconnecting converters under unbalanced conditions

In the hybrid AC/DC microgrid of Fig. 5.6, there are two types of interconnecting converters, i.e. N interfacing converters for N DG sources connected to the AC side and K parallel interlinking converters for interconnecting AC and DC sides. For simplicity of the analysis in this section, K parallel interlinking converters are treated as one unit. In general, the reference current of an interconnecting converter can be generated by a

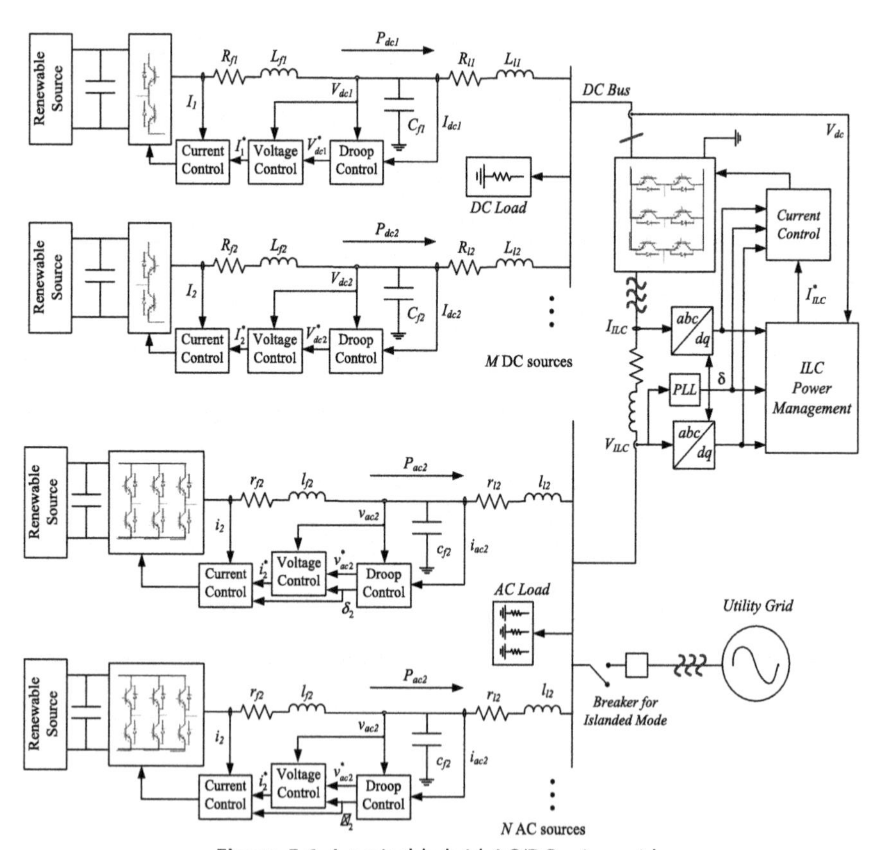

Figure 5.6 A typical hybrid AC/DC microgrid.

flexible combination of positive/negative and active/reactive current components under unbalanced conditions, as follows

$$i = i_p^+ + i_p^- + i_q^+ + i_q^-$$

where we have:

$$\begin{aligned} i_p^+ &= k_p \frac{P^*}{(V^+)^2} v^+ = K_p^+ v^+, \quad i_p^- = (1 - k_p) \frac{P^*}{(V^-)^2} v^- = K_p^- v^- \\ i_q^+ &= k_q \frac{Q^*}{(V^+)^2} v_\perp^+ = K_q^+ v_\perp^+, \quad i_q^- = (1 - k_q) \frac{Q^*}{(V^-)^2} v_\perp^- = K_q^- v_\perp^- \end{aligned} \tag{5.2}$$

where i and v are the current and voltage of the converter and $v_\perp$ is the orthogonal voltage vector (90° leading from v). The vectors with superscripts "+"/"-" and subscripts "p"/"q" denote the positive/negative and active/reactive components, respectively. The setpoints for the average active and reactive powers of the converter are represented by P^* and Q^*. Furthermore, k_p and k_q are two controlling parameters for the flexibility of the reference current generation strategy that sets the proportions between the positive and negative sequences of, respectively, the active and reactive current components under unbalanced conditions. In addition to the proper operation controllability by this setting, it thus offers valuable flexible ancillary services. Also, the positive- and negative- sequence voltage vectors can be written for any unbalanced condition in the $\alpha\beta$ frame as:

$$\begin{aligned} v^+ &= \begin{bmatrix} v_\alpha^+ \\ v_\beta^+ \end{bmatrix} = \begin{bmatrix} V^+ \cos(\omega t + \delta^+) \\ V^+ \sin(\omega t + \delta^+) \end{bmatrix}, \quad v^- = \begin{bmatrix} v_\alpha^- \\ v_\beta^- \end{bmatrix} \\ &= \begin{bmatrix} V^- \cos(\omega t + \delta^-) \\ -V^- \sin(\omega t + \delta^-) \end{bmatrix} \end{aligned} \tag{5.3}$$

Many efforts have been made in the current literature to propose the proper operation of the converters under unbalanced conditions, e.g. [9–11], [13,14,17,19]. These studies mainly focus on the following five areas. For a glance at the state of the art, the recent accomplishments in each area are introduced here:

- ***Quality of injected current:*** The most recent work on the quality of the injected current under unbalanced conditions [19] presents a model-based control design to improve the dynamic performance of the grid-connected converters (GCCs) and enhances the fault current transients.
- ***Reduction of oscillations on active power and dc-link voltage:*** Ref. [13] proposes a series of control strategies and corresponding circuit

configurations which utilize the zero sequence components to enhance the power controllability and eliminate active power oscillations.

- ***Controlled oscillations on output power:*** The flexible control of oscillations on the output active and reactive powers, different strategies based on symmetric-sequence components are proposed in [14], yielding adaptive controllability that can manage multiple objectives and constraints. It is shown that active and reactive power oscillations can be independently regulated with two individually adaptable parameters. In a range of variation for these parameters, the amplitudes of oscillating power can be also slightly controlled, as well as the peak values of the output currents. For instance, the oscillating active power can be limited below a certain amount while the output currents are controlled to be as balanced as possible.
- ***Flexible voltage support:*** The strategies in this category (e.g., the methods proposed in [3,4,9,10]) can themselves be divided into smaller groups. Overviews on different voltage support strategies in grid-connected converters are studied in [3,11], and [17]. The latest effort in this category [10] proposes to regulate the phase voltage magnitudes within the pre-defined boundaries under the unbalanced grid faults, using Eq. (5.1) and by properly compensating the undesirable effect of the zero-sequence voltage in the regulation scheme. Due to the importance of flexible voltage support under unbalanced conditions, this category is discussed in detail in the following section of this chapter.

This section of the chapter focuses on the control schemes with analytical expressions capable of finding the optimal values for the reference parameters (k_p, k_q, P^* and Q^*) under unbalanced voltage condition to achieve the following objectives:

- Minimizing the oscillations on the active and reactive powers,
- Boosting and balancing phase voltages of the PCC,
- Minimizing inverter fault current, and
- Maximizing active and reactive power delivery.

To fully accomplish these objectives, the expressions of the boosted phase voltages, maximum oscillations on instantaneous active/reactive powers, and the maximum phase currents under the unbalanced conditions must be found. The maximum allowable support (MAS) scheme obtains the maximum allowable active or reactive powers which the converter can deliver to the grid under unbalanced conditions (to support either the grid voltage or frequency) without exceeding the phase current limit, I_{limit} [8,9]. The mathematical formulas of the MAS control schemes under various conditions are presented in the coming sub-sections.

3.1 Minimum oscillation on active and reactive powers

The ability to analytically calculate the maximum power oscillations, i.e. $\tilde{p}_{max}$ and $\tilde{q}_{max}$, in terms of the scalar parameters (i.e. P, Q, k_p, k_q, and n) is very useful for proper controlling of the interconnecting converters under generic unbalanced voltage condition. Therefore, the expressions of $\tilde{p}_{max}$ and $\tilde{q}_{max}$ is obtained as follows [9]:

$$\begin{aligned} \tilde{p}_{max} &= \sqrt{P^2[k_p n + (1 - k_p)n^{-1}]^2 + Q^2[k_q n - (1 - k_q)n^{-1}]^2} \\ \tilde{q}_{max} &= \sqrt{Q^2[k_q n + (1 - k_q)n^{-1}]^2 + P^2[k_p n - (1 - k_p)n^{-1}]^2} \end{aligned} \tag{5.4}$$

By taking the derivative of Eq. (5.3) in terms of k_p and k_q, and finding the extreme point by enforcing the derivatives to be zero, the following expression can easily be obtained for the minimum oscillations on the instantaneous active and reactive powers:

$$\begin{aligned} &\tilde{p}_{max} \textit{ is minimum when } \begin{cases} k_p = 1/(1 - n^2) \\ k_q = 1/(1 + n^2) \end{cases}, \\ &\tilde{q}_{max} \textit{ is minimum when } \begin{cases} k_p = 1/(1 + n^2) \\ k_q = 1/(1 - n^2) \end{cases} \end{aligned} \tag{5.5}$$

These strategies are called minimum oscillation on the active power (or MOP) strategy and minimum oscillation on the reactive (or MOQ) strategy. In some efforts, the k_p and k_q values are only allowed to be between 0 and 1 [13]. For this case, they can be limited to $k_p = 1$ and $k_q = 1$ since Eq. (5.4) may give k_p or k_q values greater than 1. In this case (i.e., $k_p = 1$ or $k_q = 1$), only positive sequence in one current component (active or reactive) will be injected while the other component will contain both positive and negative sequences based on $1/(1 + n^2)$. However, k_p (or k_q) can be set to be $1/(1-n^2)$ in the applications when minimizing the active (or reactive) power oscillations is critical. The MOP strategy can be improved further in terms of ancillary performance. P^* or Q^* can then be obtained by using other MAS expressions, presented in the next sub-sections. Therefore, the objectives of both strategies, i.e., MOP and MAS, can simultaneously be accomplished. Similarly, MOQ can also be combined with MAS strategies to achieve double objectives.

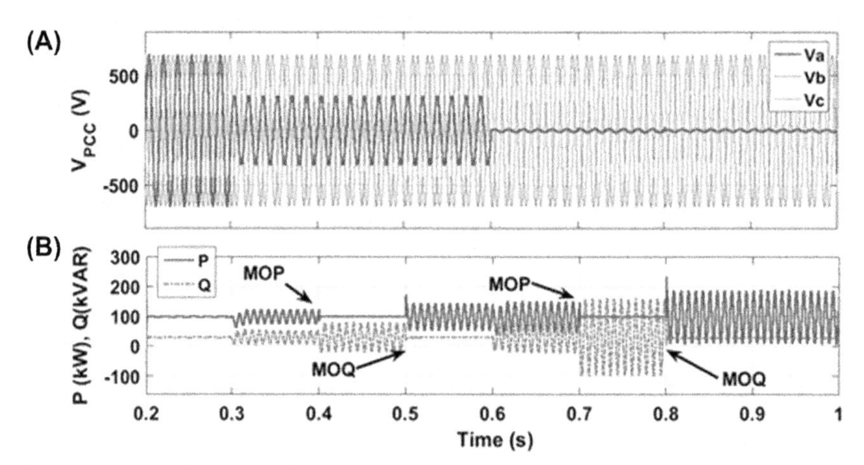

Figure 5.7 Simulation results of MOP and MOQ Strategies: (A) PCC voltage, and (B) active/reactive powers.

Fig. 5.7 shows the performance evaluation of MOP and MOQ strategies. k_p and k_q both are set to be 1 in the normal operation condition, and consequently, only positive sequence current is injected. During $t_1 = 0.3$ s and $t_2 = 0.6$ s, a medium voltage dip happens where $V_a = 0.5$ pu., and a full one-phase fault is experimented after $t_2 = 0.6$ s. Here, P and Q are set to be 100 kW and 30 kVAR. k_p and k_q are obtained according to Eq. (5.4). According to Fig. 5.7, the oscillations on the active and reactive powers are eliminated after applying the MOP and MOQ strategies, respectively.

3.2 Minimum fault current

By applying Eq. (5.2) to Eq. (5.1), the injected current under the fault can be rewritten in the $\alpha\beta$ frame:

$$\begin{bmatrix} i_\alpha \\ i_\beta \end{bmatrix} = \begin{bmatrix} i^+_{p,\alpha} \\ i^+_{p,\beta} \end{bmatrix} + \begin{bmatrix} i^-_{p,\alpha} \\ i^-_{p,\beta} \end{bmatrix} + \begin{bmatrix} i^+_{q,\alpha} \\ i^+_{q,\beta} \end{bmatrix} + \begin{bmatrix} i^-_{q,\alpha} \\ i^-_{q,\beta} \end{bmatrix} =$$

$$= \begin{bmatrix} \left[K_p^+ V^+ - K_p^- V^-\right] \cos(\omega t) & - \left[K_q^+ V^+ + K_q^- V^-\right] sin(\omega t) \\ \left[K_p^+ V^+ + K_p^- V^-\right] sin(\omega t) & + \left[K_q^+ V^+ - K_q^- V^-\right] cos(\omega t) \end{bmatrix} \tag{5.6}$$

The $\alpha\beta$ currents of Eq. (5.5) are transformed into the *abc* currents by the transformation matrix. Therefore, the maximum currents in each phase can be obtained as:

$$\begin{bmatrix} I_{\max-a}^2 \\ I_{\max-b}^2 \\ I_{\max-c}^2 \end{bmatrix} = \begin{bmatrix} (K_1)^2 + (K_2)^2 \\ \left(-\frac{1}{2}K_1 + \frac{\sqrt{3}}{2}K_4\right)^2 + \left(\frac{1}{2}K_2 + \frac{\sqrt{3}}{2}K_3\right)^2 \\ \left(-\frac{1}{2}K_1 - \frac{\sqrt{3}}{2}K_4\right)^2 + \left(\frac{1}{2}K_2 - \frac{\sqrt{3}}{2}K_3\right)^2 \end{bmatrix}$$

where

$$\begin{cases} K_1 = K_p^+ V^+ - K_p^- V^- = \frac{P}{V^-}((n+1)k_p - 1) \\ K_2 = K_q^+ V^+ + K_q^- V^- = \frac{Q}{V^-}((n-1)k_q + 1) \end{cases}$$
$$\begin{cases} K_3 = K_p^+ V^+ + K_p^- V^- = \frac{P}{V^-}((n-1)k_p + 1) \\ K_4 = K_q^+ V^+ - K_q^- V^- = \frac{Q}{V^-}((n+1)k_q - 1) \end{cases} \tag{5.7}$$

To obtain the minimum fault currents, Eq. (5.6) should be considered. According to Eq. (5.6), the maximum phase current can be one of the three expressions. For the minimum phase current (MFC) strategy, the reference values for the active and reactive powers can be determined from the operating mode controllers (either *PV* or *PQ* modes) or the grid code requirements. Also, k_q can be calculated from the voltage support method presented in the next section. Therefore, the minimum point of each expression of Eq. (5.6) should be obtained by taking their derivatives with respect to k_p:

$$I_{a-\min} \Rightarrow k_{p,a} = \frac{1}{n+1}, \quad I_{b-\min} \Rightarrow k_{p,b} = \frac{P \times (2-n) + \sqrt{3}nQ \times (2k_q - 1)}{2P \times (n^2 - n + 1)}$$
$$I_{c-\min} \Rightarrow k_{p,c} = \frac{P \times (2-n) - \sqrt{3}nQ \times (2k_q - 1)}{2P \times (n^2 - n + 1)} \tag{5.8}$$

However, only Eq. (5.7) cannot fully ensure that the obtained k_p will always minimize $I_{\max}$ in Eq. (5.6) because these equations minimize only their corresponding currents, i.e., $I_{\max\text{-}a}$, $I_{\max\text{-}b}$, and $I_{\max\text{-}c}$. Therefore, it is suggested that, in addition to the three k_ps obtained in Eq. (5.7), three other k_ps should also be considered to find the optimum value, $k_{p,opt}$. Two of these k_ps are the intersections of the magnitude curve of $I_a(k_p)$ with the magnitude curves of $I_b(k_p)$ and $I_c(k_p)$. Therefore, the equations of I_a and I_b, as well as I_a and I_c are taken to be equal to find $k_{p,ab}$ and $k_{p,ac}$, respectively.

$$I_a(k_p) = I_b(k_p) \Rightarrow k_{p,ab} = \frac{-b+\sqrt{b^2-4ac}}{2a} \begin{cases} a = 3nP^2 \\ b = -3nP^2 + \sqrt{3}nPQ(2k_q - 1) \\ c = 3nk_qQ^2(1-k_q) - \sqrt{3}nPQk_q \end{cases}$$

$$I_a(k_p) = I_c(k_p) \Rightarrow k_{p,ac} = \frac{-b+\sqrt{b^2-4ac}}{2a} \begin{cases} a = 3nP^2 \\ b = -3nP^2 - \sqrt{3}nPQ(2k_q - 1) \\ c = 3nk_qQ^2(1-k_q) + \sqrt{3}nPQk_q \end{cases} \tag{5.9}$$

Since k_p is bounded to 1, $k_{p,1} = 1$ should also be considered to find the minimum phase currents and optimal k_p value, i.e. $I_{\max,opt}$ and $k_{p,opt}$. Therefore, the three-phase currents are calculated for six possible k_ps to find the minimum $I_{\max}$. The $k_{p,opt}$, can be calculated under any operating and fault condition for any P, Q, V^+, and V^- values.

Fig. 5.8 shows the results of the MFC strategy. In this case, $P = 100$ W and $Q = 175$ VAr are injected by the interconnecting converter. A phase-to-ground fault occurs at $t = 2.6$ s. In this test case, k_q is 0.8 after the fault

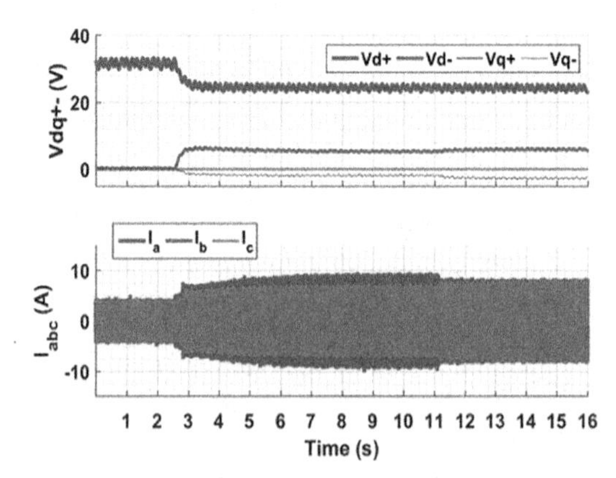

Figure 5.8 Experimental results of MFC strategy: (A) faulted voltage and (B) currents.

occurrence. Then, the phase currents increase up to 9.5 A. At $t = 11$ s, the MFC strategy is activated. The activation changes the value of the k_p from 1 to 0.79 to minimize the maximum phase current while maintaining the same values for P, Q, and k_q. Using the MFC strategy, the optimum k_p value is obtained by having three known parameters (i.e., $P = 100$ W, $Q = 175$ VAr, and $k_q = 0.8$) and the fault characteristics. The MFC strategy reduces the phase currents to 8 A, as Fig. 5.8 illustrates.

3.3 Maximum allowable active and reactive power injection

Applying the maximum allowable active power (MAP) strategy provides the maximum active power to the grid for assisting the frequency stability while simultaneously respects the phase current limits. The required formulas of P_{MAP} should be calculated such that they guarantee none of the phase currents exceeds the pre-set limits, $I_{\max}$ under the abnormal condition. In this strategy, the reference value for Q is determined either by the V-Q droop control or grid code requirements. By using the $I_{\max}$ Eq. (5.6), three possible active power values can be obtained as:

$$\begin{bmatrix} P_1 \\ P_2 \\ P_3 \end{bmatrix} = \begin{bmatrix} \dfrac{V^- \sqrt{I_{\max}^2 - K_2^2}}{k_p n + k_p - 1} \\ \left(-b + \sqrt{b^2 - 4ac} \right)/2a \\ \left(b + \sqrt{b^2 - 4ac} \right)/2a \end{bmatrix}, \text{ where}$$
$$\begin{cases} a = \dfrac{3}{V^-}(k_p(n-1)+1)^2 + \dfrac{1}{V^-}(k_p(n+1)-1)^2 \\ b = 2\sqrt{3}\left[\dfrac{K_4}{V^-}(k_p(n+1)-1) - \dfrac{K_2}{V^-}(k_p(n-1)+1) \right] \\ c = K_2^2 + 3K_4^2 - 4I_{\max}^2 \end{cases} \tag{5.10}$$

To have all the three-phase currents of Eq. (5.6) lower than the pre-set $I_{\max}$ value, P^*_{MAP} should comply with $P^*_{MAP} = \min(P_1, P_2, P_3)$.

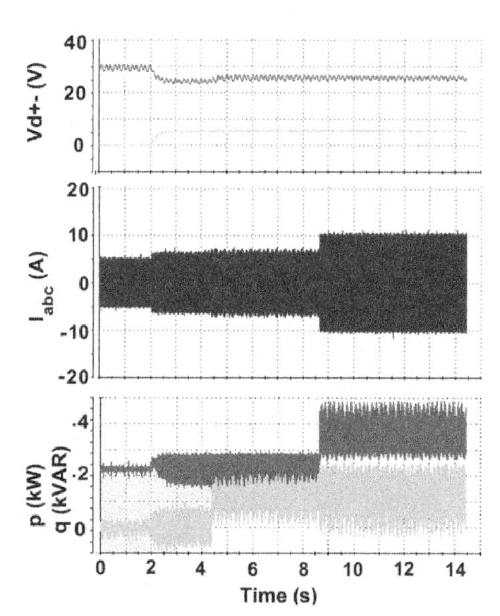

Figure 5.9 Experimental results of MAP: (A) faulted voltage, (B) phase currents, and (C) active and reactive powers.

Fig. 5.9 shows a test case where the grid code requirement is provided for the Q injection during the low-voltage. This test investigates the application of the MAP strategy. In this test case, $P = 225$ W is initially injected by the converter, as indicated in Fig. 5.9C. Similarly, a phase-to-ground fault occurs at $t = 2$ s. This fault causes the PCC voltage to drop from 30 to 25 V. At $t = 4.3$ s, the reactive power that is imposed by the grid code requirement should be injected. This injection causes the PCC voltage to increase to 26 V, as shown in Fig. 5.9A. At $t = 8.6$ s, the MAP strategy is activated to provide the maximum allowable active power considering I_{limit}=10 A. As shown in Fig. 5.9A, the *abc* currents are restricted to 10 A under the unbalanced fault by applying the MAP strategy.

Similarly, the maximum allowable reactive power (i.e., MAQ strategy) aims to provide the reference value of Q^*_{MAQ} such that all phase currents remain constrained with the current limit. The reference value of P is determined by other controllers in the GCC (e.g., by maximum power point tracking), and the Q^*_{MAQ} expressions can be obtained by using the $I_{\max}$ Eq. (5.6):

$$\begin{bmatrix} Q_1 \\ Q_2 \\ Q_3 \end{bmatrix} = \begin{bmatrix} \dfrac{V^- \sqrt{I_{\max}^2 - K_1^2}}{k_q n - k_q + 1} \\ \left(-b + \sqrt{b^2 - 4ac} \right)/2a \\ \left(b + \sqrt{b^2 - 4ac} \right)/2a \end{bmatrix}, \text{ where}$$

$$\begin{cases} a = \dfrac{3}{V^-}(k_q(n+1) - 1)^2 + \dfrac{1}{V^-}(k_q(n-1)+1)^2 \\ b = 2\sqrt{3}\left[\dfrac{K_3}{V^-}(k_q(n-1)+1) - \dfrac{K_1}{V^-}(k_q(n+1)-1)\right] \\ c = K_1^2 + 3K_3^2 - 4I_{\max}^2 \end{cases} \tag{5.11}$$

To satisfy min $(I_a, I_b, I_c) \leq I_{\max}$, Q^*_{MAQ} can be obtained as $Q^*_{MAQ} = \min(Q_1, Q_2, Q_3)$.

Fig. 5.10 shows the results of applying the MAQ strategy. Here, $P = 150$ W is injected. A similar phase-to-ground fault occurs in $t = 1.5$ s. Under the fault, the voltage profile drops from 30 to 25 V (%16.6 voltage drop). At $t = 4.3$ s, the MAQ scheme is triggered with the predefined

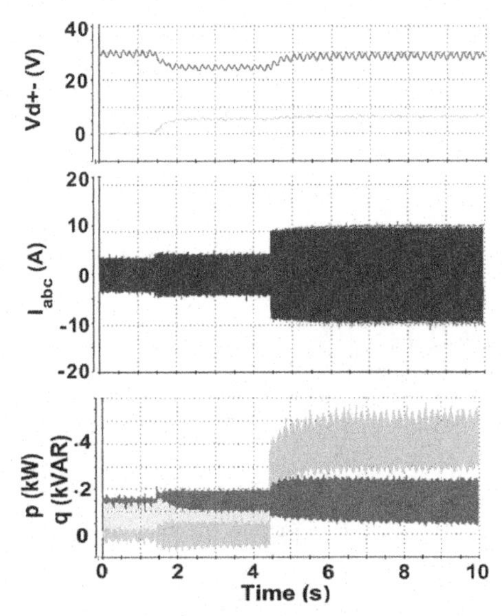

Figure 5.10 Experimental results of MAQ: (A) faulted voltage, (B) phase currents, and (C) active and reactive powers.

I_{limit}=10 A, as shown in Fig. 5.5. In practical applications, this scheme can automatically and immediately be triggered after the fault. However, it is manually configured in this test to show the results before and after applying the MAQ. According to Fig. 5.10A, the PCC voltage is increased to 29 V after applying the MAQ. Also, the injected currents are limited to 10 A, as indicated in Fig. 5.10B.

3.4 Voltage support and maximum allowable active power injection

Supporting the PCC voltage is another application of interconnecting converters that can be realized under unbalanced conditions. The three-phase voltages can be regulated at the desired range between $V_{\min}$ and $V_{\max}$. The principle goal in the voltage support is to avoid the over-voltage and under-voltage at the PCC whenever possible. However, a proper solution can be found in this range to satisfy other objectives as well. The following strategy considers the maximum active power injection and voltage support simultaneously. The PCC voltage support scheme (VSS) can be extracted as a function of the grid voltage and the injected positive and negative currents, as follows:

$$
\begin{aligned}
v = v^{+} + v^{-} &= \begin{bmatrix} v_{\alpha}^{+} + v_{\alpha}^{-} \\ v_{\beta}^{+} + v_{\beta}^{-} \end{bmatrix} \\
&= \begin{bmatrix} v_{g\alpha}^{+} + v_{g\alpha}^{-} + L_g \dfrac{d\left(i_{\alpha}^{+} + i_{\alpha}^{-}\right)}{dt} + R_g\left(i_{\alpha}^{+} + i_{\alpha}^{-}\right) \\ v_{g\beta}^{+} + v_{g\beta}^{-} + L_g \dfrac{d\left(i_{\beta}^{+} + i_{\beta}^{-}\right)}{dt} + R_g\left(i_{\beta}^{+} + i_{\beta}^{-}\right) \end{bmatrix}
\end{aligned}
\tag{5.12}
$$

Applying Eqs. (5.6) and (5.2) to Eq. (5.12) gives the following:

$$
\begin{aligned}
\begin{bmatrix} (V^{+} - V^{-})\cos(\omega t) \\ (V^{+} + V^{-})\sin(\omega t) \end{bmatrix} &= \begin{bmatrix} \left(V_g^{+} - V_g^{-}\right)\cos(\omega t - \delta) \\ \left(V_g^{+} + V_g^{-}\right)\sin(\omega t - \delta) \end{bmatrix} \\
&+ \begin{bmatrix} \omega L_g(-K_1 \sin(\omega t) + K_2 \cos(\omega t)) \\ \omega L_g(-K_3 \sin(\omega t) + K_4 \cos(\omega t)) \end{bmatrix} \\
&+ \begin{bmatrix} R_g(K_1 \cos(\omega t) + K_2 \sin(\omega t)) \\ R_g(K_3 \cos(\omega t) + K_4 \sin(\omega t)) \end{bmatrix}
\end{aligned}
\tag{5.13}
$$

In practical applications, δ is small and can be neglected for the simplicity in the analytical solution. Then, the positive and negative components of Eq. (5.13) can be separated as

$$\begin{bmatrix} V^+ \\ V^- \end{bmatrix} - \begin{bmatrix} V_g^+ \\ V_g^- \end{bmatrix} = \begin{bmatrix} \omega L_g I_q^+ \\ -\omega L_g I_q^- \end{bmatrix} + \begin{bmatrix} R_g I_p^+ \\ R_g I_p^- \end{bmatrix} \tag{5.14}$$

The maximum and minimum phase voltages can be determined simply by Ref. [13]:

$$V_{abc-\min} = \min(V_a, V_b, V_c) = \sqrt{(V^+)^2 + (V^-)^2 + 2(V^+)(V^-)\lambda_{\min}},$$
$$V_{abc-\max} = \max(V_a, V_b, V_c) = \sqrt{(V^+)^2 + (V^-)^2 + 2(V^+)(V^-)\lambda_{\max}}$$

where

$$\lambda_{\min} = \min\left(\cos(\gamma), \cos\left(\gamma - \frac{2\pi}{3}\right), \cos\left(\gamma + \frac{2\pi}{3}\right)\right), \quad \lambda_{\max} = \max\left(\cos(\gamma), \cos\left(\gamma - \frac{2\pi}{3}\right), \cos\left(\gamma + \frac{2\pi}{3}\right)\right) \tag{5.15}$$

Then the reference values for the maximum and minimum phase voltages can be determined so that the phase voltages are regulated within the explained thresholds. The reference values for V^+ and V^- can be calculated as follows:

$$V_{ref}^+ = \sqrt{\frac{\lambda_{\max} V_{\min}^2 - \lambda_{\min} V_{\max}^2 + \sqrt{\left(\lambda_{\max} V_{\min}^2 - \lambda_{\min} V_{\max}^2\right)^2 - \left(V_{\min}^2 - V_{\max}^2\right)^2}}{2(\lambda_{\max} - \lambda_{\min})}}$$

$$V_{ref}^- = \sqrt{\frac{\lambda_{\max} V_{\min}^2 - \lambda_{\min} V_{\max}^2 - \sqrt{\left(\lambda_{\max} V_{\min}^2 - \lambda_{\min} V_{\max}^2\right)^2 - \left(V_{\min}^2 - V_{\max}^2\right)^2}}{(\lambda_{\max} - \lambda_{\min})}} \tag{5.16}$$

By using Eq. (5.16), the reference values for the desired positive and negative sequences of the voltage are obtained. Then, V^+ and V^- in Eq. (5.14) are replaced with the reference values obtained from Eq. (5.16). Moreover, positive and negative sequences of the grid voltage can be estimated from the PCC measurements. Therefore, Eq. (5.14) can be rewritten as

$$I_p^+ = \frac{R}{X^2+R^2} \times \Delta V_{ref}^+, \quad I_p^- = \frac{R}{X^2+R^2} \times \Delta V_{ref}^-, \quad I_q^+ = \frac{X}{X^2+R^2} \times \Delta V_{ref}^+, \quad I_q^- = \frac{-X}{X^2+R^2} \times \Delta V_{ref}^- \tag{5.17}$$

For an inductive grid, Eq. (5.17) can be simplified as follows:

$$I_p^+ = I_p^- = 0, \quad I_q^+ = \frac{\Delta V_{ref}^+}{X}, \quad I_q^- = \frac{-\Delta V_{ref}^-}{X}. \tag{5.18}$$

To benefit from both VSS and MAP strategies, combining the VSS and MAP strategies is proposed here. Hence, the objectives of both strategies are simultaneously accomplished: (1) regulating the phase voltages within the prespecified boundaries, (2) injecting the maximum active power, and (3) respecting the predefined current limit, I_{max}. If the X/R ratio of the system is high, the active current components of Eq. (5.17), i.e., I_p^+ and I_p^-, do not contribute in regulating the voltage significantly. Consequently, the voltage support should completely be accomplished by the reactive currents, and the active current components can be used to inject the maximum allowable active power with respect to the current limitation.

Fig. 5.11 shows the test case where the performance of the VSS-MAP strategy is evaluated. Four different voltage sags occur for one phase of the grid voltage in $t = 0.1$, 0.25, 0.4, and 0.55 s, respectively. To clearly illustrate the performance of the proposed VSS method, a 0.05 delay is considered after all the fault occurrences to compare the results before and after applying the VSS strategy. Fig. 5.11 shows that by applying this strategy, all three phases are regulated in the desired range of $V_{min} = 0.9$ pu and $V_{max} = 1.1$ pu. In $t = 0.4$ s, the voltage sag in phase A is 0.25 pu (0.15 pu below V_{min}), so boosting Va with 0.15 pu by using only the positive reactive current will cause overvoltage in the other two phases. To tackle overvoltage in the other two phases, the proposed VSS strategy is applied to inject the required negative reactive current as well as the positive reactive current (see Fig. 5.11B). Fig. 5.11 demonstrates that the MAP strategy has also determined the maximum active power where the three-phase currents are under the preset limitation, i.e., $I_{max} = 200$ A. Therefore, the objectives of both strategies are simultaneously accomplished in

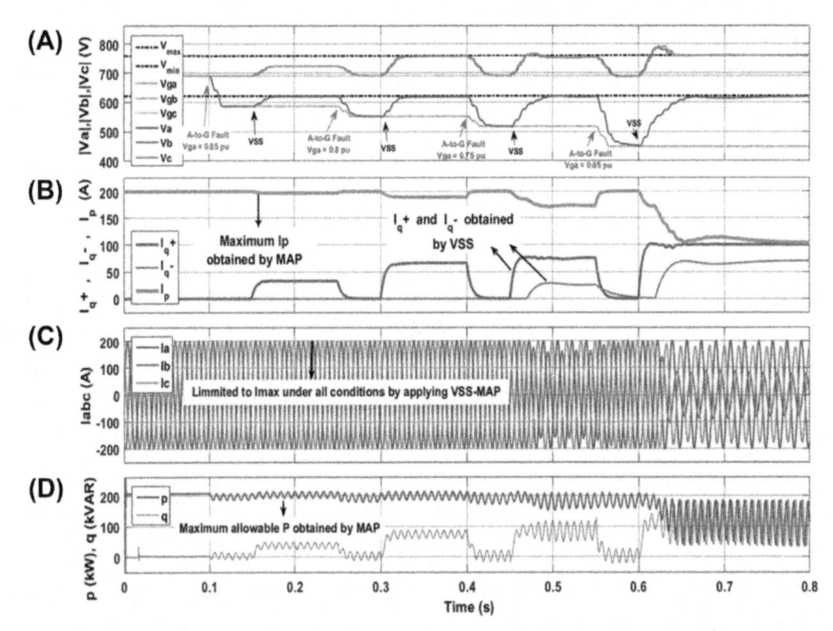

Figure 5.11 Simulation results of VSS-MAP strategy: (A) magnitude of phase voltages, (B) positive/negative and active/reactive components of currents, (C) phase currents, and (D) active/reactive powers. *MAP*, maximum allowable active power; *VSS*, voltage support scheme.

this test case: (1) the phase voltages have been regulated within the prespecified boundaries, (2) the maximum active power has been injected to support the grid, and (3) all three-phase currents have been limited to the prespecified current limitation, I_{max}.

4. Voltage support methods under unbalanced conditions

In the available literature, there are mainly four different approaches to support the voltage under unbalanced conditions by converter-interfaced units which will be discussed in this section. Such methods can also be adopted by interconnecting converters in microgrids.

4.1 Positive-sequence reactive current injection

Many efforts in the literature have followed the RCI requirement in the older version of German E.ON grid code demanding wind power plants to support grid voltage with additional reactive current, amounting to at least 2% of the rated current per each percent of the voltage drop [20,21]. In this standard, the voltage drop is defined as the magnitude of the balanced three-phase voltage sag at the low-voltage side of the interconnection transformer. Therefore, the reference value for the reactive current, I_{q+}^{ref}, can be generated by the following equation:

$$I_{q+,PSRCI}^{ref} = 2 \times \frac{V_{nom} - V_p}{V_{nom}} \times I_{\max} \tag{5.19}$$

The grid codes in Britain and Ireland require that wind farms deliver their maximum reactive current. Therefore, some efforts have been carried out [8] to present the MAQ delivery for unbalanced grid conditions. Therefore, the reference value for the reactive current can be obtained by the following equation:

$$I_{q+,MARPD}^{ref} = \sqrt{I_{\max}^2 - I_p^2} \tag{5.20}$$

where I_p is the reference value of the active current. Both Eqs. (5.19) and (5.20) only utilize the positive-sequence reactive current. Figs. 5.12 and 5.13 illustrate the voltage support performance of the PSRCI strategy by Eqs. (5.19) and (5.20), respectively. As Fig. 5.13 shows, the PSRCI strategy by Eq. (5.20) causes an overvoltage problem in unfaulted phases since it injects the maximum available reactive current. However, the performance of the PSRCI strategy is satisfactory; it boosts the positive-sequence voltage (Fig. 5.12D) without causing overvoltage (Fig. 5.12C).

4.2 Mixed-sequence reactive current injection

The MSRCI strategy imposes the injection of both positive and negative currents to support the connection voltage by a combination of boosting the positive sequence and decreasing the negative sequence of the voltage. According to Ref. [22], both positive- and negative-sequence reactive currents are injected under unbalanced faults. The reference values for the reactive currents can thus be obtained as

$$I_{q+,MSI}^{ref} = k^+ \times \frac{V_1 - V_p}{V_{nom}} \times I_{\max} \quad I_{q-,MSI}^{ref} = k^- \times \frac{V_2 - V_n}{V_{nom}} \times I_{\max} \tag{5.21}$$

where V_n is the magnitude of the negative sequence voltage at the PCC, and the recommended values for four parameters of Eq. (5.7) are $k^+ = k^- = 2.5$, $V_1 = 0.9$ pu, and $V_2 = 0.05$ pu in Ref. [22].

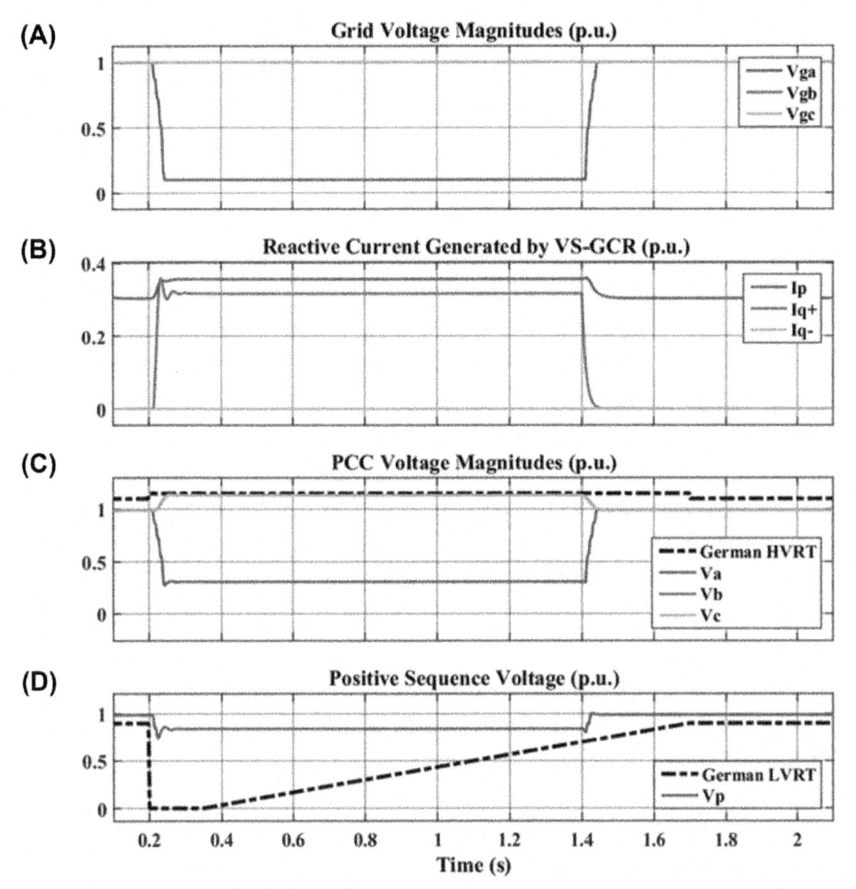

Figure 5.12 PSRCI strategy with Eq. (5.19), (A) grid voltage, (B) active/reactive currents, (C) phase voltages, and (D) positive sequence voltage. *PSRCI*, positive-sequence reactive current injection.

Similar to the previous strategy, reference [8] suggests injection of the maximum reactive power in positive and negative sequence control method using the following equations:

$$i_{q,PNSC}^{ref} = \frac{Q_{MARPD}^{PNSC}}{V^{+2} - V^{-2}} \left(\begin{bmatrix} -v_\beta^+ \\ v_\alpha^+ \end{bmatrix} - \begin{bmatrix} -v_\beta^- \\ v_\alpha^- \end{bmatrix} \right) \tag{5.22}$$

where

$$Q_{MARPD}^{PNSC} = \min \left(Q_1^{PNSC}, \ Q_2^{PNSC} \right)$$

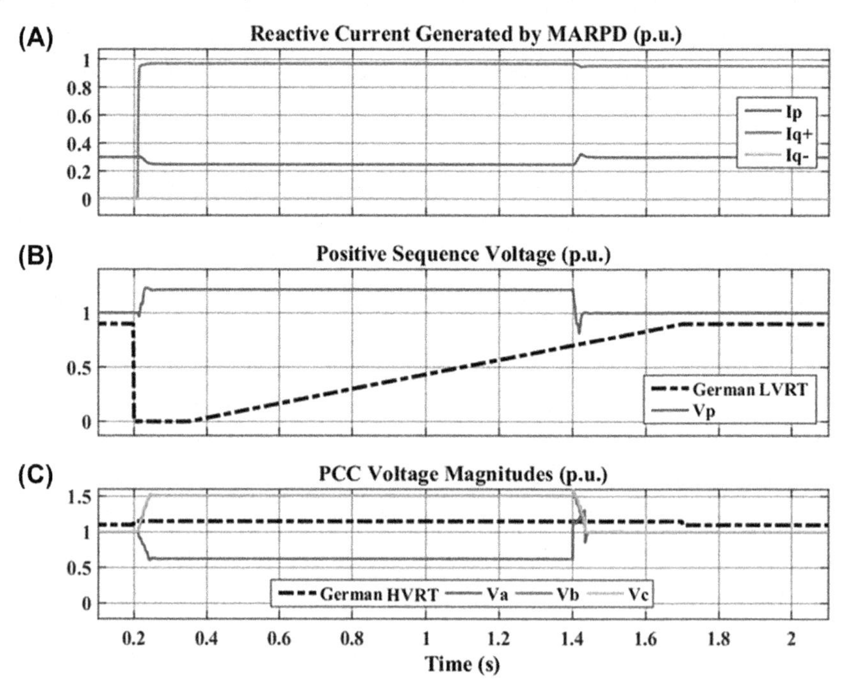

Figure 5.13 PSRCI strategy with Eq. (5.20), (A) active/reactive currents, (D) positive sequence voltage, and (C) PCC phase voltages. *PSRCI*, positive-sequence reactive current injection.

$$\begin{bmatrix} Q_1^{PNSC} \\ Q_2^{PNSC} \end{bmatrix} = \begin{bmatrix} \dfrac{\sqrt{I_{nom}^2\left[(V^+)^2-(V^-)^2\right]^2-P^2(V^++V^-)^2}}{(V^+-V^-)} \\ \dfrac{-b+\sqrt{b^2-4ac}}{2a} \end{bmatrix}$$

$$\begin{cases} a=3(V^++V^-)^2+(V^+-V^-)^2 \\ b=2\sqrt{3}P\left[(V^++V^-)^2+(V^+-V^-)^2\right] \\ c=P^2\left[(V^++V^-)^2+3(V^+-V^-)^2\right]-4I_{nom}^2\left[(V^+)^2-(V^-)^2\right]^2 \end{cases} \tag{5.23}$$

Fig. 5.14 shows the results of applying the MSRCI strategy. This voltage support strategy indicates a satisfactory support performance since it can boost the positive-sequence voltage without causing an overvoltage problem in unfaulted phases.

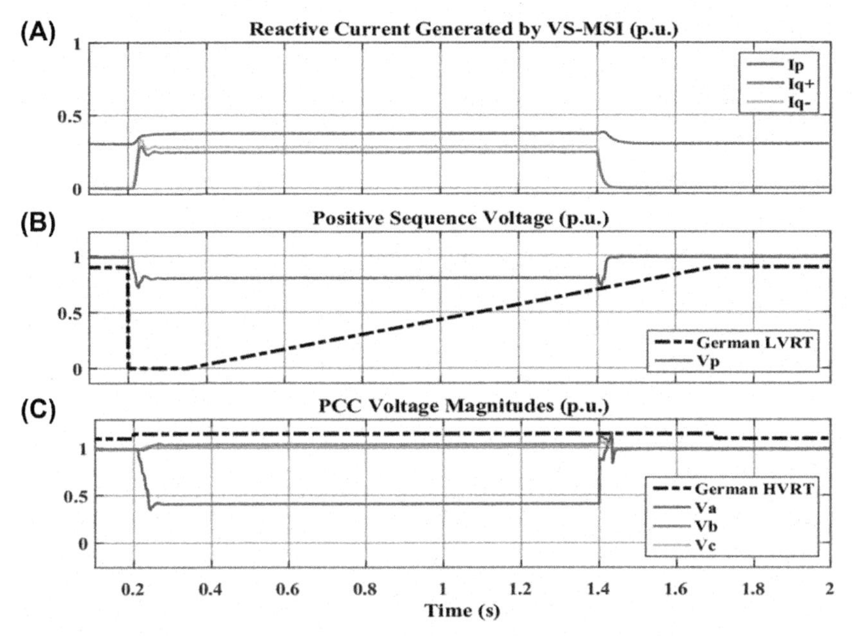

Figure 5.14 MSRCI strategy with Eq. (5.21): (A) active/reactive currents, (B) positive sequence voltage, and (C) PCC phase voltages. *MSRCI*, mixed-sequence reactive current injection; *PCC*, point of common coupling.

4.3 Dynamic regulation of phase-voltage magnitudes

The phase-voltage magnitude can be regulated using the proper expressions and reference values obtained for the positive- and negative-sequence voltages. The PCC positive- and negative-sequence voltages can be extracted as a function of the grid voltage and the injected positive/negative currents. Then, a proper set of reference values should be found for both positive- and negative-sequence reactive currents. The magnitudes of the phase voltages can be obtained in terms of the magnitudes of positive- and negative-sequence voltages by the following expressions:

$$\begin{cases} V_a = \sqrt{(V^+)^2 + (V^-)^2 + 2(V^+)(V^-)\cos(\gamma)} + (V^0)\cos(\gamma^0) \\ V_b = \sqrt{(V^+)^2 + (V^-)^2 + 2(V^+)(V^-)\cos\left(\gamma - \frac{2\pi}{3}\right)} + (V^0)\cos\left(\gamma^0 - \frac{2\pi}{3}\right) \\ V_c = \sqrt{(V^+)^2 + (V^-)^2 + 2(V^+)(V^-)\cos\left(\gamma + \frac{2\pi}{3}\right)} + (V^0)\cos\left(\gamma^0 + \frac{2\pi}{3}\right) \end{cases} \tag{5.24}$$

where $\gamma = \delta^{+} - \delta^{-}$ and $\gamma^{0} = \delta^{0} - \delta^{+}$.

Under most grid faults, the accuracy of the traditional phase-voltage magnitude regulation [12] is affected due to the existence of the zero-sequence term in Eq. (5.22). However, Shabestary and Mohamed [3,10] could dynamically regulate the phase-voltage magnitudes of the PCC within the preset limits considering the zero-sequence term. The fundamental objective was to keep the phase-voltage magnitudes between $V_{\min}^{ref}$ and $V_{\max}^{ref}$. This dynamic regulation can be obtained by proper control of the positive- and negative-sequence voltages under unbalanced fault conditions. The following equation can be written for the voltage:

$$\begin{bmatrix} \left(V^{+} - V_{g}^{+}\right)\cos(\omega t + \delta^{+}) + \left(V^{-} - V_{g}^{-}\right)\cos(\omega t + \delta^{-}) \\ \left(V^{+} - V_{g}^{+}\right)\sin(\omega t + \delta^{+}) - \left(V^{-} - V_{g}^{-}\right)\sin(\omega t + \delta^{-}) \end{bmatrix} = \begin{bmatrix} \left(L_{g}\omega I_{q}^{+} + R_{g}I_{p}^{+}\right)\cos(\omega t + \delta^{+}) + \left(R_{g}I_{p}^{-} - L_{g}\omega I_{q}^{-}\right)\cos(\omega t + \delta^{-}) \\ \left(L_{g}\omega I_{q}^{+} + R_{g}I_{p}^{+}\right)\sin(\omega t + \delta^{+}) - \left(R_{g}I_{p}^{-} - L_{g}\omega I_{q}^{-}\right)\sin(\omega t + \delta^{-}) \end{bmatrix} \tag{5.25}$$

where δ^{+} and δ^{-} are angles of the positive and negative sequences of the voltage. If the following expressions are satisfied:

$$\frac{I_{p}^{+}}{I_{q}^{+}} = \frac{R_{g}}{L_{g}\omega}, \quad \frac{I_{p}^{-}}{I_{q}^{-}} = -\frac{R_{g}}{L_{g}\omega} \tag{5.26}$$

then, the positive and negative components of Eq. (5.25) result in

$$V^{+} - V_{g}^{+} = L_{g}\omega I_{q}^{+} + R_{g}I_{p}^{+}, \quad V^{-} - V_{g}^{-} = R_{g}I_{p}^{-} - L_{g}\omega I_{q}^{-} \tag{5.27}$$

To comply with voltage limits, a combination of positive/negative and active/reactive currents (i.e., I_{p}^{+}, I_{p}^{-}, I_{q}^{+}, and I_{q}^{-}) should be injected into an inductive-resistive grid to support the grid voltage. From Eq. (5.24), the maximum and minimum phase voltages with zero-sequence consideration can be determined by the following equations:

$$\begin{aligned} V_{\min} &= \min(V_{a}, V_{b}, V_{c}) = \sqrt{(V^{+})^{2} + (V^{-})^{2} + 2(V^{+})(V^{-})\lambda_{\min} + (V^{0})\lambda_{\min}^{0}} \\ V_{\max} &= \max(V_{a}, V_{b}, V_{c}) = \sqrt{(V^{+})^{2} + (V^{-})^{2} + 2(V^{+})(V^{-})\lambda_{\max} + (V^{0})\lambda_{\max}^{0}} \end{aligned} \tag{5.28}$$

where $\lambda_{\min}$, $\lambda_{\max}$, $\lambda^0_{\min}$, and $\lambda^0_{\max}$ can be found as follows:

$$\begin{cases} \lambda_a = \cos(\gamma) \\ \lambda_b = \cos\left(\gamma - \frac{2\pi}{3}\right) \\ \lambda_c = \cos\left(\gamma + \frac{2\pi}{3}\right) \end{cases} \rightarrow \begin{cases} \lambda_{\min} = \min(\lambda_a, \lambda_b, \lambda_c) \\ \lambda_{\max} = \max(\lambda_a, \lambda_b, \lambda_c) \end{cases}$$

$$\begin{cases} if \quad \lambda_{\min} = \lambda_a \rightarrow \lambda^0_{\min} = \cos(\gamma^0) \\ if \quad \lambda_{\min} = \lambda_b \rightarrow \lambda^0_{\min} = \cos\left(\gamma^0 - \frac{2\pi}{3}\right) \\ if \quad \lambda_{\min} = \lambda_c \rightarrow \lambda^0_{\min} = \cos\left(\gamma^0 + \frac{2\pi}{3}\right) \end{cases} \quad and$$
$$\begin{cases} if \quad \lambda_{\max} = \lambda_a \rightarrow \lambda^0_{\max} = \cos(\gamma^0) \\ if \quad \lambda_{\max} = \lambda_b \rightarrow \lambda^0_{\max} = \cos\left(\gamma^0 - \frac{2\pi}{3}\right) \\ if \quad \lambda_{\max} = \lambda_c \rightarrow \lambda^0_{\max} = \cos\left(\gamma^0 + \frac{2\pi}{3}\right) \end{cases} \tag{5.29}$$

After applying proper $V^{ref}_{\min}$ and $V^{ref}_{\max}$ to Eqs. (5.28) and (5.29), the reference values of V^+_{ref} and V^-_{ref} can be solved as

$$\left(V^+_{ref}\right)^2 = \frac{-B + \sqrt{B^2 - 4A^2}}{2},$$
$$V^-_{ref} = \frac{A}{V^+_{ref}} \begin{cases} A = \frac{V_H - V_L}{2(\lambda_{\max} - \lambda_{\min})} & V_L = \left(V^{ref}_{\min} - V^0\lambda^0_{\min}\right)^2 \\ B = 2A \times \lambda_{\max} - V_H & V_H = \left(V^{ref}_{\max} - V^0\lambda^0_{\max}\right)^2 \end{cases} \tag{5.30}$$

A general solution (even applicable in grids with different X/R ratios) is to apply the results of Eq. (5.30) to Eq. (5.27) as follows:

$$I_p^+ = \frac{R_g}{X_g^2 + R_g^2} \times \Delta V_{ref}^+, \quad I_p^- = \frac{R_g}{X_g^2 + R_g^2} \times \Delta V_{ref}^-, \quad I_q^+ = \frac{X_g}{X_g^2 + R_g^2} \times \Delta V_{ref}^+, \quad I_q^- = \frac{-X_g}{X_g^2 + R_g^2} \times \Delta V_{ref}^- \tag{5.31}$$

where expressions of Eq. (5.26) are also satisfied. According to Eq. (5.31), the active components do not contribute to supporting the voltage in an inductive grid. In this case, they can be utilized to fulfill complementary objectives discussed in the previous section (i.e., MAP strategy). Fig. 5.15

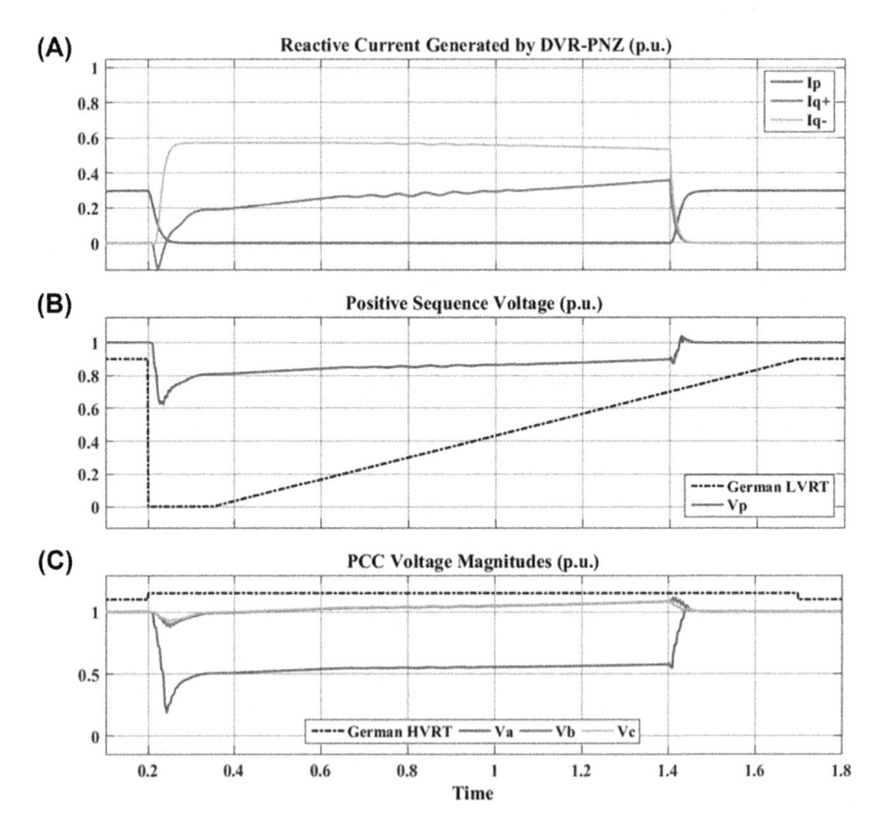

Figure 5.15 DRPVM strategy: (A) active/reactive currents, (B) positive sequence voltage, and (C) PCC phase voltage magnitudes. *DRPVM*, Dynamic Regulation of Phase-Voltage; *PCC*, point of common coupling.

shows the voltage support performance of the DRPVM strategy which regulates the phase-voltage magnitudes within the preset dynamic boundaries. This results in a proper voltage boost without overvoltage problems.

5. Summary

This chapter studied the operation of the interconnecting converters inside microgrids under unbalance conditions. The emerging grid code requirements for the operation of such converters were reviewed. The new grid codes contain requirements regarding the LVRT, HVRT, RCI, and ART. Although the newer German code has reported both positive- and negative-sequence reactive currents injection under asymmetrical grid faults, the later requirement has not explicitly been indicated in the existing grid codes yet. However, the trend of the grid code updates shows that soon there would be specific requirements for ridding through asymmetrical grid faults.

Furthermore, the optimized operation of the interconnecting converters under unbalanced conditions was studied. In this regard, different objectives were considered including minimum oscillations on the output active and reactive powers, minimum peak current, and maximum output active and reactive powers delivery. Finally, different voltage support strategies were studied and evaluated under LVRT requirements and unbalanced grid faults. The important issue in supporting the voltage while riding through unbalanced gird faults is to avoid the overvoltage on unfaulted phases. Therefore, satisfactory performance is identified as a positive-sequence voltage boost without causing an overvoltage problem.

References

[1] IEEE Recommended Practice for Monitoring Electric Power Quality, Ser., IEEE Std., June 2009, p. 1159.

[2] S. Acharya, M. S. El Moursi, A. Al-Hinai, A. S. Al-Sumaiti and H. Zeineldin, "A control strategy for voltage unbalance mitigation in an islanded microgrid considering demand side management capability," *in* IEEE Transactions on Smart Grid.

[3] M.M. Shabestary, Y.A.I. Mohamed, Asymmetrical ride-through and Grid Support in Converter-Interfaced DG Units under unbalanced Conditions, IEEE Trans. Ind. Electron. 66 (2) (2019) 1130–1141.

[4] X. Zhao, J.M. Guerrero, M. Savaghebi, J.C. Vasquez, X. Wu, K. Sun, Low-voltage ride-through operation of power converters in grid-interactive microgrids by using negative-sequence droop control, IEEE Trans. Power Electron. 32 (4) (2017).

[5] F. Shahnia, P.J. Wolfs, A. Ghosh, Voltage unbalance reduction in low voltage feeders by dynamic switching of residential customers among three phases, IEEE Trans. Smart Grid 5 (3) (2014).
[6] S. Martinenas, K. Knezovic, M. Marinelli, Management of power quality issues in low voltage networks using electric vehicles: experimental validation, IEEE Trans. Power Deliv. 32 (2) (2017).
[7] M.M. Hashempour, T. Lee, M. Savaghebi, J.M. Guerrero, Real-time supervisory control for power quality improvement of multi-area microgrids, IEEE Syst. J. 13 (1) (2019) 864–874.
[8] M.M. Shabestary, Y.A.R.I. Mohamed, An analytical method to obtain maximum allowable grid support by using grid-connected converters, IEEE Trans. Sustain. Energy 7 (4) (2016).
[9] M.M. Shabestary, Y.A.R.I. Mohamed, Analytical expressions for multi-objective optimization of converter-based DG operation during unbalanced grid conditions, IEEE Trans. Power Electron. 32 (9) (2017).
[10] M.M. Shabestary, Y. Mohamed, Advanced voltage support and active power flow control in grid-connected converters under unbalanced conditions, IEEE Trans. Power Electron. 33 (2) (2018).
[11] J. Jia, G. Yang, A.H. Nielsen, A review on grid-connected converter control for short-circuit power provision under grid unbalanced faults, in: IEEE Transactions on Power Delivery, vol. 33, 2018, 2.
[12] A. Camacho, M. Castilla, J. Miret, R. Guzman, A. Borrell, Reactive power control for distributed generation power plants to comply with voltage limits during grid faults, in: IEEE Transactions on Power Electronics, vol. 29, November 2014, pp. 6224–6234, 11.
[13] K. Ma, W. Chen, M. Liserre, F. Blaabjerg, Power controllability of a three-phase converter with an unbalanced AC source, in: IEEE Transactions on Power Electronics, vol. 30, March 2015, pp. 1591–1604, 3.
[14] F. Wang, J.L. Duarte, M.A.M. Hendrix, Pliant active and reactive power control for grid-interactive converters under unbalanced voltage dips, in: IEEE Transactions on Power Electronics, vol. 26, May 2011, pp. 1511–1521, 5.
[15] A.R.-N. VDE-, Technical Requirements for the Connection and Operation of Customer Installations to the High Voltage Network (TAB High Voltage), 4120, January 2015. Germany.
[16] M.M. Shabestary, S. Mortazavian, Y.A.R.I. Mohamed, Asymmetric low-voltage ride-through scheme and dynamic voltage regulation in distributed generation units, in: 2018 IEEE Applied Power Electronics Conference and Exposition (APEC), San Antonio, TX, 2018, pp. 1603–1608.
[17] M.M. Shabestary, S. Mortazavian, Y.A.R.I. Mohamed, Overview of Voltage Support Strategies in Grid-Connected VSCs under Unbalanced Grid Faults Considering LVRT and HVRT Requirements, SEGE, 2018.
[18] M. Tsili, S. Papathanassiou, A review of grid code technical requirements for wind farms, in: IET Renewable Power Generation, vol. 3, 2009, 3.
[19] S. Mortazavian, M.M. Shabestary, Y.A.R.I. Mohamed, Analysis and dynamic performance improvement of grid-connected voltage–source converters under unbalanced network conditions, in: IEEE Transactions on Power Electronics, vol. 32, October 2017, pp. 8134–8149, 10.
[20] M. Mirhosseini, J. Pou, V.G. Agelidis, Single- and two-stage inverter-based grid-connected photovoltaic power plants with ride-through capability under grid faults, in: IEEE Transactions on Sustainable Energy, vol. 6, July 2015, pp. 1150–1159, 3.

[21] T. Kauffmann, U. Karaagac, I. Kocar, S. Jensen, J. Mahseredjian, E. Farantatos, An accurate type III wind turbine generator short circuit model for protection applications, in: IEEE Transactions on Power Delivery, 2017, 99.
[22] K. Sharifabadi, L. Harnefors, H. Nee, S. Norrga, R. Teodorescu, Design, Control, and Application of Modular Multilevel Converters for HVDC Transmission Systems, John Wiley & Sons, Ltd., 2016.

CHAPTER 6
Microgrid protection

1. Introduction

There are numerous technical challenges regarding control, stability, power management, and protection of microgrid systems, especially when they are affected by the fault contribution caused by large-scale integration of and inverter-based and directly coupled distributed energy resources (DERs) [1]. The reported droop-based control strategies almost use inner current controllers, which can considerably enhance the stability margins and help to realize more convenient overcurrent (OC) and overload (OL) protection schemes based on the adoption of the reference points [2]. Consequently, this loop has two main functions: (1) decreasing output power and preventing OL protection scheme from tripping, and (2) improving the fault ride-through (FRT) capability without tripping out of inverter-interfaced distributed energy resources (IIDERs) [3]. The reported direct voltage control (DVC) methods almost have sufficient robustness [4]; by the way, they cannot intrinsically cope with limiting the fault current during network transients (such as symmetrical or unsymmetrical faults) and may consequently lead to damaged power electronic components of converters and tripping out of DER unit. Because the DVC-based IIDERs lack the internal current controller, they may be prone to dynamic OL conditions [5]. For the sake of handling his dilemma, in Ref. [6], an OL/OC protection method has been proposed for off-grid [7] and on-grid operation modes [8,9] as an add-on for the hierarchical controller. There are also some reported centralized and communication-assisted microgrid protection strategies [10]. However, there are two main basic challenges when utilizing these techniques:

(1) The necessity of having backup system for communication system of the Hierarchical droop-controlled (HDRC) in case that the network is faced with an interruption or failure (like the load-flow calculation method presented in Refs. [4,11] based on the radial basis function neural networks [RBFNNs]).

Microgrids and Methods of Analysis
ISBN 978-0-12-816172-2
https://doi.org/10.1016/B978-0-12-816172-2.00006-7

(2) Finding direct analytical solutions to be more adaptive with recent HDRC and DVC schemes, particularly the system has nonlinear and unbalanced loads.

For the sake of covering the mentioned themes, this chapter presents a generalized OC/OL protection methods for both DVC and HDRC structures that are superior and advantageous to the conventional fault detection methods by taking the advantages of both the classification capability of artificial neural networks (ANNs) and the transient phenomenon detection features of transient monitoring function (TMF) [12] to perform the following tasks: (1) detecting the fault, (2) discriminating between transient and permanent faults, and (3) detecting the type of faults. The presented OC protection scheme can effectively limit the output current of DERs by limiting their output power after clearance of the fault and can restore the system to its normal situation. Finally, to verify the performance and prove the effectiveness of the protection strategy, software simulations are carried out in MATLAB/Simulink time domain environment, and the results are experimentally verified by using OPAL-RT real-time digital simulator (RTDS).

2. Control system

2.1 Hierarchical droop-based control structure

In Ref. [4], an improved HDRC scheme has been presented including a harmonic voltage compensator, a self-tuning filter, a harmonic virtual impedance, and power calculation based on RBFNN-based harmonic power flow (Fig. 6.1A and B).

2.2 Direct voltage control scheme

In Ref. [7], the details of a control system for DVC-based IIDERs have been presented, which include local controllers, a power maangement system, and an open-loop frequency control scheme based on a global positioning system (GPS) (Fig. 6.1C). By the way, controlling the frequency of the system and synchronization based on GPS has some difficulties that sometimes make its implementation impossible or cumbersome. On the other hand, the hierarchical control structure (shown in Fig. 6.1A) can have not the previous drawbacks of the previously-reported centralized and droop-based controllers as they only need a low-bandwidth communication for sending the DERs' and loads' instantaneous values of the active/reactive powers to the Power management system (PMS) and LCs for providing the acceptable power-sharing (either

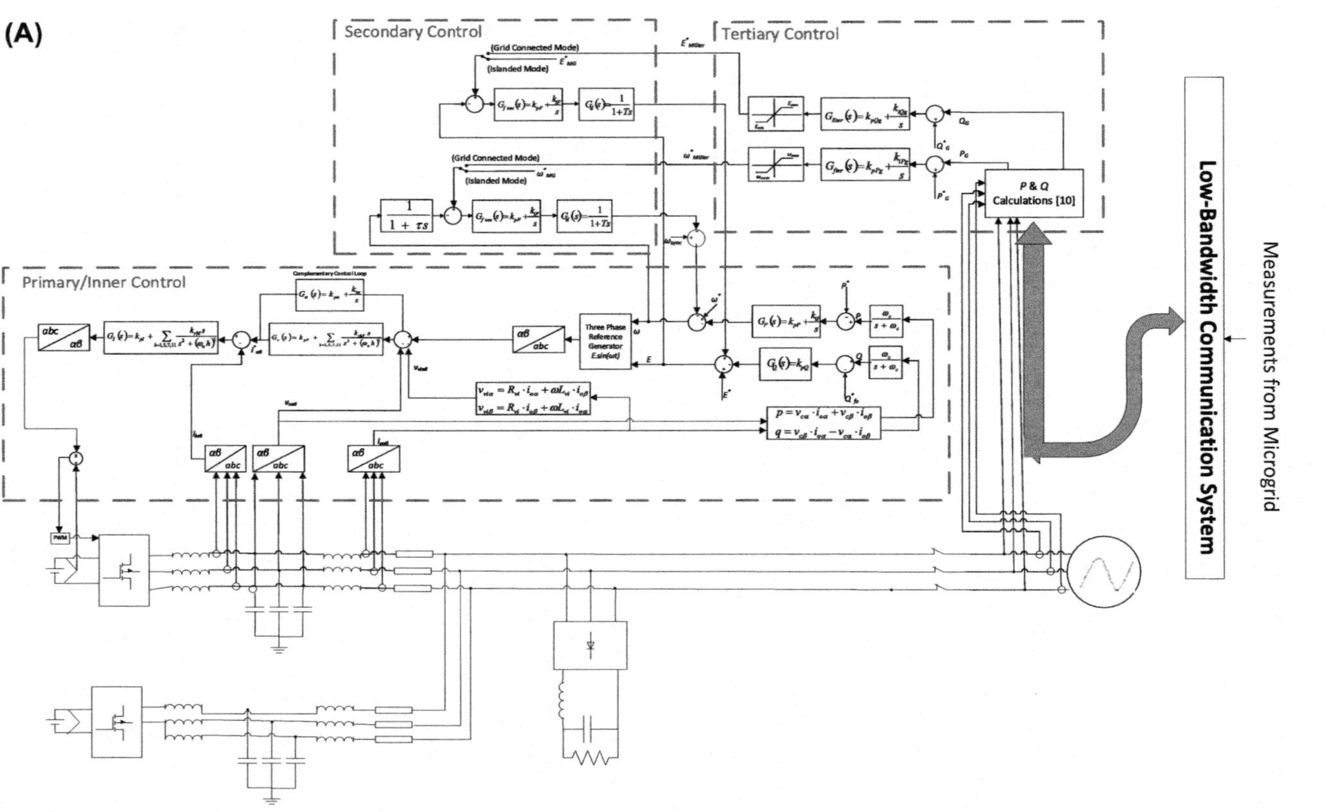

Figure 6.1 (A) and (B) Improved HDRC structure [4,11], and (C) Control system of DVC-DER unit [7]. *(C) IEEE from H.R. Baghaee, M. Mirsalim, G.B. Gharehpetian, H.A. Talebi, Nonlinear load sharing and voltage compensation of microgrids based on harmonic power-flow calculations using radial basis function neural networks, IEEE Sys. J. 99 (January 2017) 1–11, doi: 10.1109/JSYST.2016.2645165 and A.H. Etemadi, E.J. Davison, R. Iravani, A generalized decentralized robust control of islanded microgrids, IEEE Trans. Power Syst. 29 (6) (October 2014) 3102–3113.)*

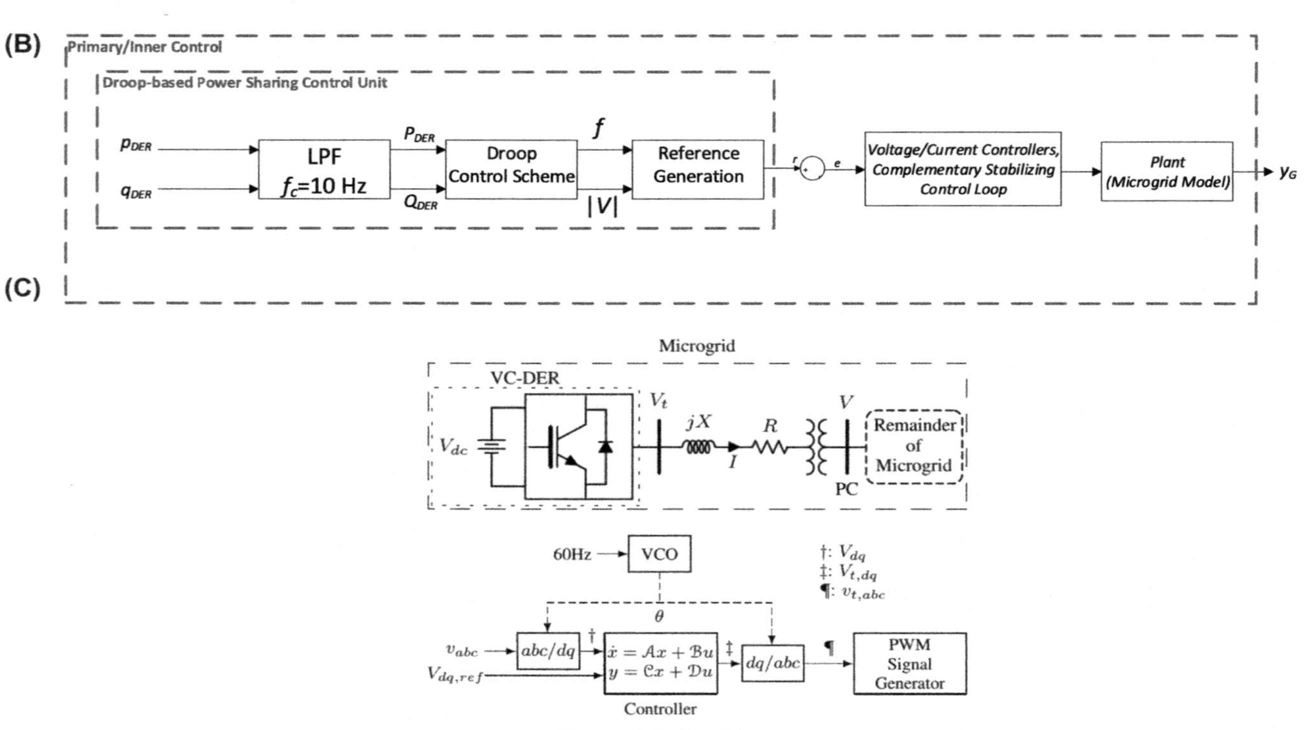

Figure 6.1 Cont'd

based on market signal or DER unit cost functions) [4,11], based on the following load flow equations [4,11]:

$$P_{DER,i} = \sum_{j=1}^{N} |V_i||V_j||Y_{ij}|\cos(\delta_i - \delta_k - \theta_{ij}) \tag{6.1}$$

$$Q_{DER,i} = \sum_{j=1}^{N} |V_i||V_j||Y_{ij}|\sin(\delta_i - \delta_k - \theta_{ij}) \tag{6.2}$$

where, δ_i and V_i are phase angle and voltage magnitude of ith bus, $P_{DER,i}$ and $Q_{DER,i}$ are net injected active and reactive powers to the ith bus, N is the number of buses, Y_{ij} and θ_{ij} are magnitude and argument of the ijth element of the Y_{Bus} matrix, and YB are elements of the Y_{BUS} matrix corresponding to the DC buses.

3. Fault model of inverter-interfaced distributed energy resource units

In this part of the analysis, first, a fault model is developed for the IIDER units. Then, the effects of each control system (either HDRC or DVC) are investigated.

3.1 HDRC-based inverter-interfaced distributed energy resource units

As pointed out in Refs. [3,4,11] and using the model of the HDRC system shown in Fig. 6.1A, the capacitor voltage can be expressed as

$$V_c(s) = G(s)V_c^*(s) - Z_o(s)I_o(s) \tag{6.3}$$

where

$$G(s) = \frac{G_V(s)G_I(s)}{sC_f(1 + G_{st}(s)G_V(s)) - G_v(s)G_I(s)} \tag{6.4}$$

$$Z_o(s) = -\frac{(VI(s)G_I(s) - 1)(1 + G_{st}(s)G_V(s))}{sC_f(1 + G_{st}(s)G_V(s)) + G_v(s)G_I(s)} \tag{6.5}$$

3.2 Inverter current limiting

A complementary stabilizing control loop has been proposed in Ref. [11], for enhancing the microgrid stability margins when faced with small- and large-signal disturbances, and improving its FRT capability. So, to decrease the fault level of the IIDERs, in the mentioned HDRC scheme, the transfer function of the stabilizing loop ($G_{st}(s)$) affects the frequency response of $G(s)$ and $Z_o(s)$ defined in Eqs. (6.4) and (6.5). In Ref. [3], based on a current-limiting function given by Eq. (6.6) and motivated by the latched limit (LL) and instantaneous saturation limit (ISL) strategies presented in Refs. [13] and [14], a current control method has been proposed for synchronous reference frame (SYRF), natural reference frame (NARF), and stationary reference frame (STRF) [15].

$$i_{L,j}^{\prime ref} = = \frac{i_{L,j}^{ref} \cdot i_{ith}}{\sqrt{2} I_{L,j}^{ref}} u\left(I_{L,j}^{ref} - \frac{i_{th}}{\sqrt{2}} \right) + u\left(\frac{i_{th}}{\sqrt{2}} - I_{L,j}^{ref} \right); j = a, b, c \tag{6.6}$$

for NARF and

$$i_{L,j}^{\prime ref} = = \frac{i_{L,j}^{ref} \cdot i_{ith}}{\sqrt{2} I_{L}^{ref}} u\left(I_{L}^{ref} - \frac{i_{th}}{\sqrt{2}} \right) + u\left(\frac{i_{th}}{\sqrt{2}} - I_{L}^{ref} \right); j = d(\alpha), q(\beta), 0(\gamma) \tag{6.7}$$

for STRF and SYRF, where $u(t)$ is the unit step function; ith is the inverter rated peak current, here 2 pu; and I_L^{ref} is the maximum value of i_L^{ref} for all phases a, b, and c. This current limiting strategy (CLS) performs as an added-on feature for the HDRC described in Section 2.1, which can affect time delay assignment and fault current coefficient of the protection system [3].

3.3 Direct voltage control—based inverter-interfaced distributed energy resource units

As mentioned, the DVC-based IIDER units lack the current controller. So, for the sake of deriving their fault model, the complementary stabilizing controller and virtual impedance terms are eliminated, and the current controller transfer function in Eqs. (6.4) and (6.5) is set to 1. So, we have

$$G(s) = \frac{G_V(s)}{sC_f - G_v(s)} \tag{6.8}$$

$$Z_o(s) = \frac{1}{sC_f + G_v(s)} \tag{6.9}$$

4. Proposed fault detection method

Some of the previous approaches have used voltage/currents symmetrical components for detecting the faults. In these approaches, the control system may become unstable when it detects these components, and based on the topology of the converter, current-limiting strategy, and the reference frame, the system can experience undesirable situations [12]. If we decide to use the LL and ISL strategies, because total harmonic distortion–based fault detection schemes have some limitations, we briefly describe the TMF method presented in Ref. [12] and use it for accurate, fast, and authentic transient detection. The TMF algorithm exploits moving data windows of prefault, during the fault and postfault operating signals (e.g., current waveforms of voltage source converters [VSCs]) (Fig. 6.2A). We can increase the length of the moving data window to improve the accuracy; however, when the window length increases, the processing speed may be decreased. This algorithm exploits the orthogonal least square (OLS) algorithm to estimates VSC current waveform fundamental frequency component [16]. In normal conditions and by using the actual signal and its reconstruction, the estimated fundamental frequency component match is then formed. Because the moving data window consists of either prefault and postfault data, the reconstructed signal is slightly different from the original VSC current if a fault occurs in the microgrid (Fig. 6.2A). Besides, owing to fault clearance, the OLS signal model will not include the added components of the fault current. As a consequence, the algorithm can discriminate the normal and fault conditions.

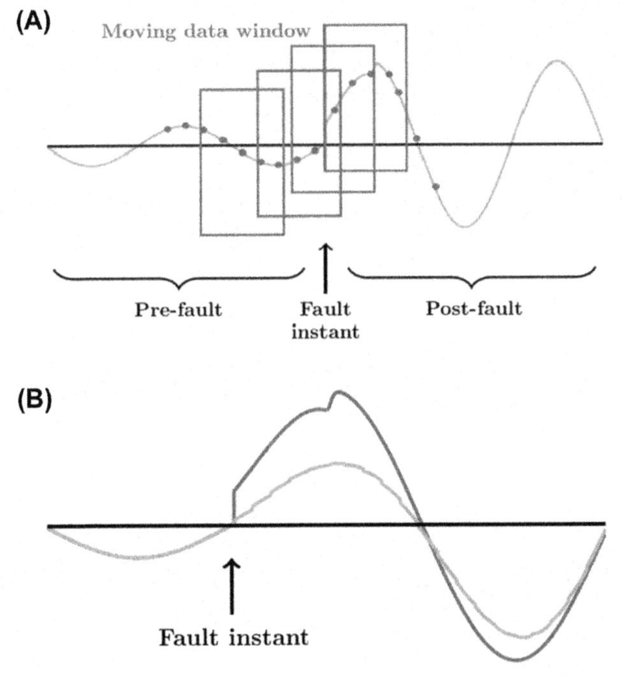

Figure 6.2 (A) The VSC current waveform moving data window, and (B) its actual and reconstructed signals [12]. *VSC*, voltage source converter. *(IEEE from I. Sadeghkhani et al., Transient monitoring function–based fault detection for inverter-interfaced microgrids, IEEE Trans. Smart Grid 99 (September 2016) 1–11.)*

4.1 Transient monitoring function

It is supposed that the VSC current can be written as

$$i(t) = \sum_{k=1}^{2N} c_k s_k(t) \tag{6.10}$$

One option for the signals $s_k(t)$ is

$$S = \begin{bmatrix} s_1 \\ s_2 \\ s_3 \\ s_4 \\ \vdots \end{bmatrix} = \begin{bmatrix} \cos(\omega_0 t) \\ \sin(\omega_0 t) \\ \cos(2\omega_0 t) \\ \sin(2\omega_0 t) \\ \vdots \end{bmatrix} \begin{matrix} \left.\right\} \textit{fundamental frequency} \\ \left.\right\} 2^{nd} \ \textit{harmonic} \\ \left.\right\} \textit{higher} - \textit{order harmonics} \end{matrix} \tag{6.11}$$

Thus, we can rewrite Eq. (6.10) as

$$\begin{aligned} i(t) &= c_1 \cos(\omega_0 t) + c_2 \sin(\omega_0 t) + c_3 \cos(2\omega_0 t) \\ &\quad + c_4 \sin(2\omega_0 t) + \cdots \end{aligned} \tag{6.12}$$

To form the discrete model of Eq. (6.10), the current is sampled with sampling period of T_s, and the current measurement vector $\boldsymbol{M}$ is then obtained as

$$m_k = \sum_{k=1}^{2N} c_k s_k(kT_s) \tag{6.13}$$

$$\boldsymbol{M} = [\, i(t_0 + T_s) \quad i(t_0 + 2T_s) \quad \dots \quad i(t_0 + KT_s) \,]^T \tag{6.14}$$

When we use the OLS algorithm, Eq. (6.13) can be reexpressed in the vector form by using the estimation of the VSC current fundamental frequency component i as

$$\boldsymbol{M} = \boldsymbol{S}_1 \boldsymbol{C}_1 \tag{6.15}$$

where $\boldsymbol{C_1} = [c_1, \ c_2]^T$ and S_1 is the discrete model of the fundamental frequency components of Eq. (6.11) that is given by

$$\boldsymbol{S}_1 = \begin{bmatrix} \cos(\omega_0 T_s) & \cos(\omega_0 2T_s) & \dots & \cos(\omega_0 KT_s) \\ \sin(\omega_0 T_s) & \sin(\omega_0 2T_s) & \dots & \sin(\omega_0 KT_s) \end{bmatrix}^T \tag{6.16}$$

The OLS solution $\widehat{C}_1$ for estimation of coefficients is given by

$$\widehat{C}_1 = \left(S_1^T S_1\right)^{-1} S_1^T M \tag{6.17}$$

Based on Eq. (6.16), the reconstruction of the current samples $\widetilde{M}$ is done from the estimated $\widehat{C}_1$ as

$$\widetilde{M} = S_1 \widehat{C}_1 = S_1 \left(S_1^T S_1\right)^{-1} S_1^T M \tag{6.18}$$

In the abnormal situations of the network that may experience a system transient, $\widetilde{M}$ goes far from $\boldsymbol{M}$, and we can define R as

$$R = \widetilde{M} - M = \left[S_1 \left(S_1^T S_1\right)^{-1} S_1^T - I\right] M \tag{6.19}$$

where $\boldsymbol{R} = [r_1, r_2, \ldots, r_k]^T$. The TMF is defined over one cycle as the sum of the absolute values of r_k as

$$TMF = \sum_{k=1}^{K} (r_k)^n \tag{6.20}$$

Unlike Ref. [16], in this chapter, we set the exponent n to be 2 for the sake of providing more emphasis on the high changes $\widetilde{M}$ from $\boldsymbol{M}$.

4.2 Proposed fault detection scheme

If a fault occurs in the microgrid, the VSC current will be changed, and its reconstructed samples depart from their actual values that increase the TMF value. Note that the data window consists of both pre- and postfault information. The maximum TMF value for phases a, b, and c is given by

$$J = \max(TMF_a, TMF_b, TMF_c) \tag{6.21}$$

In the proposed algorithm, we define a threshold for the index J that is Jth. If J becomes higher than Jth, the proposed algorithm detects the fault. The key part of this algorithm is defining a suitable threshold value because a good threshold can help it to discriminate the fault from other transient events such as load changes and switchings. Here, $i_{L,\mathrm{abc}}^{ref}$ is used in NARF because, in a three-wire VSC configuration, the TMF values of output current cannot be discriminated clearly in SYRF or STRF. For the four-wire VSC configuration, the $i_{o,abc}$ is used.

4.3 Artificial neural network–based single-phase auto reclosing scheme

To follow the signal harmonic content in an online manner, an n-input adaptive linear neuron (ADALINE) is used.

$$v(t) = A_{dc}e^{-\beta t} + \sum_{n=1}^{N} V_n \sin(n\omega t + \varphi_n) \tag{6.22}$$

The discrete form of the signal $v(t)$ is given by [16]

$$v(t) = A_{dc}(1 - \beta k T_s) + \sum_{n=1}^{N} (A_n \sin(n\omega t(k)) + B_n \cos(n\omega t(k))) \tag{6.23}$$

In Eq. (6.22), the first part consists of the first two terms of the Taylor expansion of the decaying DC component. Also, $t(k)$ is k^{th} sampling time, and T_s equals $2\pi/\omega K$. The ADALINE-ANN input vector is selected as Eq. (6.24) for the sake of harmonic extraction.

$$\begin{aligned} X(k) = {} & [\sin(\omega tk)\cos(\omega tk)\sin(2\omega tk)\cos(2\omega tk)\ldots \\ & \sin(n\omega tk)\cos(n\omega tk)1 - kT_s]^T \end{aligned} \tag{6.24}$$

We define the $W(k)$ as the input signal Fourier transform coefficients. We should select the weighting factors so that we have the minimum difference between the ADALINE-ANN output and the reference signal. By using the following weighting factor updating the law and least square error minimization, we use the Widrow–Hoff learning rule for training as [17]

$$W(k+1) = W(k) + \alpha \frac{e(k)X(k)}{X^T(k)X(k)} \tag{6.25}$$

where $W(k)$ and $W(k+1)$ are weighting factor vectors in kth and $(k+1)$th sampling period. We can reach to the perfect learning if and only if the tracking error $e(k)$ goes to zero so that we have

$$y(k) = y_d(k) = v(k) = W_0^T X(k) \tag{6.26}$$

where W_0 is defined as

$$W_0 = [A_1 \quad B_1 \quad \ldots \quad A_N \quad B_N \quad A_{dc} \quad \beta A_{dc}] \tag{6.27}$$

The harmonic content of the signal is achieved after obtaining the weighting factor's vector as follows:

$$V_n = \sqrt{w(2n)^2 + w(2n-1)^2} \tag{6.28}$$

$$\varphi_n = \cos^{-1}\left(\frac{w(2n-1)^2}{V_n}\right); \quad n = 1, 2, 3, \ldots, N \tag{6.29}$$

This scheme is proposed to estimate the harmonics in the voltage waveforms of the microgrid in an online manner. Because it requires computation burden and more easily implementable, the proposed method performs more accurately than the Furrier analysis method proposed in Ref. [18]. Besides, as mentioned in Ref. [19], the proposed strategy has less convergence time for harmonic estimation, which is less than half of a cycle using 32 samples in each cycle, in contrast with the Fourier transform method that needs at least one cycle. The sampling rate of the proposed method makes it attractive for practical implementation because it needs only a low-cost A/D chip.

In case that a fault occurs in the grid because of the changes in the voltage waveforms, we have two nonzero coefficients in Eq. (6.22). The hth voltage harmonic and its magnitude can be calculated as

$$V_h = \frac{\sqrt{2}}{N} \sum_{k=0}^{N-1} \left(v_k \cos\left(\frac{k2\pi h}{N}\right) - j v_k \cos\left(\frac{k2\pi h}{N}\right) \right); \quad j = \sqrt{-1} \tag{6.30}$$

$$V_h = \left| \underline{V}_h \right| = \sqrt{\mathrm{Re}^2\left\{ \underline{V}_h \right\} + \mathrm{Im}^2\left\{ \underline{V}_h \right\}} \tag{6.31}$$

We define two proposed harmonic indices (HIs) for discrimination of transient and permanent faults (by calculating HI_1), and specifying the type of fault (by calculating HI_2) as

$$HI_1 = \frac{1}{\sum_{h=2}^{5} \frac{h^2 V_1}{V_h}} \times 100\% \tag{6.32}$$

$$HI_2 = 100 \times \sum_{h=1}^{5} V_h \tag{6.33}$$

We have performed so many simulation studies to obtain the average value of the aforementioned indices for different kinds of faults and transients. HI_2 roughly takes the following values for different types of faults:

- Single phase to ground: (a–g): 8, (b–g): 9, (c–g):13.5
- Single line to line (a–b or a–c): 15–16.
- Three-phase fault: 12–12.5
- Double line to ground (a–c–g or a–b–g): 16–17
- Commutation error, misfire error, and fire-through error of VSCs: 3.5–5.5

In this methodology, classification features of ADALINE-ANN and TMF transient detection features are used for detecting the fault, discrimination between transient and permanent faults, and improving the system performance for realizing an interesting, attractive, and practical adaptive single-phase auto reclosing scheme. The most brilliant feature of the proposed method is that it does not need a massive training data set and can be trained online.

The proposed fault detection method does not require a high-bandwidth communication system and only needs locally measured signals and can be practically implemented with the highest possible reliability using a high-performance field-programmable gate array (FPGA) or digital signal processing chips with redundant power supply and redundant fail-safe central processing unit (CPU).

5. Overcurrent/overload protection scheme

5.1 Overload protection

The control structure of the DVC-based IIDERs does not include the inner current control loop. The three-phase instantaneous complex power of DER unit is calculated as

$$S = 3VI^* = 3V\left|\frac{V_{VSC} - V_{PCC}}{R_f + jX_f}\right|^* = 3V\left|\frac{\Delta V}{|R_f + jX_f|}\right|^* \tag{6.34}$$

Using the maximum generation capacity of the DER unit, the maximum magnitude value of the voltage difference $\Delta V_{\max}$ is given by

$$\Delta V_{\max} = \frac{S_{\max}|R + jX|}{3|V|} \tag{6.35}$$

If $|\Delta V| > \Delta V_{\max}$, the protection scheme detects that an OL has occurred. So, the VSC output voltage should be adjusted so that we have the least difference between the point of common coupling (PCC) voltage and VSC output voltage (for example, in SYRF):

$$V_{VSC,d} = V_{PCC,d} + \Delta V \max \cos\left(\tan^{-1}\left(\frac{V_{VSC,q} - V_{PCC,q}}{V_{VSC,d} - V_{PCC,d}}\right)\right) \tag{6.36}$$

$$V_{VSC,q} = V_{PCC,q} + \Delta V \max \sin\left(\tan^{-1}\left(\frac{V_{VSC,q} - V_{PCC,q}}{V_{VSC,d} - V_{PCC,d}}\right)\right) \tag{6.37}$$

Based on the fault model presented in Ref. [3], we have

$$V_{VSC,d} = \left(G(s)i_{L,\,d(\alpha)}^{\prime ref} - I_{o,d(\alpha)} + \omega C_f v_{o,q(\beta)}\right)\frac{1}{sC_f}$$

$$+\Delta V\max\cos\left(\tan^{-1}\left(\frac{V_{VSC,q} - V_{PCC,q}}{V_{VSC,d} - V_{PCC,d}}\right)\right) \tag{6.38}$$

$$V_{VSC,q} = \left(G(s)i_{L,\,q(\beta)}^{\prime ref} - I_{o,q(\beta)} + \omega C_f v_{o,d(\alpha)}\right)\frac{1}{sC_f}$$

$$+\Delta V\max\sin\left(\tan^{-1}\left(\frac{V_{VSC,q} - V_{PCC,q}}{V_{VSC,d} - V_{PCC,d}}\right)\right) \tag{6.39}$$

5.2 Overcurrent protection

The proposed fault detection method detects the OC condition. The output VSC voltage unit is set as follows [8]:

$$V_{VSC} = V_{PCC} + \left(R_f + jX_f\right)\frac{I_{pf}}{c_{trans}} \tag{6.40}$$

I_{pf} can be determined either by RBFNN-based power flow algorithm [4] or by PMS [7]. The constant c_{tran} (here, $c_{tran} = 3$) can be used for compensating the decaying DC component of the instantaneous current. The difference between the fault current during the fault (if the voltage regulation is done by IIDER), I_{rs} and prefault current is given by Eq. (6.41) [8].

$$\Delta I = \frac{\sqrt{\left(\mathrm{Re}\{I_{rs}\} - \mathrm{Re}\{I_{pf}\}\right)^2 + \left(\mathrm{Im}\{I_{rs}\} - \mathrm{Im}\{I_{pf}\}\right)^2}}{|I_{rs}|} \tag{6.41}$$

if the aforementioned conditions for discriminating between permanent and transient fault are satisfied and ΔI is greater than its predefined threshold value, the OC protection operates.

6. Simulation results

A multi-DER microgrid [7] (Fig. 6.3A) is simulated in MATLAB/Simulink environment, and the results are verified by OPAL-RT RTDS (Fig. 6.3B and C) [20]. The LCs and the rest of the microgrid systems are simulated using CPU1 and CPU2, respectively (Fig. 6.3A). Here, the microgrid is simulated with the proposed hierarchical control scheme that exploits DVC-based control of Ref. [7] and [8] in its inner voltage control loop.

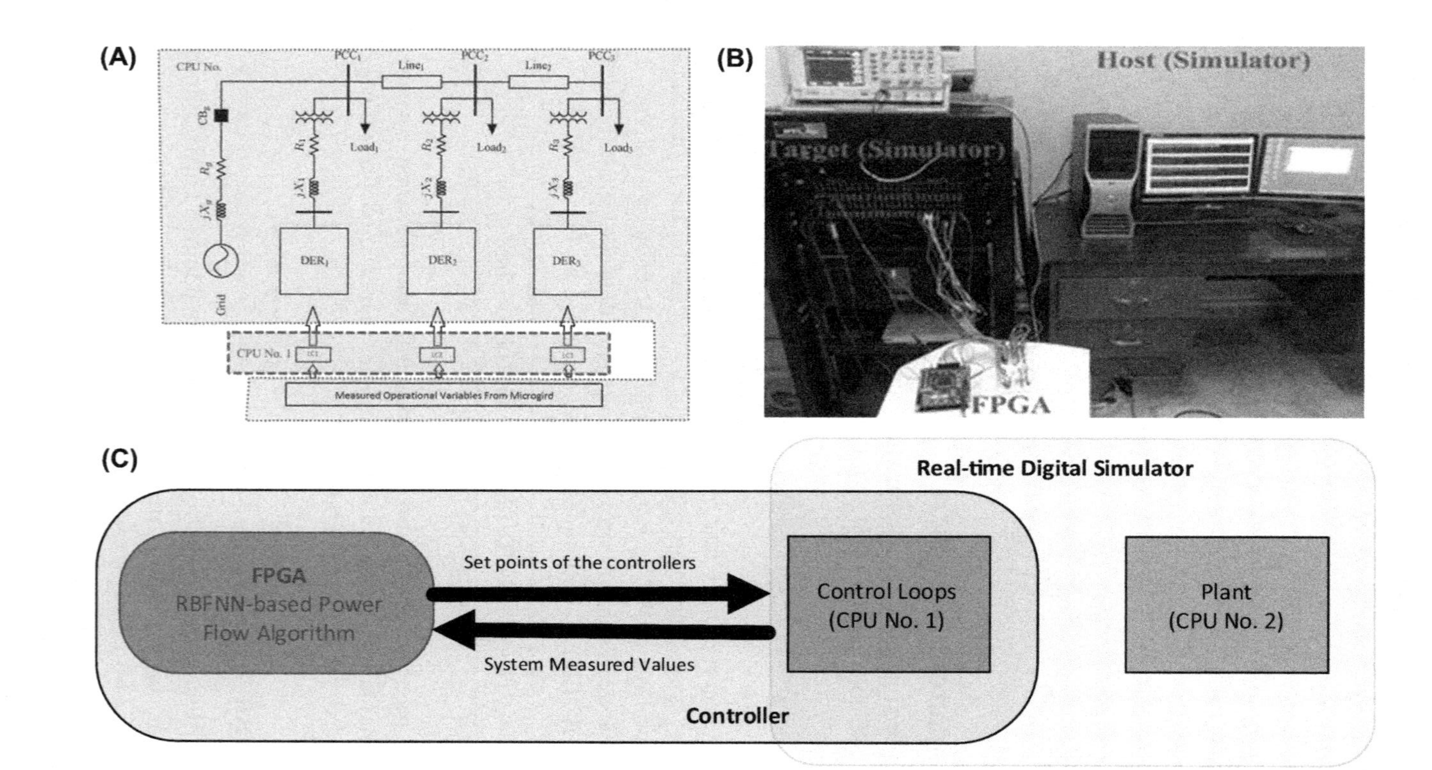

Figure 6.3 (A) The study system single-line diagram of study system [7], and (B) OPAL-RT eMEGAsim real-time digital simulator, and (C) the controller-hardware-in-the-loop real-time simulation block diagram.

6.1 Case 1: motor starting

In this section, offline time domain simulation of starting of a three-phase 132-kW, 400-V, 50-Hz, 4-pole induction motor [7], at $t = 0.8$ s in bus 2 of an autonomous microgrid, is performed with direct online (DOL) starting method. The results for hierarchical controller of Ref. [11] and the proposed hierarchical scheme with current-limiting and DVC-based control strategy of Refs. [7] and [8], for microgrid bus voltages and active and reactive powers, respectively, are illustrated in Figs. 6.4–6.6. Based on Fig. 6.4, on account of their inner current control loop of HDRC structure, after a momentarily droop, the voltage immediately returns to its nominal value, and thus, the HDRC structure has less contribution to the fault than the DVC-based one.

6.2 Case 2: overload protection

For the sake of performance evaluation of the proposed strategy, the microgrid is tested with an OL at bus 2, at $t = 1.5$ s so that the power setpoints change from 0.70, 0.53, and 0.35 p.u. apparent power at 0.95 lagging power factor, respectively for DER1, DER2, and DER3. Also, we increase the load of bus 1 at $t = 1.5$ s by 20% of its initial value and then return it to its nominal value at $t = 2$ s, and the results are presented in Fig. 6.7. The OL protection scheme limits the apparent output power at its maximum capacity, 0.75 p.u., because DER3 reaches its maximum generation capacity. Thus, as DER3 fails to supply the load, we have a decrement in the voltage to 0.9 p.u. while the DER2 output power does not affect. We can see that the proposed OL protection scheme limits the output power of DERs effectively. At $t = 2.5$ s, the new setpoints of active and reactive powers and voltage magnitude and phase angles are calculated based on the RBFNN-based power flow of Refs. [4,11], and assigned by PMS.

6.3 Case 3: fault response of isolated inverter-interfaced distributed energy resource for different reference frames

In the next simulation, we perform different tests to investigate the fault response of an isolated IIDER. The output voltage, actual inductor current, output current, and limited inductor current waveforms have been illustrated in Fig. 6.8 in line–line to ground fault (a–b–g) in bus 2, respectively. As we can observe, the presented current-limiting strategy results in having less fault current, better voltage profile, improved FRT capability, and higher power quality.

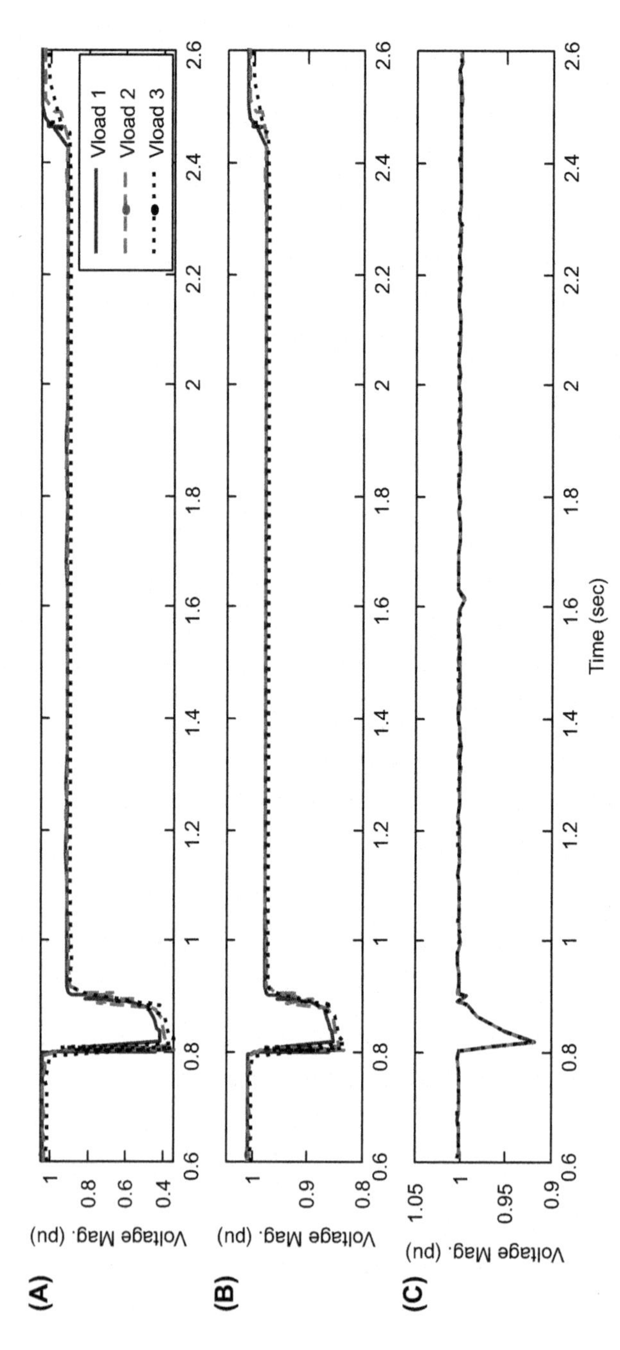

Figure 6.4 Motor starting: microgrid voltages for (A) hierarchical droop-based control scheme of Ref. [11], (B) the proposed hierarchical scheme with CLS, and (C) DVC-based control strategy of Ref. [7]. *DVC*, direct voltage control.

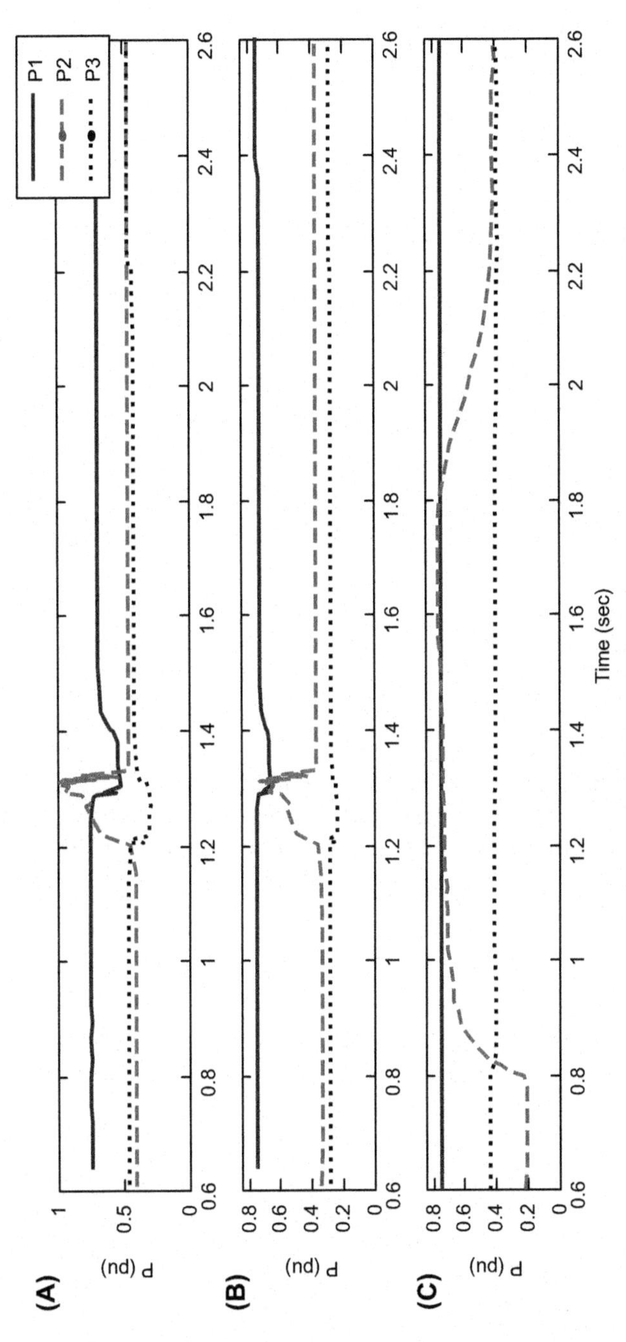

Figure 6.5 Motor starting: microgrid active powers for (A) hierarchical droop-based control scheme of Ref. [11], (B) the proposed hierarchical scheme with CLS, and (C) DVC-based control strategy of Ref [7]. *DVC*, direct voltage control.

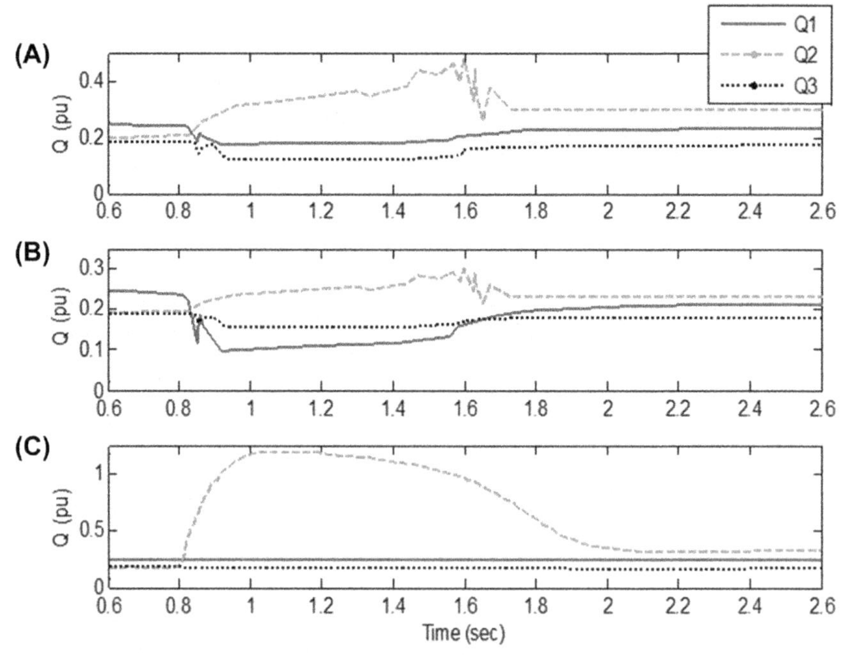

Figure 6.6 Motor starting: microgrid reactive powers for (A) hierarchical droop-based control scheme of Ref. [11], (B) the proposed hierarchical scheme with CLS, and (C) DVC-based control strategy of Ref. [7]. DVC, direct voltage control.

6.4 Case 4: real-time verification using real-time digital simulator—Three-phase fault, overcurrent protection, and improvement of fault ride-through capability

Hardware-in-the-loop (HIL) simulation is a type of real-time simulation. You use HIL simulation to test your controller design. HIL simulation shows how your controller responds in real time to realistic virtual stimuli. You can also use HIL to determine if your physical system (plant) model is valid. In HIL simulation, you use a real-time computer as a virtual representation of your plant model and a real version of your controller. The figure shows a typical HIL simulation setup. The desktop computer (development hardware) contains the real-time capable model of the controller and plant. The development hardware also contains an interface with which to control the virtual input to the plant. The controller hardware contains the controller software that is generated from the controller model. The real-time processor (target hardware) contains code for the physical system that is generated from the plant model.

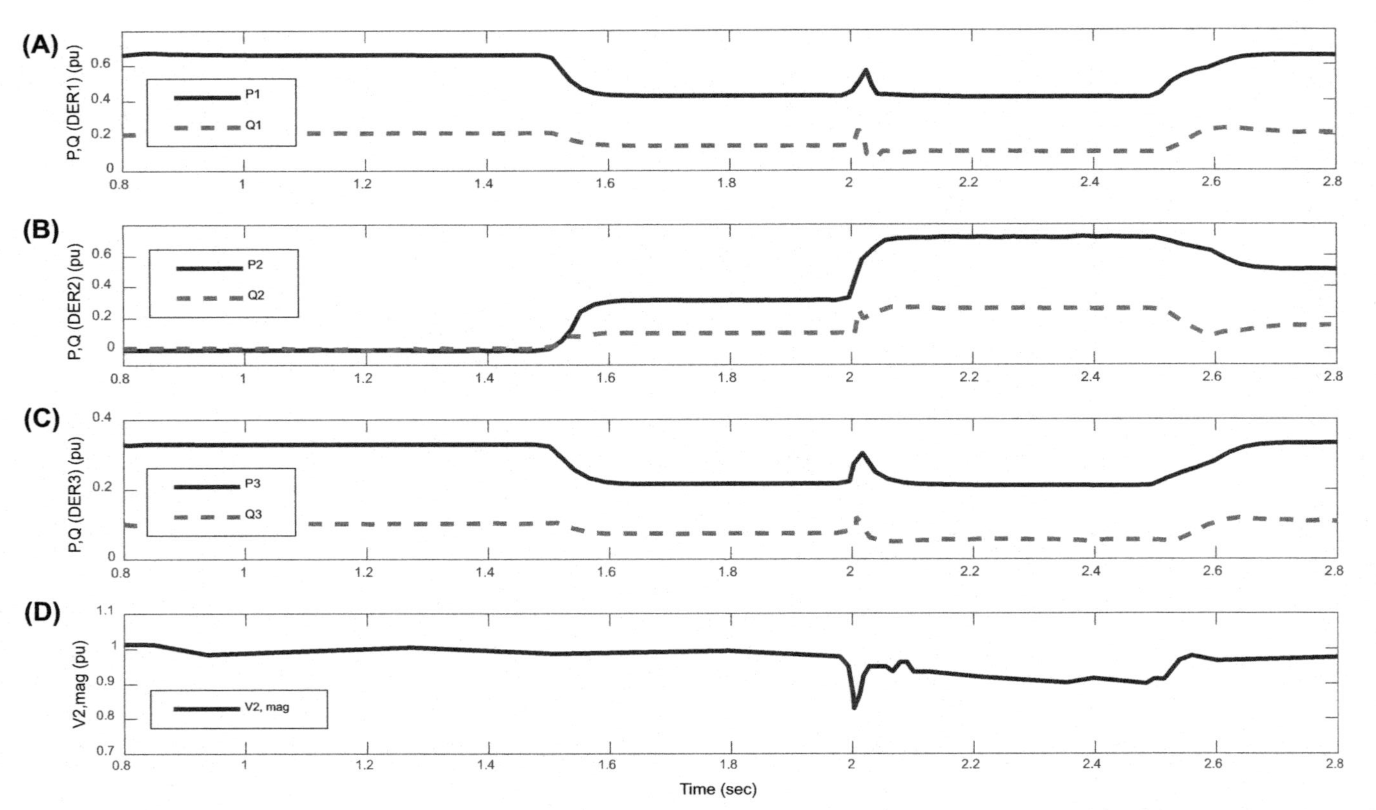

Figure 6.7 Overload protection: active and reactive powers for (A) hierarchical droop-based control scheme of Ref. [11], (B) the proposed hierarchical scheme with CLS, (C) DVC-based control strategy of Ref. [7], and (D) variation of voltage magnitude of PCC2. *DVC*, direct voltage control; *PCC*, point of common coupling.

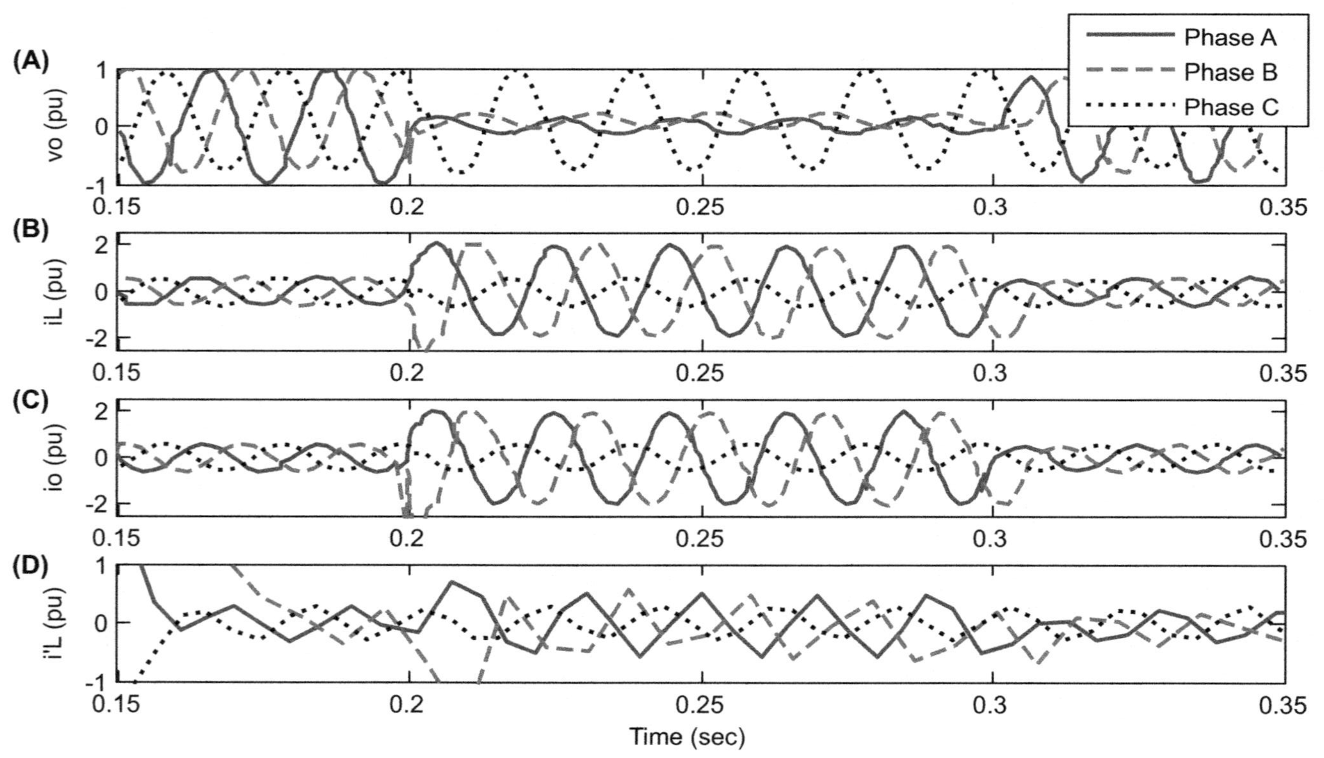

Figure 6.8 Microgrid in line to line fault (*a–b–g*): (A) output voltage, (B) inductor current, (C) output current, and (D) limited inductor currents.

Fig. 6.3C shows the basic real-time controller hardware-in-the-loop (CHIL) block diagram so that the CPU No. 1 and No. 2 are in charge of simulating the control loops and the plant, respectively, and a power flow algorithm–supported RBFNN implemented on FPGA determines the initial conditions required for the setpoints of the primary and secondary controllers in a very real-time manner.

We can do the experimental real-time verification for other cases; however, in this part, we use eMEGAsim OPAL-RT RTDS for verification of the results by simulating a three-phase fault (as a large-signal disturbance) that occurs in the middle of the line connecting DER1 and DER2 (that happens at the moment $t = 0.8$ s and clears after six cycles) [20]. The setpoints of active and reactive powers and voltage magnitude and phase angles are calculated based on the RBFNN-based power flow of Refs. [4,11] (implemented on a Xilinx Virtex 7 FPGA), and assigned by PMS. The CPU No. 1 and No. 2 are in charge of simulating the control system (either hierarchical droop-based or DVC-based scheme) and the plant, respectively (Figs. 6.9 and 6.10). It can be observed that after clearing of the fault, without any problem, power sharing continues, and voltage becomes stable. However, we have less voltage droop and better power sharing curves when the proposed CLS is used.

Furthermore, to prove the effectiveness of the proposed index for discriminating permanent and transient faults (occurred in the previous location), the system is simulated for two types of three-phase faults: (1) a transient fault (that its secondary arch flash is quenched after $t = 0.76$ s), and (2) the permanents fault in the previous location with an impedance of $Z = 5 + 10\,\Omega$ (Fig. 6.11A and B). The results prove that the proposed indices can discriminate permanent and transient faults authentically.

6.5 Discussions

In this chapter, we aimed to present a generalized fault model for HDRC-based and DVC-based IIDER units and offer a new OC/OL protection strategy. In this way, first, we developed analytical models for the investigation of the fault response of different IIDERs in different reference frames. Also, the study system was simulated for different scenarios, including motor starting and three-phase fault. The analytical fault model and equivalent phase networks, along with the effectiveness of the proposed OC/OL protection strategy and TMF/ANN-based fault detection method, were evaluated and verified for DVC- and HDRC-based IIDERs by

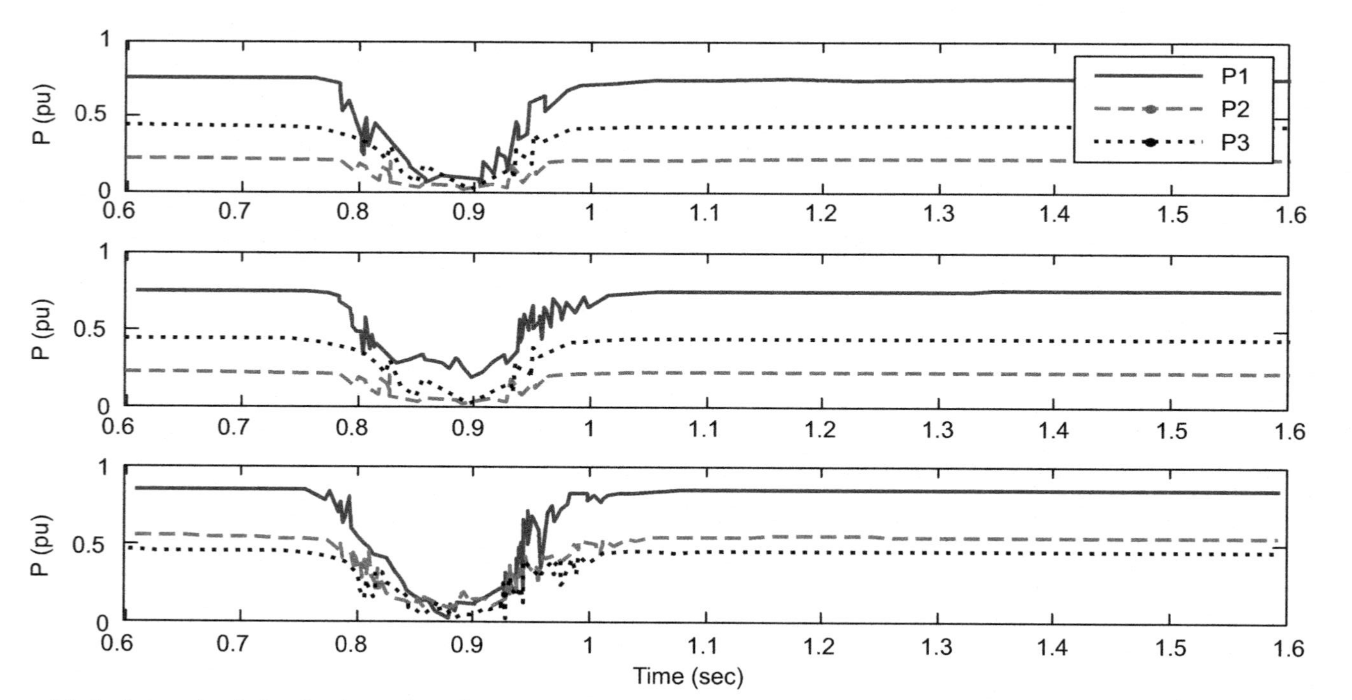

Figure 6.9 Real-time simulation for three-phase fault: microgrid active powers for hierarchical droop-based control scheme of Ref. [11], the proposed hierarchical scheme with current-limiting and DVC-based control strategy [7].

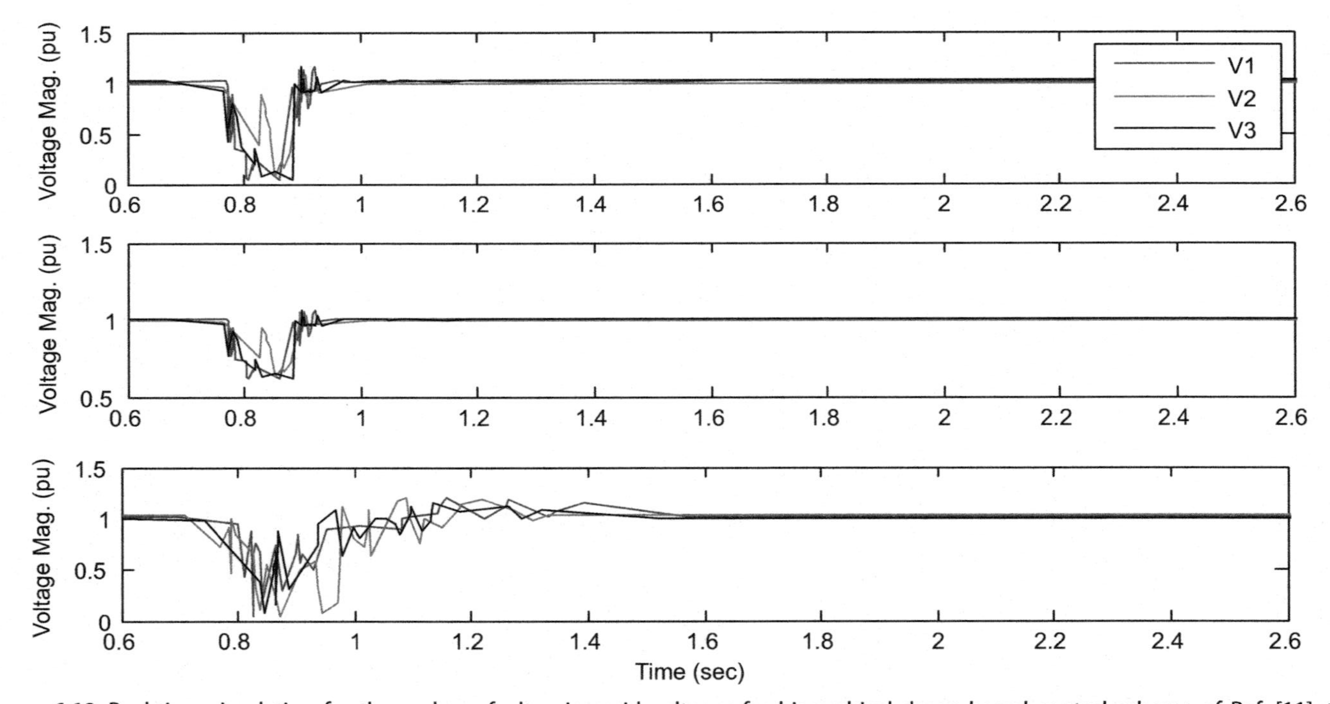

Figure 6.10 Real-time simulation for three-phase fault: microgrid voltages for hierarchical droop-based control scheme of Ref. [11], the proposed hierarchical scheme with current-limiting and DVC-based control strategy [7].

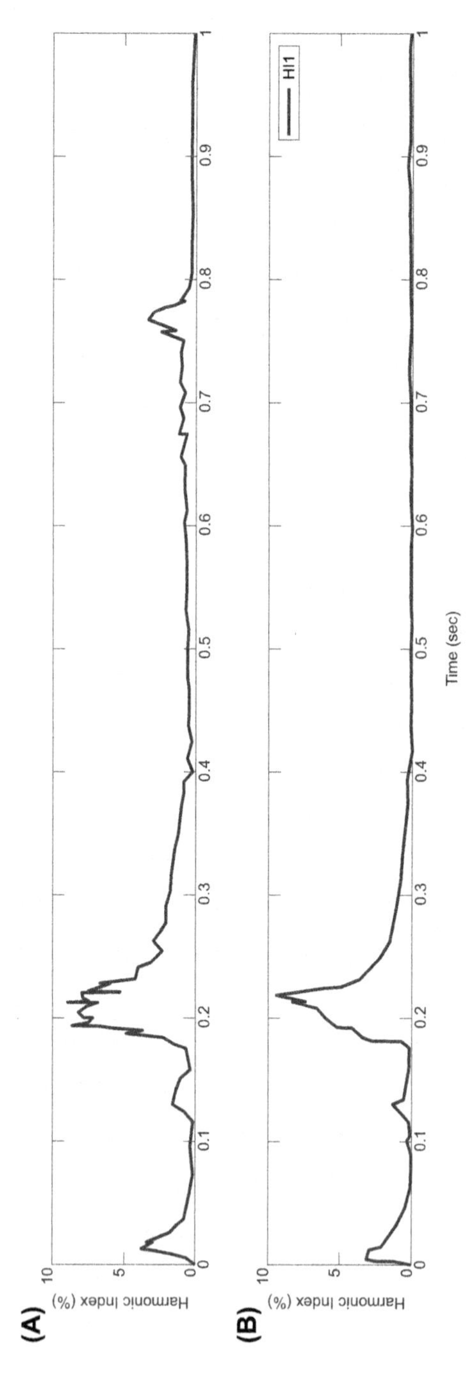

Figure 6.11 Real-time simulation for three-phase fault: microgrid: Value of HI1 index for (A) transient fault and (B) permanent fault. *HI*, harmonic index.

performing offline digital time domain simulations in MATLAB/Simulink software environment and experimental verification by CHIL real-time tests using OPAL-RT RTDS. The obtained results and experimental verifications once again proved the efficiency, authenticity, and accuracy of the proposed fault detection and protection strategies and the system stability for small- and large-signal disturbances.

7. Pareto-optimal solution for coordination of overcurrent relays in interconnected networks and Multi–distributed energy resource microgrids

OC relays are used as both main and backup protective relays in sub-transmission and distribution and upstream transmission networks, respectively [22–24]. The backup protection provides the backup to the main protection whenever it fails in operation or is cut out for repairs. The backup protection is the second line of defense, which isolates the faulty section of the system with a predefined time interval concerning operation time (OT) of the main relay in case the main protection fails to function properly. The failure of the primary protection occurs because of the failure of the DC supply circuit, current or voltage supply to relay circuit, relay protective circuit, or the circuit breaker. Usually, the backup relay OT is selected more than the sum of OTs for the main circuit breakers, main relay, and overshoot time of the backup relay [25].

There are four essential parameters which should be adjusted for OC relays, i.e., pick up current (defined as the current for which the relay initiates its operation), current setting (that is expressed in the percentage ratio of relay pick up current to the rated secondary current of CT), plug setting multiplier (PSM) (that is referred as the ratio of fault current in the relay to its pick up current, traditionally from 50% to 200% with 25% steps), and time setting multiplier (TSM) (calibrated from 0 to 1 in steps of 0.05 s) (which its ratings are traditionally from 0.05 to 1 with 0.05 steps) [26–30]. However, using the current commercially available digital relays, it is possible to adjust these parameters in the steps of 0.01 [31]. In this chapter, we use the directional OC (DOC) relays normally used in interconnected networks. The coordination constraint between main and backup relays is expressed as follows (see Fig. 6.12):

$$\Delta t_{mb} = t_b - t_m - CTI \geq 0 \tag{6.42}$$

where CTI is the coordination time interval between main and backup relays. In the meshed and interconnected systems, dissimilar to the weakly meshed or radial structured distribution networks that each DOC relay acts as the back of its downstream relays, multiple relays act as a backup of one

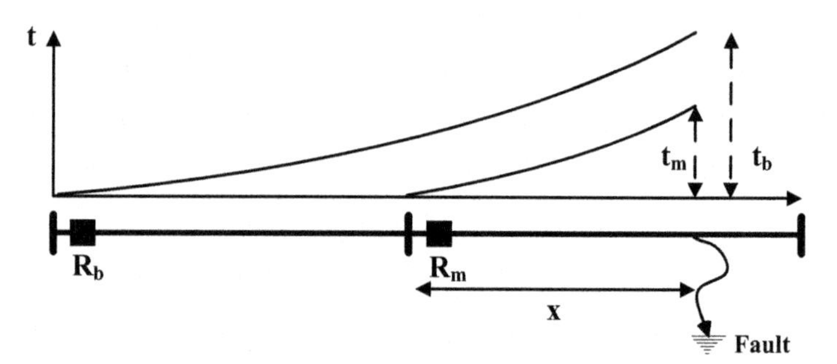

Figure 6.12 Main and backup relay coordination constraint [1].

specified relay and vice versa; each relay may act as the backup of one or more relays. The problem of relay coordination is almost computationally intensive, and in this regard, using the numerical and heuristic optimization tools is always beneficial, especially in large and interconnected power systems. When optimization algorithms are exploited for solving the problem of relay coordination, defining an appropriate objective function(s) (OF(s)) is essential. Among these algorithms, genetic algorithm (GA) has been used before to solve this problem [22,32]; however, in some cases, the reported GA-based strategies have caused miscoordination so that the main relay OTs are higher than the backup ones leading to the operation of backup relay sooner than the main relay, which is not acceptable [31,33–35]. Also, in the reported GA-based algorithms, the TSM has been modeled as a continuous variable that has very small steps even for new modern digital relays that may not be optimal. Some researches are presented in Refs. [31,33,34]. It should be noted that the OT of the main and backup relays shall not be so large that it results in long toleration of the fault by the equipment, which may lead to damaging the power system equipment [35,36]. To reduce the tripping times for a selective fault clearing, an optimal relay coordination algorithm has been proposed in Ref. [37], based on a convex optimization model.

Because TSM ranges continuously, the discrete TSM may result in a near-optimal or subsolutions. Different optimization algorithms have been used to solve the over-current relay coordination (OCRC) problem and then compared from the accuracy and the efficiency points of view in Ref. [38]. By the way, when the OCRC problems are solved by the single-objective optimization (SOO) algorithm, the selection of weighting factors of becomes challenging [39,40]. In these cases, some tools such as the analytical hierarchy process may be beneficial [41]. Also, the interior point method (IPM) algorithm and its combination with heuristic optimization tools have been proposed for solving the OCRC problem in Ref. [43].

In this section, we exploit multiobjective particle swarm optimization (MOPSO) and fuzzy decision-making tool (FDMT) to address the OCRC problem in interconnected power systems and microgrids, and solve the issues of large OT of OC relays and miscoordination by defining suitable OFs in multiobjective optimization (MOO) algorithm along with good operational constraints and obtaining Pareto-optimal solution for both continuous and discrete variables. Moreover, we address the problem of weighting factor selection and scaling of OFs by using FDMT in the MOPSO to find the Pareto-optimal solution. The proposed algorithm increases the flexibility of the solution by including the coefficients of OC relay characteristic curves and PSMs and providing the possibility of choosing the best curve for main and backup relays. Finally, based on IEC-6090 recommendations and fault calculations presented in the previous sections for the microgrids consisting of DERs [3], we extend the proposed strategy for addressing OCRC problem in microgrids.

7.1 Problem formulation

To solve the OCRC problem, we need to take care of the high-speed operation of the relays, the selectivity (that force a relay to only operate if a fault occurs in its corresponding protection zone), and moreover, the reliability (which can assign the backup relay appropriately). All of these issues can affect the dependability of the protection system, and at a higher level, the reliability and security of the whole power network [42–44]. Usually, the overall operation speed of all relays is expressed as [22]

$$OF = \sum_{i=1}^{n} t_i\left(x_{close-in}\right) \tag{6.43}$$

For proper selection of PSM, we choose the minimum current that can be detected by DOC relay as fault current, which shall be greater than the multiplication of a certainty factor (usually 1.3) and the maximum continuous load current, mathematically expressed as

$$1.3 I_{L\max} \leq I_p \leq I_{F\min} \tag{6.44}$$

Based on the aforementioned explanations, the problem can be mathematically formulated as the following SOO problem.

$$\begin{aligned}
&\text{Minimize } F = \sum_{i=1}^{n} t_i \\
&st. \quad \Delta t_j = t_{b_j} - t_{m_j} - CTI > 0 \\
&\quad 1 \leq j \leq mb \\
&\quad I_{p\min} \leq I_p \leq I_{p\max} \\
&\quad t_{i\min} \leq t_{i,op} \leq t_{i\max} \\
&\quad TSM_{\min} \leq TSM \leq TSM_{\max} \\
&\quad PSM_{\min} \leq PSM \leq PSM_{\max}
\end{aligned} \tag{6.45}$$

where the $PSM_{\min}$ and $PSM_{\max}$ values are given by [38,39]

$$PSM_{\min} = \max\left(0.5, \min\left(1.25\frac{I_{L,\max,i}}{CTR_i}, \frac{I_{f,\max,i}}{3CTR_i}\right)\right) \tag{6.46}$$

$$PSM_{\max} = \min\left(2, \frac{2I_{f,\min,i}}{3CTR_i}\right) \tag{6.47}$$

where CTR_i is the current transformer ratio (CTR) of ith current transformer (CT). Depending on fault current and other settings, the OT of OC relay is obtained from its characteristic curve as

$$t = \left(\frac{A}{\left(I/I_p\right)^B - C} + D\right) \times TSM \tag{6.48}$$

and in a more convenient form, we have [22,46,47]

$$t = \frac{a}{PSM^b - 1} \times TSM \tag{6.49}$$

The optimization problem of Eq. (6.45) can be solved using direct mathematical/analytical and/or intelligent metaheuristic search algorithms. The implementation of the first strategy is sometimes hard or cumbersome, if not possible, while the second method is easier to be implemented. Because PSM, TSS, and the coefficients of the characteristic curves are set in small values in digital DOC relays, the problem can be continuous, or the OCRC problem is formulated as a mixed-integer optimization problem. A robust and powerful algorithm for solving this problem is particle swarm optimization (PSO). Motivated by Solati-Alkaran et al. [22], an OF including the operational optimization constraints is considered as penalty value as

$$OF = \sum_{i=1}^{n} a_i f(t_i) + \sum_{j=1}^{mb} b_j g(\Delta t_j) \tag{6.50}$$

where a_i and b_j are the constants.

7.2 Proposed method

7.2.1 Challenges for time setting multiplier: continuous or discrete

The assumption of the TSM as continuous variable and its rounding to the nearest discrete value (that has been done in the earlier researches) can increase the possible states and consequently the search space, which can affect the convergence of the algorithm and sometimes may result in divergence in large test systems. Besides, in this conditions, we cannot evaluate the mathematical relation of optimization variables and OF(s), and moreover, the algorithm may find the local minimum instead of the global one (Fig. 6.13A).

7.2.2 Optimization algorithm

Multiple-criteria decision-making (MCDM) and MOO are exploited to solve the engineering and mathematical optimization problems, including several OFs that all should be optimized simultaneously [48], which can be expressed in the general form (with d variables and M objective functions) as

$$f(x) = [f_1(x), f_2(x), \ldots, f_M(x)]^T \tag{6.51}$$

s.t.

$$\begin{cases} g_j(x) \leq 0; & i = 1, 2, \ldots J \\ h_k(x) = 0; & i = 1, 2, \ldots \infty K \end{cases} \tag{6.52}$$

where $x = [x_1,\ x_2\ \ldots\ x_d]^T$. A feasible region Ω and feasible solution are defined by constraints given by Eq. (6.51) and any point x inside Ω, respectively. The set Ω is mapped by vector function $f(x)$ (representing all possible objective function(s) values) into the set Λ [48]. In the MOO, we use the concept of Pareto dominance formulated first by an Italian scientist named Vilfredo Pareto. The Pareto optimal set can consist of many solutions that can have the potential to be optimal so that the most proper solution vector will be determined based on some criteria based on them [49–51]. In the proposed methodology, based on the concept of MCDM, the FDMT is

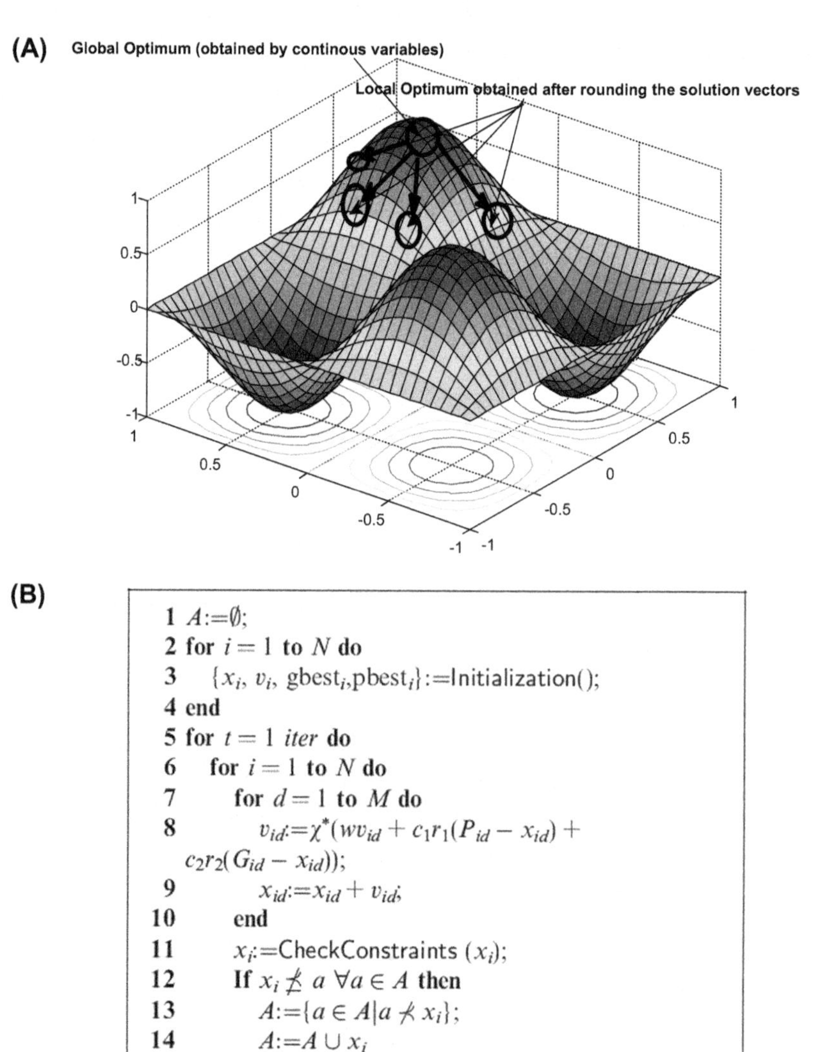


```
1 A:=∅;
2 for i = 1 to N do
3     {x_i, v_i, gbest_i, pbest_i}:=Initialization();
4 end
5 for t = 1 iter do
6     for i = 1 to N do
7         for d = 1 to M do
8             v_id:=χ*(wv_id + c1r1(P_id − x_id) +
      c2r2(G_id − x_id));
9             x_id:=x_id + v_id;
10        end
11        x_i:=CheckConstraints (x_i);
12        If x_i ⋠ a ∀a ∈ A then
13            A:={a ∈ A|a ⊀ x_i};
14            A:=A ∪ x_i
15        end
16    end
17    If x_i ⪯ pbest_i ∨ (x_i ⊀ pbest_i ∧ pbest_i ⊀ x_i)
18        pbest_i:=x_i;
19    end
20    gbest_i:=GlobalGuide (x_i;A);
21 end
```

Figure 6.13 (A) Concept of global and local optimums, (B) pseudocode of the proposed archive-based MOPSO algorithm [49]. *MOPSO*, multiobjective particle swarm optimization.

exploited as the criteria based on the defining a linear membership function (MF) (μ_i) for each OF (that should be minimized) as

$$\underset{i=1,\ldots,j}{\mu_i^n} = \begin{cases} 1 & F_i^n \leq \min(F_i) \\ \dfrac{\max(F_i) - F_i^n}{\max(F_i) - \min(F_i)} & \min(F_i) \leq F_i^n \leq \max(F_i) \\ 0 & F_i^n \geq \max(F_i) \end{cases} \tag{6.53}$$

If the OFs should be maximized, the MF (μi) is expressed as

$$\underset{i=j+1,\ldots,n_{obj}}{\mu_i^n} = \begin{cases} 0 & F_i^n \leq \min(F_i) \\ \dfrac{F_i^n - \min(F_i)}{\max(F_i) - \min(F_i)} & \min(F_i) \leq F_i^n \leq \max(F_i) \\ 1 & F_i^n \geq \max(F_i) \end{cases} \tag{6.54}$$

For the sake of evaluating the optimality degree of the Pareto-optimal solutions, using the MFs defined in Eqs. (6.53) and (6.54), based on the fuzzification of first-order linear equations, the best OF values are uniformly distributed. The most preferred solution is then chosen as

$$\mu_{opt,I} = \max\left\{ \sup \frac{\sum_{i=1}^{p} w_i \cdot \mu_i^n}{\sum_{n=1}^{M_P} \sum_{i=1}^{p} w_i \cdot \mu_i^n} \right\} \tag{6.55}$$

where

$$\omega_i \geq 0, \quad \sum_{i=1}^{p} w_i = 1 \tag{6.56}$$

where M_p is the number of Pareto-optimal solutions, and w_i is the weight value associated with the ith objective function. We offer the best compromise solution (best trade-off I) as the solution with the minimum MF. We can also offer another criterion for the selection of the best compromise solution. A pseudocode of the whole MOPSO algorithm has been illustrated in Fig. 6.13B. A more detailed description of the MOPSO algorithm along with nomenclature and definition of related variables has been provided in Ref. [49].

7.2.3 Objective functions

The OF of OCRC problem has been defined in Ref. [31] as

$$OF = \alpha_1 \sum_{i=1}^{n} (t_i)^2 + \alpha_2 \sum_{j=1}^{mb} \left(\Delta t_j - \beta\left(\Delta t_j - \left|\Delta t_j\right|\right)\right)^2 \tag{6.57}$$

The second term of Eq. (6.57) is optimization constraint added to OF as penalty term, i.e., $g(\Delta_{tj}) = (\Delta_{tj} - \beta(\Delta_{tj} - |\Delta_{tj}|))^2$, and thus, the main function of the OCRC problem of Ref. [7] is $f(t_i) = (t_i)^2$. However, if we add the square of OT to OF, the algorithm goes far from its optimality. As an example, if the ith and jth relays OT are 0.04 and 0.814, respectively, the OF value ($f(t_i) = (t_i)^2$) is 0.82, the same as when they are, respectively, chosen as 0.1 and 0.9. In this section, we handle this dilemma by defining the first OF as $f(t_i) = t_i$. In this chapter, for considering the penalty function for both positive and negative Δt values, the constraints consist of both $(|\Delta t_j| - \Delta t_j)$ and $(|\Delta t_j| + \Delta t_j)$ terms. We have some constraints that are not satisfied and can increase the protection scheme costs directly due to the requirement for alternate backup. Thus, we use the sign function sgn(x) in the definition.

$$\text{sgn}(x) = \begin{cases} 1 & \Delta t_j < 0 \\ 0 & \Delta t_j \geq 0 \end{cases} \tag{6.58}$$

To cover the mentioned themes, the proposed OFs are defined as a three-dimensional (3D) MOO as

$$F_1 = \sum_{i=1}^{n} (t_i)^2$$

$$F_2 = \sum_{j=1}^{mb} \left(\Delta t_j - \beta\left(\Delta t_j - \left|\Delta t_j\right|\right)\right)^2$$

$$F_3 = \left\{ \begin{array}{l} a_0 \sum_{j=0}^{m} BC_j + a_1 \dfrac{\sum_{i=1}^{n} t_i}{n} + a_2 \dfrac{\sum_{j=1}^{mb}\left(\left|\Delta t_j\right| - \Delta t_j\right)}{\max\left(1, \sum_{j=1}^{mb} \text{sgn}\left(\left|\Delta t_j\right| - \Delta t_j\right)\right)} + \\ \\ a_3 \dfrac{\sum_{j=1}^{mb}\left(\left|\Delta t_j\right| + \Delta t_j\right)}{mb - \sum_{j=1}^{mb} \text{sgn}\left(\left|\Delta t_j\right| - \Delta t_j\right)} + a_4 \dfrac{\sum_{j=1}^{mb} \text{sgn}\left(\left|\Delta t_j\right| - \Delta t_j\right)}{mb} \end{array} \right\} \tag{6.59}$$

where for primary-backup combination j, BC_j is defined as either 0 or 1 depending on $\Delta t_j > \varepsilon$ or $\Delta t_j < \varepsilon$, respectively.

For decreasing the dependency of coefficients $a_1, \ldots, a_4$ on the number of OCRC constraints and also determining thresholds of coefficients, each term of F_3 is normalized concerning the number of nonzero terms [22]. Anywhere required, the weighting factor coefficients of F_3 are chosen the same as Ref. [22].

7.2.4 Software implementation and coefficients of characteristic curves

In Refs. [31] and [36], optimization variables just include TSM of OC relays in the OCRC problem. By the way, some other researchers like Ezzeddine et al. [52] have considered the characteristic curves; however, all variables are discrete and the modern digital relays capabilities (like the standard form of characteristic curves given by Eq. (6.48) [53]) have not been considered. By using this trick, the proposed strategy has no limitation/restriction for using any type of OC relay characteristic curves, considering the allowable range of PSM and TSM and the related characteristic curve vectors for the OC relays selected based on the vendor catalog.

Besides, in the presented MOPSO/FDMT algorithm, we can automatically adjust the relayed coefficients using FDMT instead of setting weighting factors by trial and error. In the proposed strategy, we use the accurate analytical method of Ref. [46] to obtain critical fault point and $IF_{\min}$ and $IL_{\max}$, dissimilar to other researches that have used the six-pair fault current method of Ref. [54]. In the proposed algorithm, after selecting the OC relay, the related type, OT, and functions are coded in the PSO module. It should be further noted that the pair (a, b) is equal to (0.0144, 0.02) for IEC normally inverse, (13.5, 1) for IEC very inverse, and (80, 2) for IEEE extremely inverse types. Also, for any other user-defined relay, these values can be arbitrarily selected. In each iteration, each generation contains solution vectors as $X = [TSM, PSM, Type]$, where $TSM = [TSM_1, \ldots, TSM_n]$, $PSM = [PSM_1, \ldots, PSM_n]$, and Type $= [Type_1, \ldots Type_n]$ include details of all OC relays based on Eq. (6.49).

7.3 Simulation results and discussion

Finally, the proposed MOPSO/FDMT-based OCRC algorithm is tested on 8-bus [31] and IEEE 30-bus test power systems [55], to validate its accuracy, efficiency, effectiveness, and authenticity. To have a fair comparison with Ref. [30], for providing the same condition and having a fair comparison, in the first case, the relay operation characteristics and PSM are assumed the same as Ref. [30]. In the second scenario, additional optimization variables such as coefficients of characteristic curves and PSM are considered. The software program has been implemented and run using MATLAB 8.1 R2013 version software with Intel Corei7, 2.4 GHz CPU, 4 GB of RAM, 500 GB hard disc computer.

7.3.1 Case 1: 8-bus system—comparison with Refs. [22] and [31].

The data of the 8-bus test system have been gotten from Ref. [21]. The proposed MOPSO/FDMT-based strategy is here compared with the SOO algorithms of Refs. [22] and [31]. The basic difference that discriminates the proposed strategy with the reported methods in the procedure of obtaining weighting factors that are obtained by FDMT in the proposed strategy (that can reflect the effect of all OFs and find the best tradeoff solutions), and by trial and error in Ref. [22], and by FIS in Ref. [31]. As depicted in Fig. 6.14,

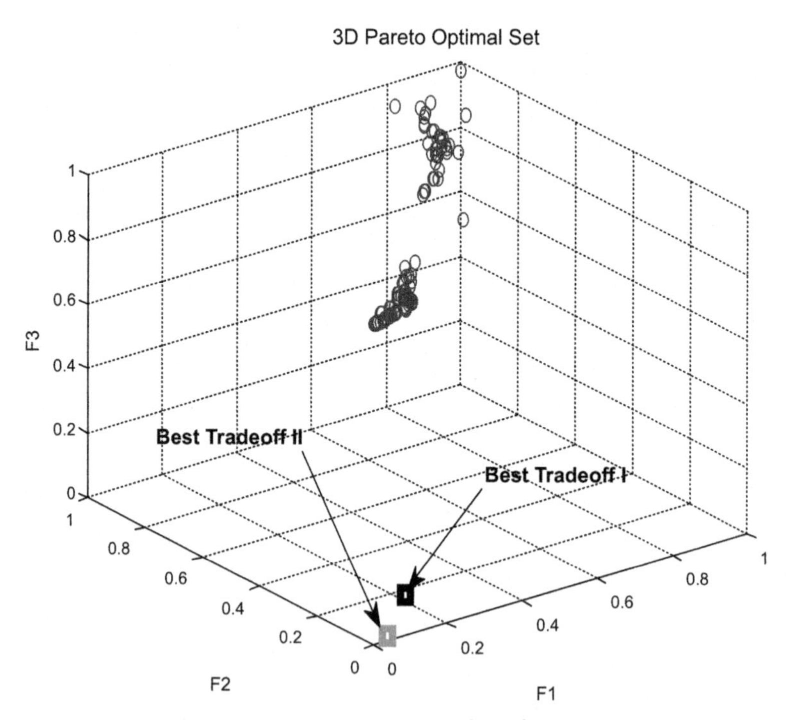

Figure 6.14 3D Pareto-optimal set for case 1.

the MOPSO evaluates the OFs individually and determines the best trade-off solution using the Pareto-optimal set. The obtained results for the proposed algorithm and its comparison with the methods of Refs. [22] and [31] are presented in Table 6.1A and B. As it can be seen, the proposed MOPSO/FDMT-based algorithm can decrease the TSM and OT of OC relays considerably in comparison with the algorithms presented in Refs. [22] and [31].

7.3.2 Case 2: IEEE 30-bus system—comparison with Refs. [21] and [31].

The data of the 8-bus test system have been gotten from Ref. [55]. The proposed MOPSO/FDMT-based strategy is here compared with the SOO algorithms of Refs. [21] and [31] (Table 6.2). Like IEC standard [40], the parameter D is here set to zero. The TSM and PSM, values of Δt and characteristic curve parameters calculated by the proposed MOPSO/FDMT-based strategy and other reported SOO methods [22,31] are presented in Table 6.3A,B, and C. As it can be seen, the proposed MOPSO/FDMT-based algorithm can decrease the TSM and OT of OC relays considerably in comparison with the algorithms presented in Refs. [22] and [31] that can prove that the proposed strategy can solve the OCRC problem and it is applicable for a wide range of characteristic curves.

7.3.3 Case 3: Wood and Woollenberg 6-bus system– comparison with Ref. [43].

In this part of the analysis, the proposed strategy is compared for a 6-bus power test system [43,56], with the SOO algorithms such as differential evolution algorithm (DEA), IPM-branch, and bound (IPM-BBM), IPM-GA, and IPM-DEA algorithms [43]. It is assumed that the TSM and PSM are assumed to continuously changed from 0.1 to 1.1 and discretely from 0.5 to 2 (with 0.25 resolution), respectively. We compared the obtained results for the proposed strategy with SOO algorithms of Refs. [22,31], by doing more than 100 simulations, and the results are presented in Table 6.4A and B. It is evident that based on the data presented in Table 6.4A and B, using the proposed strategy, there is no mis-coordination among the pairs of primary and backup relays, and the OTs of the OC relays are less than ones obtained by other reported SOO algorithms.

Table 6.1A Simulation results for case 1 (Part A).

Relay No.	Proposed method of Refs. [12,31] [12,31]			Proposed method of Refs. [12,31]			Proposed method		
	PSM (A)	TSM (s)	OT (s)	PSM (A)	TSM (s)	OT (s)	PSM (A)	TSM (s)	OT (s)
1	500	0.15	0.5894	500	0.1	0.3929	500	0.1	0.1058
2	800	0.35	1.2173	800	0.3	1.0434	800	0.25	0.2447
3	600	0.25	0.9621	600	0.25	0.9621	600	0.2	0.2052
4	800	0.1	0.6636	800	0.1	0.6636	800	0.1	0.0979
5	550	0.1	0.766	550	0.05	0.383	550	0.05	0.0520
6	550	0.25	0.7623	550	0.2	0.6099	550	0.15	0.1561
7	650	0.2	0.7059	650	0.2	0.7059	650	0.15	0.1518
8	550	0.25	0.7625	550	0.2	0.61	550	0.15	0.1561
9	540	0.05	0.3479	540	0.05	0.3479	540	0.05	0.0522
10	550	0.15	0.6883	550	0.15	0.6883	550	0.1	0.1041
11	650	0.25	0.9834	650	0.25	0.9834	650	0.2	0.2025
12	550	0.4	1.1847	550	0.4	1.1847	550	0.35	0.3643
13	600	0.1	0.4665	600	0.1	0.4665	600	0.1	0.1026
14	800	0.15	0.5944	800	0.15	0.5944	800	0.1	0.0979
Sum		2.75	10.694	—	2.5	9.636	—	2.05	2.0931

Results for PSM, TSM, and ΔT of the relays for SOO methods of Refs. [12,22,31] and the proposed MOPSO/FDMT-based algorithm. *OT*, operation time; *PSM*, plug setting multiplier; *TSM*, timesetting multiplier.

Table 6.1B Simulation results for case 1 (Part B).

Main relay	Backup relay	Main relay close-in fault current (kA)	Backup relay close-in fault current (kA)	Δt proposed in [31] (s)	Δt proposed in [22] (s)	Δt of proposed method(s)
8	7	4.9618	1.5209	0.4673	0.6198	0.3338
2	7	5.3623	1.5281	0.0028	0.1767	0.0020
2	1	5.3623	0.8049	0.6290	0.0541	0.4493
3	2	3.3345	3.3345	0.2675	0.0332	0.1911
4	3	2.2343	2.2343	0.2057	0.2057	0.1469
5	4	1.3529	1.3529	0.1885	0.5715	0.1346
6	14	4.965	1.5229	0.4825	0.635	0.3446
14	1	4.2327	0.7941	1.3181	0.5473	0.9415
1	6	2.6825	2.6825	0.0532	0.0411	0.0380
9	10	1.4437	1.4437	0.3173	0.3173	0.2266
10	11	2.3347	2.3347	0.2195	0.2195	0.1568
11	12	3.4808	3.4808	0.0492	0.0492	0.0351
12	14	5.3651	1.5294	0.0489	0.0489	0.0349
12	13	5.3651	0.8056	0.8135	0.8135	0.5811
13	8	2.4907	2.4907	0.2292	0.0101	0.1637
7	13	4.2326	0.8098	1.4085	1.2523	1.0061

Results of close-in fault currents for and ΔT each pair of OC relay for SOO methods of Refs. [22] and [31] and the proposed MOPSO/FDMT-based algorithm. *FDMT*, fuzzy decision-making tool; *MOPSO*, multiobjective particle swarm optimization; *SOO*, single-objective optimization.

Table 6.2 Case 2: Δt for each pair of relays in the second scenario.

Main relay	Backup relay	Δt proposed in Ref. [31] (s)	Δt proposed in Ref. [22] (s)	Δt of the proposed method (s)	Main relay	Backup relay	Δt proposed in Ref. [31] (s)	Δt proposed in Ref. [22] (s)	Δt of the proposed method (s)
21	1	0.3566	0.1161	0.0775	29	10	0.1688	0.0121	0.0077
22	1	−0.014	0.0281	0.0161	31	9	0.9071	0.0017	0.0010
22	2	−0.048	0.0395	0.0275	29	11	0.168	0.1034	0.0817
21	3	0.3377	0.0029	0.0020	31	10	0.876	0.0404	0.0198
2	23	0.1307	0.1876	0.1338	30	11	0.4193	0.2208	0.1488
2	5	0.0277	0.0854	0.0615	10	12	0.0304	0.1073	0.0569
25	21	0.0275	0.0507	0.0237	32	30	−0.178	0.0126	0.0070
2	13	0.0718	0.1113	0.0642	11	32	0.2525	0.3939	0.2145
33	21	0.0692	0.0819	0.0376	12	31	−0.203	0.1386	0.0706
3	4	0.0035	0.3203	0.2443	11	14	0.2132	0.4679	0.3345
24	22	−0.219	0.0117	0.0071	34	31	0.0151	0.2493	0.1300
25	23	0.0583	0.0872	0.0497	12	14	−0.26	0.0231	0.0119
33	23	0.1001	0.1186	0.0664	34	32	0.0000	0.0606	0.0289
4	29	−0.021	0.0301	0.0139	13	15	0.0000	0.0033	0.0023
9	24	−0.165	0.099	0.0497	35	33	−0.014	0.001	0.0006
5	6	0.1193	0.0089	0.0043	14	35	−0.044	0.0885	0.0499
26	25	−0.017	−0.0113	−0.0051	15	34	0.0023	0.0052	0.0039
33	5	0.0000	0.0167	0.0088	14	16	−0.024	0.0653	0.0390
25	13	0.0000	0.0112	0.0060	36	34	0.0976	0.0143	0.0082
6	7	−0.027	0.0259	0.0167	15	16	0.0213	0.0842	0.0408
27	26	0.0506	−0.3191	−0.1980	36	35	0.0966	0.1166	0.0595
7	28	−0.036	0.0177	0.0101	16	17	0.0975	0.1985	0.1190
8	27	0.3351	0.0046	0.0022	16	18	−0.012	0.0475	0.0360

29	8	0.0813	0.1057	0.0714	37	36	0.8487	0.0446	0.0205
28	9	0.0472	0.032	0.0169	37	17	1.045	0.3121	0.1428
30	8	0.3167	0.2178	0.1145	18	19	0.0055	0.0323	0.0170
28	10	0.0000	0.0675	0.0316	18	20	0.0107	0.0053	0.0043
31	8	0.7737	0.1309	0.1079	19	38	2.6075	0.1094	0.0710
28	11	0.0000	0.159	0.1052	20	39	0.2813	0.547	0.3031
30	9	0.4513	0.0884						

Table 6.3A Case 2: PSM, TSM, OT, and the coefficients of characteristic curves for the relays in the second scenario (Part A).

	Proposed method of Ref. [31]					
Relay No.	**PSM (A)**	**TSM (s)**	**OT (s)**	**A**	**B**	**C**
1	132.5025	1.36	0.0006	146.6626	2.6862	0.8699
2	617.565	1.26	0.0555	112.9513	2.4943	0.2738
3	128.7975	1.6	0.223	110.3803	1.4208	0.7861
4	86.19	0.57	0.8413	146.494	1.1429	0.5938
5	136.37	0.58	0.1132	134.4898	1.5744	0.4241
6	78.65	1.13	0.23	60.0673	1.4766	0.5716
7	242.19	0.19	0.17	161.4479	2.1584	0.6536
8	106.145	0.72	0.2361	154.0273	1.3002	0.2628
9	188.5	0.88	0.078	117.4913	1.762	0.4378
10	406.64	1.64	0.0873	156.9227	2.3419	0.2573
11	290.355	1.91	0.0148	30.168	2.1916	0.8045
12	76.635	1.4	0.1702	177.4172	1.6713	0.5555
13	159.25	1.31	0.2277	127.3411	1.6508	0.502
14	214.695	1.61	0.0479	73.5127	2.1578	0.3288
15	72.735	1.59	0.2092	55.905	1.5199	0.4092
16	78.39	1.99	0.0409	142.4735	2.0509	0.1829
17	47.645	0.99	0.0002	33.9021	2.6001	0.7824
18	106.73	0.76	0.1123	105.8932	2.0803	0.6544
19	178.1	0.69	0.0119	133.6827	2.6559	0.3091
20	202.3125	0.09	0.0412	49.0794	1.4426	0.3664
21	411.71	0.11	0.059	65.4844	1.9975	0.756
22	257.595	0.1	0.0581	41.207	1.6209	0.9225
23	18.655	1.46	0.001	80.9993	1.9044	0.5534
24	129.285	1.97	0.021	67.8589	2.2023	0.8789
25	170.4625	0.42	0.1831	47.6603	1.7069	0.3383
26	78.65	0.8	0.272	35.3231	1.2101	0.4417
27	121.095	0.64	0.5036	163.6328	1.4794	0.3648
28	159.2175	0.85	0.0577	55.0352	2.5033	0.6107
29	113.1	0.99	0.4981	66.605	1.5234	0.4082
30	203.32	0.07	0.0231	126.1495	1.9704	0.3807
31	145.1775	0.68	0.0175	133.539	2.2172	0.3538
32	51.09	0.64	0.0876	155.3271	1.5213	0.5549
33	159.25	0.9	0.0722	171.5927	2.7692	0.5531
34	107.3475	1.47	0.0595	124.7478	2.3069	0.7895
35	60.6125	1.3	0.0572	63.5611	1.631	0.8231
36	78.39	2	0.0123	71.2508	2.7448	0.4333
37	133.4125	1.82	0.5934	139.7214	1.8598	0.5799
38	80.925	1.18	0.0589	62.8565	2.8621	0.2354
39	42.51	0.26	0.0215	94.2032	2.0096	0.7216
Sum	–	39.93	5.5681	–	–	–

OT, operation time; *PSM*, plug setting multiplier; *TSM*, time setting multiplier.

Table 6.3B Case 2: PSM, TSM, OT, and the coefficients of characteristic curves for the relays in the second scenario (Part B).

Proposed method of [22]					
PSM (A)	TSM (s)	OT (s)	A	B	C
176.67	1.71	0.0861	93.4511	1.6808	0.6859
308.7825	0.55	0.0742	126.3971	1.782	0.5201
85.865	0.7	0.1965	140.753	1.2184	0.6834
129.285	0.68	0.2081	87.38	1.5614	0.5763
204.555	1.6	0.105	65.9812	1.8458	0.8797
78.65	0.75	0.2956	112.8085	1.4686	0.5487
80.73	0.81	0.2473	141.6921	1.7511	0.7212
212.29	0.15	0.0754	80.9713	1.2587	0.3516
150.8	1.23	0.2039	150.3858	1.5827	0.8071
406.64	0.97	0.1683	74.0219	1.7756	0.6316
145.1775	1.51	0.0774	94.7212	1.6841	0.7251
25.545	0.24	0.3001	76.3254	0.7547	0.4801
191.1	1.77	0.079	152.5927	2.1333	0.4498
107.3475	0.65	0.1295	142.6291	1.5241	0.872
36.3675	0.1	0.2304	185.7382	0.9402	0.4843
52.26	0.83	0.1338	79.3001	1.314	0.4796
95.29	0.54	0.0077	158.4595	2.3707	0.4812
53.365	0.44	0.1583	129.673	1.5281	0.7206
71.24	0.57	0.0262	55.2575	1.6547	0.9007
242.775	0.38	0.0534	76.1376	2.0574	0.6346
205.855	0.52	0.0392	81.257	2.2516	0.724
128.7975	0.3	0.1025	84.7053	1.6557	0.2245
18.655	0.89	0.003	96.9041	1.666	0.6986
86.19	0.33	0.1375	107.3112	1.2669	0.372
68.185	0.5	0.1155	125.4209	1.7169	0.3199
31.46	0.27	0.2209	99.2452	1.0095	0.4943
242.19	1.53	0.0594	100.6542	2.6956	0.3155
159.2175	0.34	0.1055	116.185	2.2151	0.6008
226.2	0.27	0.2603	28.3074	1.3478	0.3246
609.96	0.02	0.0716	30.7511	1.1471	0.4464
96.785	0.14	0.0141	39.0118	1.3975	0.2929
51.09	1.57	0.2025	137.8084	1.5084	0.4985
191.1	0.14	0.1364	88.3953	1.746	0.8212
71.565	1.35	0.2125	46.1327	1.4615	0.519
36.3675	1.33	0.1104	173.3709	1.5374	0.5967
26.13	0.18	0.2747	98.7698	0.9272	0.6374
53.365	0.41	0.318	164.7845	1.2836	0.2596
80.925	0.38	0.1068	21.8593	1.7491	0.7283
42.51	0.43	0.0398	117.4362	2.0405	0.4853
—	27.08	5.3868	—	—	—

OT, operation time; *PSM*, plug setting multiplier; *TSM*, time setting multiplier.

Table 6.3C Case 2: PSM, TSM, OT, and the coefficients of characteristic curves for the relays in the second Scenario (Part C).

The proposed method					
PSM (A)	**TSM (s)**	**OT (s)**	**A**	**B**	**C**
177.297	1.6537	0.0168	90.0199	1.7555	0.6600
308.853	0.5192	0.0008	119.5992	1.9682	0.5113
85.995	0.6919	0.3321	130.7755	1.2594	0.6750
129.979	0.6127	0.0118	81.1767	1.7147	0.5578
204.592	1.5400	0.0021	65.1586	2.0233	0.8479
79.019	0.6959	0.0707	111.1511	1.6016	0.5223
80.876	0.6889	0.0216	141.3052	1.9155	0.6863
212.687	0.1167	0.0037	76.7642	1.4526	0.3225
151.280	1.1534	0.0282	144.7093	1.7307	0.8032
406.891	0.9370	0.0013	73.1781	1.8160	0.6209
145.766	1.3601	0.0195	83.0958	1.7395	0.6940
26.050	0.1744	0.6426	65.1111	0.8890	0.4697
191.205	1.7663	0.0017	147.2540	2.2731	0.4146
107.772	0.6164	0.0465	131.5480	1.5949	0.8574
36.891	0.0767	0.3224	182.3584	1.0477	0.4482
53.028	0.8043	0.2702	73.8981	1.3588	0.4774
95.709	0.4781	0.0011	150.3276	2.4248	0.4360
54.056	0.3915	0.0786	128.4819	1.6198	0.6757
71.939	0.5193	0.0096	49.9889	1.8485	0.8594
243.316	0.2383	0.0002	74.0676	2.0946	0.6320
206.273	0.3933	0.0001	74.7267	2.3262	0.7203
128.846	0.2792	0.0044	72.6815	1.7369	0.2233
19.147	0.8889	0.4995	83.8969	1.6972	0.6806
86.267	0.2407	0.0383	102.9129	1.4523	0.3340
68.904	0.3713	0.0254	122.0516	1.7686	0.2736
31.800	0.1581	0.3206	88.1389	1.0935	0.4732
242.849	1.3901	0.0000	87.0611	2.8645	0.2694
159.995	0.2477	0.0002	105.7327	2.3581	0.5597
226.339	0.2467	0.0041	27.1604	1.3653	0.3133
610.254	0.0131	0.0001	25.8490	1.2050	0.3967
97.222	0.0782	0.0019	25.0178	1.5158	0.2872
51.205	1.5363	0.2506	133.5339	1.7044	0.4749
191.500	0.0821	0.0005	86.3632	1.8107	0.7726
71.880	1.3215	0.0876	39.5143	1.4951	0.4769
36.481	1.2942	0.6132	167.4728	1.6318	0.5943
26.635	0.0534	0.1885	95.2973	1.0109	0.6124
53.916	0.3168	0.1856	160.1831	1.4073	0.2516
81.635	0.3082	0.0015	13.1563	1.7937	0.7022
43.255	0.2978	0.0115	110.9480	2.1132	0.4779
—	23.8826	4.0173	—	—	—

OT, operation time; *PSM*, plug setting multiplier; *TSM*, time setting multiplier.

Table 6.4A TSM and PSM for the relays for case 3 (Part A).

	DE		IPM-GA		IPM-DE		IPM-BBM		IPM-IPM	
Relay No.	**TSM**	**PSM**	**TSM**	**PSM**	**TSM**	**PSM**	**TSM**	**TSM**	**TSM**	**PSM**
1	0.1061	2	0.1019	2	0.102	2	0.1019	2	0.1019	2
2	0.1001	1.75	0.1102	1.25	0.1001	1.5	0.1	1.5	0.1081	1.25
3	0.1499	1.5	0.1443	1.5	0.1331	1.75	0.1313	1.75	0.1314	1.75
4	0.1	0.5	0.1	0.5	0.1001	0.5	0.1	0.5	0.1	0.5
5	0.1	2	0.1092	1.75	0.1001	2	0.1	2	0.107	1.75
6	0.1008	2	0.1098	1.75	0.1	2	0.1	2	0.1	2
7	0.1124	2	0.1208	1.75	0.1123	2	0.1074	2	0.1076	2
8	0.218	0.75	0.1921	1	0.1736	1.25	0.1734	1.25	0.1741	1.25
9	0.1001	1.75	0.1	1.5	0.1001	1.5	0.1	1.5	0.1	1.5
10	0.1001	1.75	0.1082	1.25	0.1	1.5	0.1054	1.25	0.1055	1.25
11	0.102	1.25	0.1037	1.25	0.1	1.5	0.1015	1.25	0.1015	1.25
12	0.1001	1.5	0.1022	1.25	0.1	1.5	0.1	1.5	0.1022	1.25
13	0.1109	2	0.1095	2	0.1088	2	0.1087	2	0.1089	2
14	0.1242	2	0.1213	2	0.1196	2	0.1195	2	0.1199	2
15	0.1227	2	0.1604	1.25	0.1435	1.5	0.1434	1.5	0.157	1.25
16	0.1508	2	0.1579	1.75	0.1507	2	0.1507	2	0.1444	2
17	0.1001	1.5	0.1068	1.25	0.1	1.5	0.1	1.25	0.106	1.25
18	0.1314	2	0.129	2	0.1277	2	0.1277	2	0.1279	2
19	0.1001	2	0.1	2	0.1	2	0.1	2	0.1024	1.75
20	0.1229	2	0.1279	1.75	0.1181	2	0.1181	2	0.1167	2
21	0.1168	2	0.1275	1.75	0.1146	2	0.1146	2	0.1147	2
22	0.1378	1.5	0.1401	1.5	0.1257	1.75	0.1257	1.75	0.1371	1.5
$\sum_{i=1}^{n} t_{op,i}$	7.8086		7.7024		7.6499		7.5487		7.5482	

DE, differential evolution; *GA*, genetic algorithm; *IPM*, interior point method; *IPM-BBM*, IPM-bound; *PSM*, plug setting multiplier; *TSM*, time setting multiplier.

Table 6.4B TSM and PSM for the relays for case 3 (Part B).

	Fuzzy GA		Proposed method	
Relay No.	**TSM**	**PSM**	**TSM**	**PSM**
1	0.102	2	0.101	2
2	0.1079	1.25	0.1079	1.25
3	0.131	1.75	0.128	1.75
4	0.1	0.5	0.1	0.5
5	0.1069	1.75	0.1067	1.75
6	0.1	2	0.1	2
7	0.1075	2	0.1072	2
8	0.1686	1.25	0.1683	1.25
9	0.1	1.5	0.1	1.5
10	0.1046	1.25	0.1046	1.25
11	0.1014	1.25	0.1012	1.25
12	0.1022	1.25	0.1022	1.25
13	0.1072	2	0.107	2
14	0.1198	2	0.1195	2
15	0.1571	1.25	0.157	1.25
16	0.1443	2	0.1443	2
17	0.1059	1.25	0.1055	1.25
18	0.1251	2	0.1252	2
19	0.1024	1.75	0.1024	1.75
20	0.116	2	0.116	2
21	0.1129	2	0.1129	2
22	0.1371	1.5	0.1371	1.5
$\sum_{i=1}^{n} t_{op,i}$	7.5059		7.4504	

PSM, plug setting multiplier; *TSM*, time setting multiplier.

7.3.4 *Case 4: OCRC problem in microgrids—comparison with Refs. [61,62].*

In this section, we extend the proposed MOPSO/FDMT-based strategy for solving the OCRC problem in microgrids. There are some basic assumptions that should be considered:

(1) The calculation of the fault contribution is done based on the contribution of DERs [57–62], IEC60909 recommendations [63–67], and the fault model presented in Section 3.

(2) As we discussed in Section 3, the hierarchical droop-controlled DERs [3,8,11] have less fault contribution compared with directly voltage-controlled DERs [2,3,8], as they have inner current control loops [3,8,63].

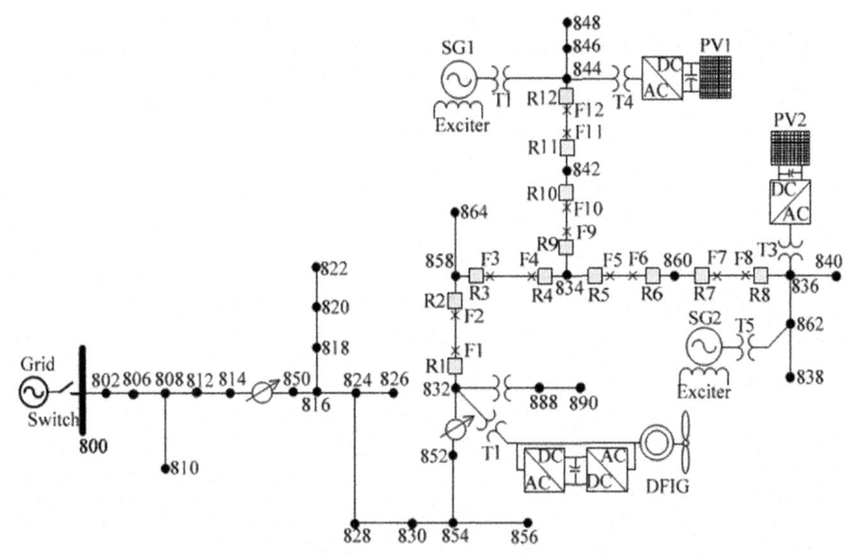

Figure 6.15 Modified IEEE 34-bus microgrid schematic diagram [61]. *(IEEE from H. Muda, P. Jena, Superimposed adaptive sequence current based microgrid protection: a new technique, IEEE Trans. Power Deliv. 32 (2) (2017) 757–767.)*

(3) The current-limiting strategy discussed in Chapter 4 for hierarchical droop-controlled and directly voltage-controlled DER units and the recommendations for wind and solar DER units provided in Refs. [2,3] are applied.

(4) Before performing fault analysis, the initial values of the microgrid system are obtained by the three-phase AC/DC power flow method discussed in Chapter 3.

Based on the aforementioned assumptions, the performance of the proposed strategy is evaluated by testing on a modified 24.9 kV, 12 MVA, 50 Hz IEEE 34-bus microgrid system shown in Fig. 6.15 [61]. The ratios of CTs at relay location are 300/1 for R1 and R2, 250/1 for R3 and R4, 100/1 for R5 and R6, 50/1 for R7 and R8, 150/1 for R9, and R10, and 100/1 for R11 and R12, respectively. We consider four different scenarios: (1) the microgrid operates in islanded mode consisting of only IIDER units (or inverter-based distributed generation [IBDG]), (2) the microgrid operated in an islanded mode having IIDERs as well as synchronous-based distributed generation/distributed energy resource (SBDG/SBDER) units, (3) the grid-connected mode, and (4) the connection/disconnection of DER unit depends on the situation.

In this case, Eq. (6.49) is rewritten as

$$t = \frac{a}{PSM^b - 1} \times TSM = f(PSM) \times TSM \tag{6.60}$$

The TSM of the relay is calculated as

$$TSM^c = \frac{f(PSM^r)}{f(PSM^s)} \times TSM^r \tag{6.61}$$

where superscript "*r*" indicates the previous operating mode of the microgrid and superscript "*c*" denotes the present operating mode. The upper and lower TMS limits are set to 1 and 0.05, respectively. The OT limit of both the primary and backup relays is considered to be 1 s, with the desired CTI of 0.3 s [61]. The results are presented in Tables 6.5A,B, and 6.6. As it can be seen, the proposed algorithm can effectively coordinate the OC relays in a multi-DER microgrid using the mentioned assumptions. Compared with synchronous-based DERs, the power electronic converters of inverter-based DERs can decrease the microgrid fault level considerably that can affect the *TMS*, *OT*, and *CTI* (Δt). Also, the contribution of the upstream grid (in grid-connected mode) and connection/disconnection of DERs affect the fault level. However, the control structure of DER affects the FRT capability and plug-and-play (P&P) functionality of DERs and consequently the design of the microgrid protection system.

7.3.5 Application of directional OC relays and OCRC problem in microgrids — comparison with [61].

Finally, based on the assumption of Section 7.3.4, we evaluate the performance of the proposed strategy for directional OC relays. In this regard, the proposed OCRC algorithm has been tested to coordinate directional OC relays in the test microgrid of Fig. 6.5 [62]. The results for the protection of feeders, IIDERs and SBDGs and the OT (for islanded and grid-connected modes) are provided in Tables 6.7 and 6.8. The obtained simulation results can prove the effectiveness of the proposed strategy for coordinating DOC relays compared with other the reported techniques.

8. Conclusion

In this chapter, we proposed a generalized protection scheme for HDRC-based and DVC-based IIDERs. Also, we added the benefits of a current-limiting strategy for hierarchical droop-based strategy in different

Table 6.5A Results for relay settings, fault currents and operating times for scenarios 1, 22, and 3 (Part A).

		The DORC-based protection [45]											
		Scenario 1				Scenario 2				Scenario 3			
FP	PR/ BR	I_p, A (Mag)	TSM (s)	OT (s)	Δt (CTIs) [45] (s)	I_p, A (Mag)	TSM (s)	OT (s)	Δt (CTIs) [45] (s)	I_p, A (Mag)	TSM (s)	OT (s)	Δt (CTIs) [45] (s)
F1	R1	0.160	0.060	0.276	–	0.218	0.055	0.230	–	0.531	0.053	0.276	–
F2	R2	0.150	0.050	0.230	–	0.169	0.053	0.275	–	0.531	0.050	0.230	–
	/R4	0.330	0.057	0.608	0.378	0.362	0.050	0.725	0.450	0.574	0.053	0.608	0.378
F3	R3	0.300	0.052	0.215	–	0.323	0.061	0.211	–	0.562	0.061	0.215	–
	/R1	0.160	0.060	0.560	0.345	0.218	0.055	0.601	0.390	0.531	0.053	0.560	0.315
F4	R4	0.330	0.057	0.191	–	0.362	0.050	0.216	–	0.547	0.066	0.191	–
	/R6	0.300	0.050	0.501	0.310	0.325	0.057	0.532	0.316	0.252	0.063	0.501	0.335
	/R10	0.475	0.051	0.604	0.413	0.513	0.051	0.665	0.439	0.491	0.064	0.604	0.347
F5	R5	0.290	0.060	0.279	–	0.312	0.051	0.189	–	0.252	0.071	0.279	–
	/R3	0.300	0.052	0.594	0.315	0.323	0.061	0.495	0.306	0.562	0.061	0.594	0.352
	/R10	0.475	0.051	0.776	0.497	0.513	0.051	0.588	0.399	0.491	0.064	0.776	0.348
	/R6	0.300	0.050	0.210	–	0.325	0.056	0.314	–	0.252	0.063	0.221	–
	/R10	0.890	0.098	0.512	0.301	0.963	0.059	0.893	0.579	0.207	0.054	0.512	0.448
F6	R6	0.300	0.050	0.183	–	0.325	0.057	0.247	–	0.252	0.063	0.183	–
	/R8	0.890	0.098	0.486	0.303	0.963	0.059	0.686	0.439	0.207	0.050	0.486	0.360
	/R5	0.290	0.060	0.362	–	0.312	0.051	0.248	–	0.252	0.071	0.362	–
	/R3	0.300	0.052	0.662	0.300	0.323	0.061	0.720	0.472	0.562	0.061	0.662	0.664
	/R10	0.475	0.051	0.864	0.502	0.513	0.051	0.929	0.682	0.491	0.064	0.864	0.573
F7	R7	0.900	0.078	0.216	–	0.970	0.050	0.225	–	0.212	0.050	0.216	–
	/R5	0.290	0.060	0.668	0.452	0.312	0.051	0.658	0.433	0.252	0.071	0.668	0.330

Continued

Table 6.5A Results for relay settings, fault currents and operating times for scenarios 1, 22, and 3 (Part A).—cont'd

		The DORC-based protection [45]											
		Scenario 1				Scenario 2				Scenario 3			
FP	PR/ BR	I_p, A (Mag)	TSM (s)	OT (s)	Δt (CTIs) [45] (s)	I_p, A (Mag)	TSM (s)	OT (s)	Δt (CTIs) [45] (s)	I_p, A (Mag)	TSM (s)	OT (s)	Δt (CTIs) [45] (s)
F8	R8	0.890	0.098	0.285	—	0.963	0.059	0.201	—	0.207	0.054	0.258	—
F9	R9	0.470	0.073	0.237	—	0.518	0.058	0.235	—	0.497	0.064	0.237	—
	/R6	0.300	0.050	0.570	0.303	0.325	0.057	0.556	0.321	0.252	0.063	0.570	0.396
	/R3	0.300	0.052	0.550	0.357	0.323	0.061	0.572	0.337	0.562	0.061	0.550	0.380
F10	R10	0.475	0.051	0.200	—	0.513	0.051	0.208	—	0.491	0.064	0.200	—
	/R12	0.861	0.130	0.518	0.318	0.930	0.051	0.511	0.303	0.294	0.052	0.518	0.316
F11	R11	0.850	0.057	0.171	—	0.920	0.050	0.198	—	0.308	0.050	0.171	—
	/R9	0.470	0.073	0.494	0.320	0.518	0.058	0.528	0.303	0.497	0.064	0.494	0.384
F12	R12	0.861	0.130	0.375	—	0.930	0.051	0.212	—	0.294	0.052	0.375	—

Br, *Breaker;* *CTI*, coordination time interval; *DORC*, directional over-current relay coordination; *FP*, feeder protection; *OT*, operation time; *PR*, protection; *TSM*, time setting multiplier.

Table 6.5B Results for relay settings, fault currents and operating times for scenarios 1, 2 and 3 (Part B).

The proposed method											
Scenario 1				Scenario 2				Scenario 3			
I_p, A (Mag)	TSM (s)	OT (s)	Δt (CTIs) [45] (s)	I_p, A (Mag)	TSM (s)	OT (s)	Δt (CTIs) [45] (s)	I_p, A (Mag)	TSM (s)	OT (s)	Δt (CTIs) [45] (s)
0.161	0.060	0.276	—	0.218	0.055	0.230	—	0.530	0.052	0.275	—
0.150	0.051	0.235	—	0.170	0.052	0.276	—	0.531	0.050	0.230	—
0.330	0.057	0.608	0.378	0.361	0.051	0.727	0.452	0.572	0.052	0.606	0.376
0.301	0.050	0.213	—	0.323	0.061	0.211	—	0.562	0.061	0.215	—
0.160	0.060	0.560	0.345	0.218	0.055	0.601	0.390	0.531	0.053	0.560	0.315
0.330	0.057	0.191	—	0.360	0.050	0.215	—	0.546	0.065	0.190	—
0.305	0.052	0.529	0.313	0.326	0.055	0.530	0.314	0.252	0.063	0.501	0.335
0.475	0.051	0.604	0.413	0.512	0.050	0.663	0.437	0.490	0.062	0.602	0.346
0.290	0.060	0.279	—	0.310	0.050	0.187	—	0.252	0.071	0.279	—
0.300	0.051	0.597	0.315	0.323	0.061	0.495	0.306	0.562	0.061	0.594	0.352
0.475	0.051	0.776	0.497	0.511	0.050	0.586	0.398	0.491	0.064	0.776	0.348
0.300	0.050	0.210	—	0.325	0.056	0.314	—	0.251	0.062	0.220	—
0.890	0.098	0.512	0.301	0.963	0.059	0.893	0.579	0.207	0.054	0.512	0.448
0.300	0.050	0.183	—	0.324	0.055	0.246	—	0.252	0.063	0.183	—
0.895	0.099	0.489	0.303	0.963	0.059	0.686	0.439	0.208	0.051	0.487	0.361
0.290	0.060	0.362	—	0.312	0.051	0.248	—	0.252	0.071	0.362	—
0.305	0.051	0.659	0.305	0.323	0.061	0.720	0.472	0.562	0.061	0.662	0.664
0.475	0.051	0.864	0.502	0.513	0.051	0.929	0.682	0.491	0.064	0.864	0.573
0.900	0.078	0.216	—	0.970	0.050	0.225	—	0.212	0.050	0.216	—
0.291	0.062	0.682	0.452	0.311	0.052	0.656	0.433	0.251	0.070	0.666	0.328

Continued

Table 6.5B Results for relay settings, fault currents and operating times for scenarios 1, 2 and 3 (Part B).—cont'd

The proposed method											
Scenario 1				Scenario 2				Scenario 3			
I_p, A (Mag)	TSM (s)	OT (s)	Δt (CTIs) [45] (s)	I_p, A (Mag)	TSM (s)	OT (s)	Δt (CTIs) [45] (s)	I_p, A (Mag)	TSM (s)	OT (s)	Δt (CTIs) [45] (s)
0.890	0.098	0.285	—	0.963	0.059	0.201	—	0.207	0.054	0.258	—
0.470	0.073	0.237	—	0.518	0.058	0.235	—	0.495	0.063	0.235	—
0.305	0.052	0.575	0.308	0.325	0.057	0.556	0.321	0.252	0.063	0.570	0.396
0.300	0.052	0.550	0.357	0.321	0.060	0.571	0.336	0.560	0.060	0.551	0.381
0.475	0.051	0.200	—	0.513	0.051	0.208	—	0.491	0.064	0.200	—
0.865	0.132	0.521	0.318	0.930	0.051	0.511	0.303	0.292	0.051	0.520	0.318
0.850	0.057	0.171	—	0.921	0.051	0.195	—	0.308	0.050	0.171	—
0.472	0.074	0.496	0.321	0.517	0.056	0.526	0.303	0.497	0.064	0.494	0.384
0.861	0.130	0.375	—	0.930	0.051	0.212	—	0.293	0.051	0.374	—

Table 6.6 Results for relay settings, fault currents and operating times for scenario four.

		The DORC-based protection				The proposed method			
FP	PR/BR	I_p, A (Mag)	TSM (s)	OT (s)	Δt (CTIs) (s)	I_p, A (Mag)	TSM (s)	OT (s)	Δt (CTIs) (s)
F1	R1	0.608	0.054	0.214	—	0.606	0.053	0.213	—
F3	R3	0.635	0.059	0.241	—	0.635	0.059	0.241	—
	/R1	0.608	0.054	0.658	0.417	0.607	0.053	0.658	0.415
F5	R5	0.283	0.053	0.201	—	0.282	0.052	0.200	—
	/R3	0.635	0.059	0.683	0.482	0.634	0.060	0.665	0.481
F7	R7	0.241	0.051	0.175	—	0.241	0.051	0.175	—
	/R5	0.283	0.053	0.503	0.328	0.283	0.053	0.503	0.328
F9	R9	0.554	0.054	0.226	—	0.554	0.054	0.226	—
	/R3	0.635	0.059	0.666	0.440	0.634	0.058	0.664	0.441
F11	R11	0.347	0.050	0.210	—	0.347	0.050	0.210	—
	/R9	0.554	0.054	0.623	0.413	0.552	0.053	0.620	0.411

Table 6.7 Result of relay setting for operation time of instantaneous directional OC relays in microgrid system shown in Fig. 6.5

(Islanded mode)

		Protection scheme of		The proposed method	
		MGFPR1 (OC)	**SGDG relay (OC)**	**MGFPR1 (OC)**	**SGDG relay (OC)**
Fault 2	AG	2.324	–	2.322	–
	AB	1.669	–	1.667	–
	ABG	1.332	–	1.331	–
Fault 3	AG	–	0.181	–	0.180
	AB	–	0.233	–	0.231
	ABG	–	0.179	–	0.177

(Grid-connected mode)

		Protection scheme of				The proposed method			
		MGFPR1 (OC)	**MGFPR2 (OC)**	**MGFPR1 (OC)**	**MGFPR2 (OC)**	**MGFPR1 (OC)**	**MGFPR2 (OC)**	**MGFPR1 (OC)**	**MGFPR2 (OC)**
Fault 2	AG	0.045	–	–	0.247	0.044	–	–	0.244
	AB	0.056	–	–	0.308	0.055	–	–	0.305
	ABG	0.046	–	–	0.242	0.044	–	–	0.240
	ABC	0.045	19.143	–	0.234	0.044	19.141	–	0.232
Fault 3	AG	–	0.065	0.179	0.383	–	0.062	0.177	0.382
	AB	–	0.062	0.222	0.358	–	0.060	0.220	0.355
	ABG	–	0.064	0.177	0.381	–	0.061	0.175	0.380

MGFPR, microgrid feeder protection; *OC*, overcurrent; *SGDG*, synchronous-based distributed generation.

Table 6.8 Result of relay setting for instantaneous directional OC relays for feeder, IIDER, and SGDER.

		OC relay time delay (msec)	I_p, A	TMS
Protection scheme of	MGFPR1	300	220	0.8
	MGFPR2	300	200	0.94
	MGFPR3	500	90	0.85
Proposed method	MGFPR1	300	222	0.8
	MGFPR2	300	203	0.93
	MGFPR3	500	92	0.86
Protection scheme of	IIDER	2000	350	0.244
	SGDG	2000	209	0.167
Proposed method	IIDER	2000	352	0.246
	SGDG	2000	208	0.166

IIDER, inverter-interfaced distributed energy resource; *OC*, overcurrent.

reference frames that can enhance the FRT capability of microgrid. We also offered a new fault detection algorithm that can take the advantages of both ADALINE-ANN and TMF and complete the proposed OC/OL protection scheme. Next, we tested the fault model for three- and four-wire inverter topologies and various types of balanced and unbalanced SC faults, verified the performance of the proposed fault detection method and OC/OL protection scheme by offline time domain simulations in MATLAB/Simulink environment, then validated by experimental investigation through real-time digital simulations using OPAL-RT RTDS. In the next part, we proposed an OC relay coordination algorithm based on MOSPSO and FDMT to address the OCRC problem in large and interconnected power systems and then developed for multi-DER microgrids, which can effectively handle both continuous and discrete values of TSMs, considers the coefficients of characteristic curves of DOC relays and related PSMs, performs optimal OCRC with less OT values, and considers that the OFs are separately and independently by using the FDMT in the proposed MOO algorithm. Finally, the proposed MOPSO/FDMT-based algorithm was tested on different test power systems such as 8-bus, IEEE 30-bus, and Wood and Woollenberg 6-bus networks. In all cases, the obtained simulation results and their comparison with other SOO OCRC methods reveal the effectiveness, efficiency, accuracy, and authenticity of the proposed MOPSO/FDMT-based relay coordination algorithm.

References

[1] H.R. Baghaee, M. Mirsalim, G.B. Gharehpetian, H.A. Talebi, A generalized descriptor-system robust H∞ control of autonomous microgrids to improve small and large signal stability considering communication delays and load nonlinearities", Int. J. Electr. Power Energy Syst. 92 (1) (November 2017) 63–82, https://doi.org/10.1016/j.ijepes.2017.04.007.

[2] T.N. Boutsika, S.A. Papathanassiou, Short-circuit calculations in networks with distributed generation, Elec. Power Syst. Res. 78 (1) (Dec. 2007) 1181–1191.

[3] H.R. Baghaee, M. Mirsalim, G.B. Gharehpetian, H.A. Talebi, A new current limiting strategy and fault model to improve fault ride-through capability of inverter interfaced DERs in autonomous microgrids, Sustain. Energy Technol. Assess. 24 (December 2017) 71–81, https://doi.org/10.1016/j.seta.2017.02.004. Part C.

[4] H.R. Baghaee, M. Mirsalim, G.B. Gharehpetian, H.A. Talebi, Nonlinear load sharing and voltage compensation of microgrids based on harmonic power-flow calculations using radial basis function neural networks, IEEE Sys. J. PP (99) (January 2017) 1–11, https://doi.org/10.1109/JSYST.2016.2645165.

[5] J.M. Guerrero, et al., Advanced control architecture for intelligent MicroGrids – Part I: decentralized and hierarchical control, IEEE Trans. Ind. App. 60 (4) (April 2013) 1254–1262.

[6] S. Dasgupta, S. Mohan, S. Sahoo, S. Panda, Lyapunov function based current controller to control active and reactive power flow from a renewable energy source to a generalized three-phase microgrid system, IEEE Trans. Ind. Elect. 60 (2) (Feb. 2013) 799–813.

[7] A.H. Etemadi, E.J. Davison, R. Iravani, A generalized decentralized robust control of islanded microgrids, IEEE Trans. Power Sys. 29 (6) (Oct.2014) 3102–3113.

[8] A.H. Etemadi, R. Iravani, Overcurrent and overload protection of directly voltage-controlled distributed resources in a microgrid, IEEE Trans. Ind. Elect. 60 (12) (June 2013) 5629–5638.

[9] H. Nikkhajoei, R.H. Lasseter, Microgrid protection, in: Proc. IEEE PES General Meeting, Tampa, Florida, USA, 24–28 June 2007, pp. 1–6.

[10] T.S. Ustun, C. Ozansoy, A. Zayegh, Fault current coefficient and time delay assignment for microgrid protection system with central protection unit, IEEE Trans. Power Sys. 28 (2) (May 2013) 598–606.

[11] H.R. Baghaee, M. Mirsalim, G.B. Gharehpetian, Power calculation using RBF neural networks to improve power sharing of hierarchical control scheme in multi-DER microgrids, IEEE J. Emer. Sel. Top. Power Electron. 4 (4) (December 2016) 1217–1225, https://doi.org/10.1109/JESTPE.2016.2581762.

[12] I. Sadeghkhani, et al., Transient monitoring function–based fault detection for inverter-interfaced microgrids, IEEE Trans. Smart Grid PP (99) (September 2016) 1–11.

[13] N. Bottrel, T.C. Green, Comparison of current limiting strategies during fault ride-through of inverters to prevent latch-up and wind-up, IEEE Trans. Power Elect. 29 (7) (July 2014) 3786–3797.

[14] C.A. Plet, M. Brucoli, J.D.F. McDonald, T.C. Green, Fault models of inverter-interfaced distributed generators: experimental verification and application to fault analysis, in: Proc. IEEE Power and Energy Society General Meeting, Detroit, Mich., USA, 2011, pp. 1–8.

[15] P. Krause, O. Wasynczuk, S. Sudhoff, Analysis of Electric Machinery and Drive Systems, second ed., Wiley, Hoboken, NJ, 2002.

[16] A.G. Phadke, J.S. Thorp, Computer Relaying for Power Systems, J. Wiley & Sons, 2009.

[17] R.E. Shatshat, M. Kazerani, M.M.A. Salama, Modular active power-line condition, IEEE Trans. Power Deliv. 16 (4) (October 2001) 700–709.
[18] N.I. Elkalashy, H.A. Darwish, A.M.I. Taalab, M.A. Izzularab, An adaptive single pole autoreclosure based on zero sequence power, Electr. Power Syst. Res. 77 (5–6) (April 2007) 438–446.
[19] Z.M. Radojevic, J.R. Shin, New one terminal algorithm for adaptive reclosing and fault distance calculation on transmission lines, IEEE Trans. Power Deliv. 21 (3) (July 2006) 1231–1237.
[20] J.-F. Cécile, L. Schoen, V. Lapointe, A. Abreu, J. Bélanger, A Distributed Real-Time Framework for Dynamic Management of Heterogeneous Co-simulations, 2006. Available online at: http://www.opal-rt.com/sites/default/files/technical_papers/scs_article.pdf.
[21] S.A. Morello, H.A.A. Hassan, B.G. Campbell, R.J. Kerestes, G.F. Reed, Upstream fault detection using second harmonic magnitudes in a grid tied microgrid setting, in: IEEE Power & Energy Society General Meeting (PESGM), IEEE, Portland, OR, USA, Aug. 2018, pp. 1–5.
[22] D. Solati-Alkaran, M.R. Vatani, M.J. Sanjari, G.B. Gharehpetian, M.S. Naderi, Optimal overcurrent relay coordination in interconnected networks by using fuzzy-based GA method, IEEE Trans. Smart Grid PP (99) (2016) 1–11.
[23] S.A. Ahmadi, H. Karami, M.J. Sanjari, H. Tarimoradi, G.B. Gharehpetian, Application of hyper-spherical search algorithm for optimal coordination of overcurrent relays considering different relay characteristics, Int. J. Elect. Power Energy Syst. 83 (1) (2016) 443–449.
[24] M. Jannati, B. Vahidi, S.H. Hosseinian, H.R. Baghaee, A new adaptive single phase auto-reclosure scheme for EHV transmission lines, in: Proc. 12th IEEE Int. Middle-East Power Syst. Conf. (MEPCON), Aswan, Egypte, March 2008, pp. 203–207.
[25] S.A. Ahmadi, H. Karami, G.B. Gharehpetian, Comprehensive coordination of combined directional overcurrent and distance relays considering miscoordination reduction, Int. J. Elect. Power Energy Syst. 92 (1) (2017) 42–52.
[26] R. Mohammadi, H.A. Abyaneh, H.M. Rudsari, S.H. Fathi, H. Rastegar, Overcurrent relays coordination considering the priority of constraints, IEEE Trans. Power Deliv. 26 (3) (2011) 1927–1938.
[27] H.A. Abyaneh, et al., A new optimal approach for coordination of overcurrent relays in interconnected power systems, IEEE Trans. Power Del. 18 (2) (2003) 430–435.
[28] H.K. Kargar, et al., Pre-processing of the optimal coordination of overcurrent, Elect. Power Syst. Res. 75 (1) (2005) 134–141.
[29] T. Amraee, Coordination of directional overcurrent relays using seeker algorithm, IEEE Trans. Power Deliv. 27 (3) (2012) 1415–1422.
[30] C.A.C. Salazara, A.C. Enríqueza, S.E. Schaeffer, Directional overcurrent relay coordination considering non-standardized time curves, Elect. Power Syst. Res. 122 (1) (2015) 42–49.
[31] F. Razavi, et al., A new comprehensive genetic algorithm method for optimal overcurrent relays coordination, Elect. Power Syst. Res. 78 (1) (2008) 713–720.
[32] C.W. So, K.K. Li, Intelligent method for protection coordination, in: Proc. IEEE Conf. Elect. Utility Der. Res. Power Tech., Hong Kong, China, Apr. 2004, pp. 378–382.
[33] R. Mohammadi, et al., Optimal relays coordination efficient method in interconnected power systems, J. Elec. Eng. 61 (2) (2010) 75–83.
[34] F. Adelnia, Z. Moravej, M. Farzinfar, A new formulation for coordination of directional overcurrent relays in interconnected networks, Int. Trans. Electr. Energy Syst. 25 (1) (2015) 120–137.

[35] P.P. Bedekar, S.R. Bhide, Optimum coordination of overcurrent relay timing using continuous genetic algorithm, Expert Sys. Appl. 38 (1) (2011) 11286–11292.
[36] C.W. So, K.K. Li, Time coordination method for power system protection by evolutionary, IEEE Trans. Ind. Appl. 36 (5) (2000) 1235–1240.
[37] T. Keil, J. Jäger, Advanced coordination method for overcurrent protection relays nonstandard tripping character, IEEE Trans. Power Deliv. 23 (1) (2008) 52–57.
[38] K.A. Saleh, et al., Optimal coordination of directional overcurrent relays using a new time–current–voltage characteristic, IEEE Trans. Power Deliv. 30 (2) (2014) 537–544.
[39] M.N. Alam, B. Das, V. Pant, A comparative study of metaheuristic optimization approaches for directional overcurrent relays coordination, Elect. Power Syst. Res. 128 (1) (2015) 39–52.
[40] Z. Moraveja, et al., Optimal coordination of directional overcurrent relays using NSGA, Elect. Power Syst. Res. 119 (1) (2015) 228–236.
[41] H.R. Baghaee, M. Abedi, Calculation of weighting factors of static security indices used in contingency ranking of power systems based on fuzzy logic and analytical hierarchy process, Int. J. Electr. Power Energy Syst. 33 (4) (2011) 855–860, https://doi.org/10.1016/j.ijepes.2010.12.012.
[42] J. Jäger, R. Lubiatowski, G. Ziegler, R. Krebs, Protection security assessment for large power systems, in: Proc. PowerTech, Bucharest, Romania, July 2009, pp. 1–6.
[43] M.N. Alam, et al., An interior point method based protection coordination scheme for directional overcurrent relays in meshed networks, Int. J. Elec. Power Energy Syst. 81 (1) (2016) 153–164.
[44] J. Jäger, T. Keil, A. Dienstbier, P. Lund, Network security assessment-An important task in distribution systems with dispersed generation, in: Proc. CIRED Elect. Dist, -Part 1, Prague, Czech Rep, 8-11 Jun. 2009, pp. 1–4.
[45] H. Gers, M. Juan, Protection of Electricity Distribution Networks, third ed., Institution of Engineering and Technology, 2011.
[46] D. Solati-Alkaran, M.R. Vatani, M.J. Sanjari, G.B. Gharehpetian, Overcurrent relays coordination in interconnected networks using accurate analytical method and based on determination fault critical, IEEE Trans. Power Deliv. 30 (2) (2015) 870–877.
[47] IEC standard for single input energizing quantity measuring relays with dependent specified time, IEC Stand. 255–4 (1976).
[48] R. Saborido, et al., Evolutionary multi-objective optimization algorithms for fuzzy portfolio selection, Appl. Soft Comput. 39 (1) (2016) 48–63.
[49] H.R. Baghaee, M. Mirsalim, G.B. Gharehpetian, H.A. Talebi, Reliability/cost based multi-objective pareto optimal design of stand-alone wind/PV/FC generation microgrid system, Energy 115 (1) (2016) 1022–1041, https://doi.org/10.1016/j.energy.2016.09.007.
[50] H.R. Baghaee, M. Mirsalim, G.B. Gharehpetian, A. Kashefi-Kaviani, Security/cost-based optimal allocation of multi-type FACTS devices using multi-objective particle swarm optimization, Simul. Int. Trans. Soc. Model. Simul. 88 (8) (2012) 999–1010, https://doi.org/10.1177/0037549712438715.
[51] H.R. Baghaee, M. Mirsalim, G.B. Gharehpetian, Multi-objective optimal power management and sizing of a reliable wind/PV microgrid with hydrogen energy storage using MOPSO, J. Intel. Fuzzy Syst. 32 (3) (2017) 1753–1773, https://doi.org/10.3233/JIFS-152372.
[52] M. Ezzeddine, et al., Novel method for optimal coordination of directional overcurrent relays considering their available discrete settings and several operation characteristics, Electr. Power Syst. Res. 81 (1) (2011) 1475–1481.

[53] ABB Group, ABB Relion Prod. Fam. Zurich, Switzerland, 2016 [Online], http://new.abb.com/substationautomation/products/protection-control/relion-product-family/relion-615-series.
[54] Applied Protective Relaying, first ed., Westinghouse Electric Corp., FL, 1979.
[55] Power System Test SCENARIO archive," [Online]. Available: http://www.ee.washington.edu/research/pstca.
[56] R.D. Zimmerman, C.E. Murillo-S_anchez, R.J. Thomas, Matpower's extensible optimal power flow architecture, in: Proc. IEEE Power and Energy Society General Meeting, 2009, pp. 1–7.
[57] A. Hooshyar, R. Iravani, Microgrid protection, Proc. IEEE 105 (7) (2017) 1032–1053, https://doi.org/10.1109/JPROC.2017.266934.
[58] H.R. Baghaee, M. Mirsalim, M.J. Sanjari, G.B. Gharehpetian, Effect of type and interconnection of DG units in the fault level of distribution networks, in: Proc. 13th Power Elect. & Motion Control. (EPE-PEMC), Poznan, Poland, Sept. 2008, pp. 313–319.
[59] M.A. Zamani, T.S. Sidhu, A. Yazdani, A protection strategy and microprocessor-based relay for low-voltage microgrids, IEEE Trans. Power Deliv. 26 (3) (2011) 1873–1883.
[60] H.R. Baghaee, M. Mirsalim, M.J. Sanjari, G.B. Gharehpetian, Fault current reduction in distribution systems with distributed generation units by a new dual functional series compensator, in: Proc. 13th Power Elect. & Motion Cont, Conf. (EPE-PEMC), Pozn., Poland, Sept. 2008, pp. 750–757.
[61] H. Muda, P. Jena, Superimposed adaptive Sequence current based microgrid protection: a new technique, IEEE Trans. Power Deliv. 32 (2) (2017) 757–767.
[62] S.F. Zarei, M. Parniani, A comprehensive digital protection scheme for low voltage microgrids with inverter-based and conventional distributed generations, IEEE Trans. Power Deliv. 32 (1) (2017) 441–452.
[63] Short-circuit Currents in Three-phase a.c. Systems—Part 0: Calculation of Short-Circuit Currents, IEC 60909-0, 2016.
[64] Short-circuit Currents in Three-phase A.C. Systems—Part 1: Factors for the Calculation of Short-Circuit Currents According to IEC, IEC 60909-1, 2002.
[65] Short-circuit Currents in Three-phase A.C. Systems-Part 2: Data of Electrical Equipment for Short-Circuit Current Calculations, IEC 60909-2, 2008.
[66] Short-circuit Currents in Three-phase AC Systems - Part 3: Currents during Two Separate Simultaneous Line-To-Earth Short Circuits and Partial Short-Circuit Currents Flowing through Earth, IEC 60909-3, 2013.
[67] Short-circuit Currents in Three-phase A.C. Systems—Part 4: Examples for the Calculation of Short-Circuit Currents, IEC 60909-4, 2000.

Further reading

[1] H.R. Baghaee, D. Mlakic, S. Nikolovski, T.D. Dragicevic, Support vector machine-based islanding and grid fault detection in active distribution networks, IEEE J. Emerg. Sel. Top. Power Electron. PP (99) (May 2019) 1–19.
[2] M. Mishra, P.K. Rout, Detection and classification of micro-grid faults based on HHT and machine learning techniques, IET Gener. Transm. Distrib. 12 (2) (January 2018) 388–397.
[3] J. Moshtagh, M. Jannati, H.R. Baghaee, E. Nasr, A novel approach for online fault detection in HVDC converters, in: 12th International Middle-East Power System Conference, IEEE, Aswan, Egypt, Mar. 2008, pp. 307–311.
[4] P.P. Bedekar, S.R. Bhide, Optimum coordination of directional overcurrent relays using the hybrid GA-NLP approach, IEEE Trans. Power Deliv. 26 (1) (2011) 109–119.

[5] Soman S.A. Lecture on Power System Protection (web) Module 4 and 5. [Online] Available: www.nptel.ac.in/courses/108101039/.
[6] H.R. Baghaee, M. Mirsalim, G.B. Gharehpetian, Real-time verification of new controller to improve small/large-signal stability and fault ride-through capability of multi-DER microgrids, IET Gener. Transm. Distrib. 10 (12) (2016) 3068–3084, https://doi.org/10.1049/iet-gtd.2016.0315.
[7] H.R. Baghaee, M. Mirsalim, G.B. Gharehpetian, et al., Eigenvalue, robustness and time delay analysis of hierarchical control scheme in multiDER microgrid to enhance small/large-signal stability using complementary loop and fuzzy logic controller, J. Circuits Syst. Comput. 26 (6) (2016) 1–30, https://doi.org/10.1142/S0218126617500992.
[8] H.R. Baghaee, M. Mirsalim, G.B. Gharehpetian, H.A. Talebi, A decentralized power management and Sliding mode control strategy for hybrid AC/DC microgrids including renewable energy resources, IEEE Trans. Ind. Inform. (2017) 1–15, https://doi.org/10.1109/TII.2017.2677943 (99).
[9] H.R. Baghaee, M. Mirsalim, G.B. Gharehpetian, H.A. Talebi, Decentralized sliding mode control of WG/PV/FC microgrids under unbalanced and nonlinear load conditions for on and off-grid modes, IEEE Syst. J. (2017) 1–12, https://doi.org/10.1109/JSYST.2017.2761792 (99).
[10] H.R. Baghaee, M. Mirsalim, G.B. Gharehpetian, H.A. Talebi, A decentralized robust mixed $H_2/H\infty$ voltage control scheme to improve small/large-signal stability and FRT capability of islanded multi-DER microgrid considering load disturbances, IEEE Syst. J. (2017) 1–11, https://doi.org/10.1109/JSYST.2017.2716351 (99).

CHAPTER 7

Optimal sizing of microgrids

1. Introduction

Renewable and sustainable energy markets can be enhanced by harnessing renewable energy sources, which are capable of reducing emissions of greenhouse gases [1,2]. As renewable energies have a limited capacity, distributed energy resources (DERs) can be used to provide end-user power alone or in a small-scale network. Microgrids are small-scale, low-voltage grids that are capable of solving local energy problems, enhancing flexibility, and are either grid-connected or islanded (autonomous) [3–6]. In remote and rural areas, electrification is primarily provided by solar panels and wind turbines [7–9], which their generation cannot accurately be predicted. Both photovoltaic (PV) and wind generation (WG) systems suffer from these drawbacks, which pose serious reliability concerns. To improve reliability, oversizing is a solution; however, it may seem to be costly. Another approach to improve system reliability and reduce costs is to utilize combination of PV and WG units, which merge their benefits [10–12] (Fig. 7.1). Also, combination of a fuel cell (FC) with electrolyzer and a hydrogen storage tank can be connected to generating units by a common DC bus. It is obvious that in DC microgrids, AC loads are supplied through DC/AC converters.

Many studies have recently been conducted on hydrogen storage, as energy storage system (ESS), along with renewable energy resources [13–15]. A supercapacitor may be used to enhance the system short-term response as a very short-term storage system, but this system can also be used for long-term storage [8].

A diesel generator, on the other hand, can reduce the cost of storage. In addition to their fuel requirement and emissions, diesel-powered systems have a host of disadvantages. A hydrogen-based storage system, however, has no emissions or fuel needs. Furthermore, since fuel prices are expected to rise and FC costs should significantly decrease in the future, hydrogen-powered systems will become economically realistic [16–18]. Alternatively,

Microgrids and Methods of Analysis
ISBN 978-0-12-816172-2
https://doi.org/10.1016/B978-0-12-816172-2.00007-9

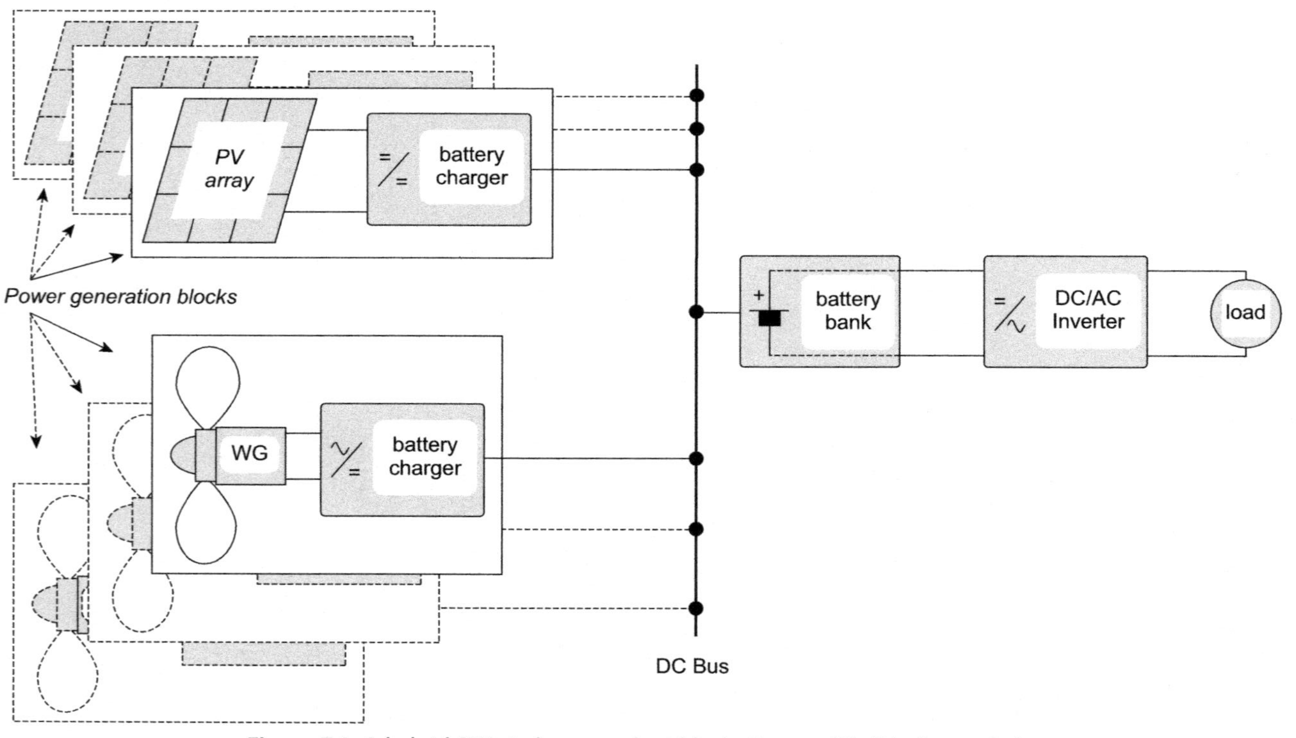

Figure 7.1 A hybrid PV/wind power plant block diagram [7]. *PV*, photovoltaic.

with hybrid PV/wind/diesel systems, surplus solar and wind energy cannot be stored over the course of the good seasons. As opposed to this, in the hydrogen-based storage system, an electrolyzer produces hydrogen, so that excess energy is converted to chemical form and stored in tanks. Using FCs, reaction of hydrogen with oxygen can generate electricity, which can be delivered to the local load in case of wind and solar generation shortage or during peak periods [19]. Furthermore, the FC system can independently be operated with high levels of reliability when hydrogen sources or networks are available [20].

As wind speed and radiation are intermittent in nature, the major challenge in designing such systems is the provision of reliable power requirement in different weather conditions while considering operation and investment costs. Therefore, it is essential to design a hybrid set that presents a reliable and economical source of power [15]. There is a wealth of literature on hybrid PV/WG systems, which provides various methods for optimal design [7,9,10,15,19,21–27].

Based on simulations, the authors of Ref. [24] have used nonlinear programming to optimally allocate and size grid-connected WG units. An optimal WG/PV combination with battery ESS (BESS) has been optimized using simple iterative search algorithms [19,28–34]. In this iterative optimization, the technical criteria called loss of power supply probability (LPSP) [34,35] and deficiency of power supply probability [28] have been used. In Ref. [25], a genetic algorithm (GA) has been employed to calculate the optimal dimensions of components. Other researchers have implemented particle swarm optimization (PSO) to optimally size a hybrid standalone power system under the assumption of a constant and reliable supply of electrical energy [21,27,36–38]. Researchers in Refs. [7,26] have evaluated the performance of genetic and also preference-inspired coevolutionary algorithms to optimally size the combination of WG, PV, and BESS under different scenarios of LPSP. However, wind turbines and solar panels have not been considered in the outage probabilities. An economic cost calculation methodology has been developed in Ref. [23] for calculating interruption costs based on user sectors and the duration of interruptions.

In many researches, optimal design of microgrids has not been considered system reliability extensively. Also, outages of generating units and their failures have been neglected even though they might have a significant impact on system reliability and cost. Any engineering system must be subjected to reliability assessment [39]. Studies on renewable energies, whether for single or multiple systems, must solely evaluate the reliability of various combinations of renewables considering their diverse configurations, element characteristics, load profiles, and their availability. Mawardi and colleagues [40] have investigated how parameters uncertainties and low reliability can hinder the application of an FC and concluded that both of them should be taken into account in various applications. In conclusion, the paper argues that stochastic optimization can result in robust results.

A lot of inspiration came from this suggestion, as it motivates the study of the effects of component reliability on stand-alone DERs. Consequently, in Ref. [10], the outage probability of PV arrays and wind turbines has been considered as a first stage toward design of a robust, economically viable, and reliable hybrid system. Further analyses have shown that the availability of AC/DC converters has a considerable effect on the hybrid system, since this is the only cutoff requirement in the reliability diagram [39]. A later study performed by Kashefi-Kaviani and colleagues [38] have addressed the problem in more detail. The objective function, minimized by PSO, has been modeled as a single-objective optimization (SOO) problem including replacements, operation and maintenance (O&M), dissatisfaction of customers, and annualized costs for investments. A cost function has been added to the reliability indices to alter them as if they were cost functions. In spite of this, reliability has not been considered a function independently.

Hybrid energy systems have also been designed and optimized using multiobjective optimization (MOO) methods [28,41]. Dufo-López and Bernal-Agustín [18] have presented a hybrid system consisting of PV, WG diesel hydrogen ESS with minimum total net present cost, CO_2 emissions, and shed load, minimized by genetic and multiobjective evolutionary algorithms. In Ref. [42], a microgrid composed of PV, FCs, and gas engines has been optimized by multiobjective linear programming, to meet thermal and electrical requirements.

Our first case study in this chapter will illustrate how a hybrid wind/PV power system can provide typical load patterns. Considering a full supply and demand, the main goal is to minimize system cost over its 20-year life span using PSO algorithm. In this study, the data collected in the North West region of Ardebil province (in Iran) and installation, replacement, and O&M costs are used. In the second case, a multiobjective PSO algorithm will be applied considering reliability and operational constraints, to reduce costs of the same system of the first case over its lifetime. The hybrid system will also be evaluated using a proposed approximate reliability method. In addition, the approximate method reduces both problem complexity and computation burden, which leads to decreased time and resources needed.

2. Optimal hybrid wind generation/photovoltaic/battery energy storage system

2.1 Modeling and simulation of wind generation/photovoltaic system

A simulation of the system, shown in Fig. 7.1, is carried out in hourly time steps for 1-year operation period. So, it is assumed that the output power of renewable generation units is constant in each operation period. A schematic diagram showing typical $V-I$ and $P-V$ curves for a given PV system consisting of N_P and N_S parallel and series modules, respectively, can be found in Fig. 7.2A. The power of PV array, $P_M^i(t)$, can be calculated using the PV nominal power at 25.6°C and 1 kW/m^2, given by the manufacturer, based on the PV maximum power output at the end of the ith day and tth hour. A typical PV array can be modeled by the following equations [7]:

$$P_M^i(t) = N_S \times N_P \times V_{OC}^i(t) \times I_{SC}^i(t) \times FF^i(t) \tag{7.1}$$

$$I_{SC}^i(t) = \left\{ I_{SC,STC} + K_I \left[T_A^i(t) - 25^\circ\mathrm{C} \right] \right\} \frac{G^i(t)}{1000} \tag{7.2}$$

$$V_{OC}^i(t) = V_{OC,STC} - K_V \cdot T_A^i(t) \tag{7.3}$$

$$T_A^i(t) = T_{Amb}^i(t) + \frac{NAOT - 20^\circ\mathrm{C}}{800} G^i(t) \tag{7.4}$$

In these equations, $I_{SC}^i(t)$ in A is short-circuit current of PV, $I_{SC,STC}$ is the same but under standard conditions, $G^i(t)$ in W/m^2 represents irradiation density, K_I in A/°C introduces the temperature coefficient of I_{SC}, $T_A^i(t)$ in °C is PV system real temperature, $V_{OC}^i(t)$ in V represents the open-circuit

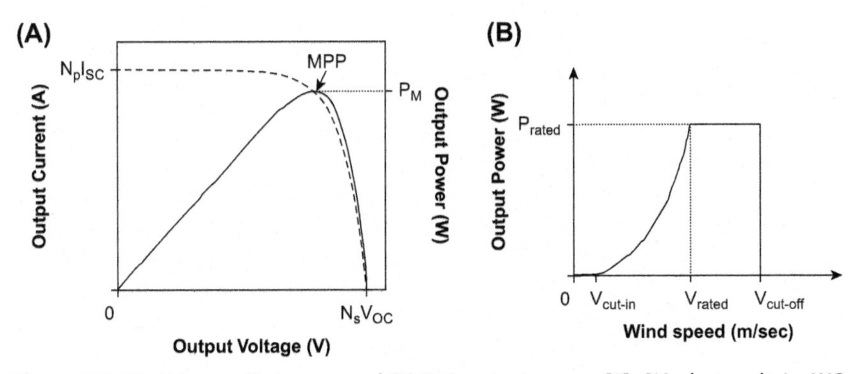

Figure 7.2 (A) PV array $V-I$ curve and (B) WG output power [7]. *PV*, photovoltaic; *WG*, wind generation.

condition voltage, $V_{OC,STC}$ is the same under standard conditions, K_V in V/°C represents the temperature coefficient of V_{OC}, $T^i_{Amb}(t)$ and $NAOT$ both in °C are the ambient and nominal operating temperature, respectively, and the fill factor is presented by $FF^i(t)$.

The number of series modules can be written, as follows:

$$N_S = \frac{V^m_{DC}}{V^m_{OC}} \tag{7.5}$$

where V^m_{DC} represents maximum input voltage of battery charger, while PV array maximum open-circuit voltage is presented by V^m_{OC}.

The transfer factor of the battery charger, n_s, is calculated using the following equation:

$$n_s = \frac{P^i_{PV}(t)}{P^i_M(t)} = n_1 \times n_2 \tag{7.6}$$

where $P^i_{PV}(t)$ represents the power delivered from PV to the battery, the maximum PV output power is presented by $P^i_M(t)$, the electronic equipment efficiency is given by n_1, and n_2 represents the conversion factor which algorithm of battery charging affects it. In maximum power point tracking (MPPT) mode, n_2 is slightly greater than one, when the battery is operating. The PV array power is measured by a battery charger. The MPPT systems have a 50% chance of increasing n_2 to about 100%. However, battery prices increase by less than the typical 10%. The cost of MPPT batteries can only result in one present increase in the total cost for a 30% increment in output energy. The MPPT systems are economically feasible in hybrid systems, according to the results shown in Ref. [1]. Thus, merely MPPT battery chargers will be discussed in this study (Fig. 7.3).

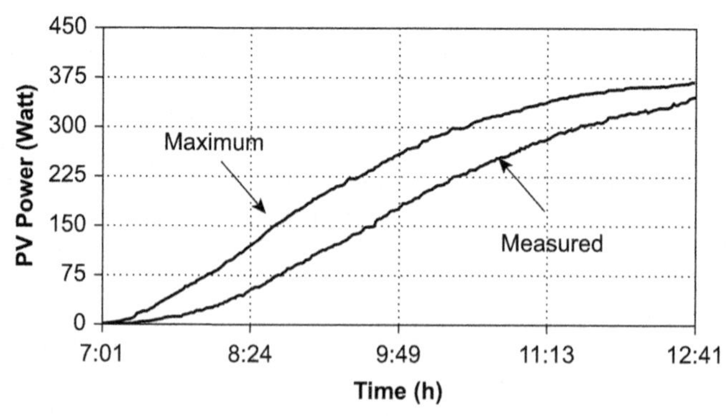

Figure 7.3 MPPT system output power deviation from maximum [7]. *MPPT*, maximum power point tracking.

Using the number of PV batteries N_{ch}^{PV} for the battery charger, one can obtain the following:

$$N_{ch}^{PV} = \frac{N_{PV} \times P_{PV}^{m}}{P_{ch}^{m}} \tag{7.7}$$

where P_{ch}^{m} and P_{pv}^{m} both in W are the battery charger nominal power and PV maximum output power under standard conditions, respectively.

Fig. 7.2B shows how output power varies according to wind speed. These curves are usually provided by manufacturers, and they show the actual power received by the battery bank through the generator. Without considering the characteristics of the battery charger, the WG can be modeled, since the MPPT system and its efficiency are usually taken into account in its output curve. A lookup table based on the curve is used in the optimization process. $P_{WG}^{i}(t)$ is the power transmitted to the battery bank in the tth hour and in the ith day, which is calculated as [7]

$$P_{WG}^{i}(t) = \begin{cases} 0 & ; v^{i}(t) \leq v_{cut-in}, v^{i}(t) \geq v_{cut-off} \\ P_{WG,rated} \times \left(\dfrac{v^{i}(t) - v_{cut-in}}{v_{rated} - v_{cut-in}} \right)^{3} & ; v_{cut-in} \leq v^{i}(t) \leq v_{rated} \\ P_{WG,rated} & ; v_{rated} \leq v^{i}(t) \leq v_{cut-off} \end{cases} \tag{7.8}$$

where $P_{WG,rated}$ represents nominal output power of WG at rated wind speed. Additionally, $v^{i}(t)$ acts as a measure of wind speed at WG hub, determined by the following equation:

$$v^{i}(t,h) = v_{ref}^{i}(t) \times \left(\frac{h}{h_{ref}} \right)^{a} \tag{7.9}$$

where a can be in the range of $1/7 - {}^{1}/_{4}$ [7]. It is possible to charge a battery bank of C_n (Ah) up to a limited level, but there is also an amount of depth of discharge (DOD) compatible with the battery bank.

$$C_{\min} = (1 - DOD) \times C_n \tag{7.10}$$

In this equation, $C_{\min}$ represents the minimum permissible capacity during discharges. To determine the charge level, we have

$$C^{i}(t) = C^{i}(t-1) + n_B \frac{P_B^{i}(t)}{V_{BUS}} \Delta t \tag{7.11}$$

$$C^{i}(24) = C^{i}(0) \tag{7.12}$$

In these equations, the battery available capacity (in Ah) is presented by $C^i(t)$ for the tth hour of day i, the DC bus voltage (in V) is presented by V_{BUS}, and the battery absorbed or delivered output power is represented by $P_B^i(t)$ ($P_B^i(t) < 0$ and $P_B^i(t) > 0$ during discharging and charging periods, respectively. Also, Δt is time step of simulations (1 h). The round-trip efficiency of the battery, i.e., n_B, is 80% and 100% for charging and discharging periods, respectively. The number of series batteries N_B^S, can be determined considering the voltage of DC bus and the voltage of battery V_B, as follows:

$$N_B^S = \frac{V_{BUS}}{V_B} \tag{7.13}$$

An optimization procedure will utilize the aforementioned equations. As long as the load demand is fully supplied, the goal is to minimize hybrid system costs. The algorithm inputs will be hourly averaged data on system components, annual hourly demand data, and hourly averaged data on temperature, wind speed, and solar irradiation. The power, generated by PV and WG units and injected to the battery, is determined as follows:

$$P_{re}^i(t) = N_{PV} \cdot P_{PV}^i(t) + N_{WG} \cdot P_{WG}^i(t);\ 1 \leq i \leq 365,\ 1 \leq t \leq 24 \tag{7.14}$$

where N_{PV} and N_{WG} represent the number of PV and WG units. $P_{Load}^i(t)$, the load demand in tth hour of day i, can be calculated using AC/DC converter input power, $P_L^i(t)$, and the converter efficiency, n_i, as follows:

$$P_L^i(t) = \frac{P_{Load}^i(t)}{n_i} \tag{7.15}$$

Based on the battery capacity, we calculate the following:

- If $P_{re}^i(t) = P_L^i(t)$, the capacity of the battery remains unaltered.
- If $P_{re}^i(t) > P_L^i(t)$, the surplus power, $P_B^i(t) = P_{re}^i(t) - P_L^i(t)$, will be applied for charging the battery, and its new capacity will be calculated by Eq. (7.11). If its SOC (state of charge) achieves 100%, the surplus power must be dissipated in a dummy resistor.
- If $P_{re}^i(t) < P_L^i(t)$, the battery provides the power shortage and its new capacity can be determined using Eq. (7.11).

The hybrid system will not be reliable whenever SOC is less than a minimum permissible value, i.e., a maximum DOD, which undermines the optimization constraint. Simulations are conducted for 1 year in 1-hour time steps.

2.2 System objective function

The proposed optimal sizing algorithm output is the WG units, PV units, and batteries number. In the optimization problem, the total cost in 20-year operation period, $J(X)$ (€), i.e., capital costs, $C_c(X)$ (€), and replacement and O&M costs, i.e., $C_m(X)$ (€), have to be minimized.

$$\min_x\{J(X)\} = \min_x\{C_c(X) + C_m(X)\} \tag{7.16}$$

Therefore, it makes sense to minimize the objective function (OF) including utility, as follows:

$$\begin{aligned} J(X) &= N_{PV}(C_{PV} + 20M_{PV}) + N_{WG}(C_{WG} + 20M_{WG}) \\ &+ N_{BAT}[C_{BAT}(y_{BAT} + 1) + M_{BAT}(20 - y_{BAT} - 1)] \\ &+ N_{ch}^{PV}\left[C_{ch}^{PV}\left(y_{ch}^{PV} + 1\right) + M_{ch}^{PV}\left(20 - y_{ch}^{PV} - 1\right)\right] \\ &+ N_{INV}[C_{INV}(y_{INV} + 1) + M_{INV}(20 - y_{INV} - 1)] \end{aligned} \tag{7.17}$$

where $X = [N_{PV}, N_{WG}, N_{BAT}]$. Constraints on this optimization problem include the following:

$$N_{PV}, N_{WG}, N_{BAT} \geq 0 \tag{7.18}$$

$$P_{Supply}(i, t, X) \geq P_{Demand}(i, t); \begin{cases} i = 1, 2, 3, ..., 365 \\ t = 1, 2, 3, ..., 24 \end{cases} \tag{7.19}$$

C_{PV}, C_{WG}, C_{BAT}, C_{ch}^{PV}, and C_{INV} represent capital costs of PV unit, WG system, battery, charger, and required converter, respectively. M_{PV}, M_{WG}, M_{BAT}, M_{ch}^{PV}, and M_{INV} represent O&M costs of mentioned devices, respectively. y_{INV}, y_{ch}^{PV}, and y_{BAT} are replacement number of inverters, chargers, and batteries in 20-year period of operation. If the inverters have been replaced five times during this operation period, we can calculate this by dividing the value of mean time between failures (MTBF) (here, 20 years) by the number of inverters changed during this time. A perfect supply of load demand is guaranteed by the last constraint. The optimization algorithm penalizes the current solution by 1010 €(here) in case of violation of this constraint. If another possible combination can be found, a new penalty is imposed.

3. Hybrid photovoltaic/wind generation/fuel cell network optimal design

3.1 Modeling

Fig. 7.4 depicts the network studied. The simulation of this network is carried out for 1 operational year. In this period, the wind speed and solar irradiation are used to calculate their generated power.

3.1.1 Photovoltaic and wind generation units

A PV panel output power is calculated using insulation data, as follows:

$$P_{PV} = \frac{G}{1000} P_{PV,rated} \times \eta_{PV,conv} \tag{7.20}$$

where G in W/m^2 represents perpendicular radiation on panel surface, $P_{PV,rated}$ implies nominal power of PV panel in $G = 1000$ W/m^2, and η_{MPPT} represents DC/DC converter efficiency. Most of PV units are working under MPPT condition; therefore, it is reasonable to assume that they work near the maximum power point [9]. As a result, this system increases the average of power extracted from PV units by about 30%, making hybrid systems economically feasible [7]. As a result, PV units are assumed to be equipped with 95% MPPT systems that provide 48 V DC on the DC bus side in the presented study. It must be mentioned that the temperature effects are not considered in this study.

Fig. 7.5 shows the output power of the WG system versus wind speed provided by the manufacturer, which indicates the power transferred between WG unit and DC bus. The curve BWC Excel R/48 [43] by Bergey Wind Power is considered in this chapter. The output voltage is 48 V DC, and its rated capacity is 7.5 kW. In this figure, the relationship between power and wind speed is given by the following equation.

$$P_{WG} = \begin{cases} 0 \quad ; v_W \leq v_{cut\ in}, v_W \geq v_{cut\ out} \\ P_{WG,\max} \times \left(\dfrac{v_W - v_{cut\ in}}{v_{rated} - v_{cut\ in}} \right)^m ; v_{cut\ in} \leq v_W \leq v_{rated} \\ P_{WG,\max} + \dfrac{P_{furl} - P_{WG,\max}}{v_{cut\ out} - v_{rated}} \times (v_W - v_{rated}); v_{rated} \leq v_W \leq v_{furl} \end{cases} \tag{7.21}$$

where $P_{WG,\max}$ and $P_{WG,furl}$ represent the output power of WG unit at rated and cutout speeds, respectively. In this chapter, $m = 3$ and v_W is the

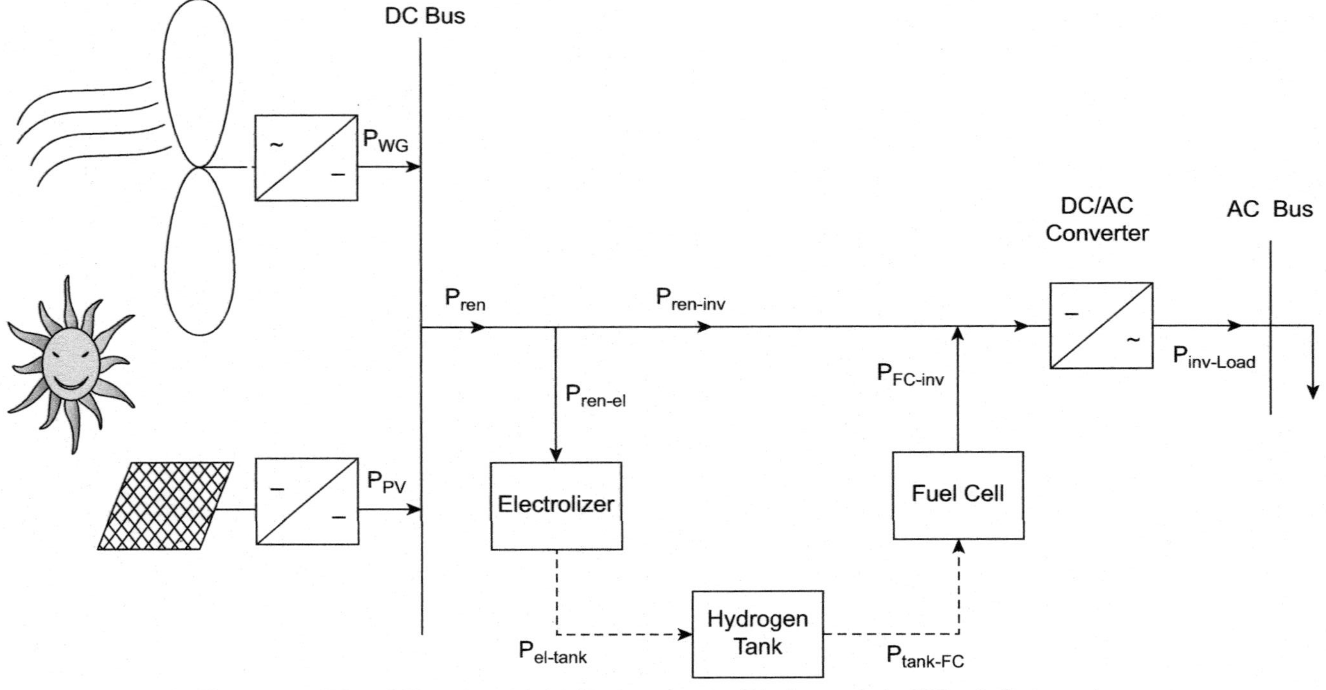

Figure 7.4 WG and PV units with hydrogen storage. *PV*, photovoltaic; *WG*, wind generation.

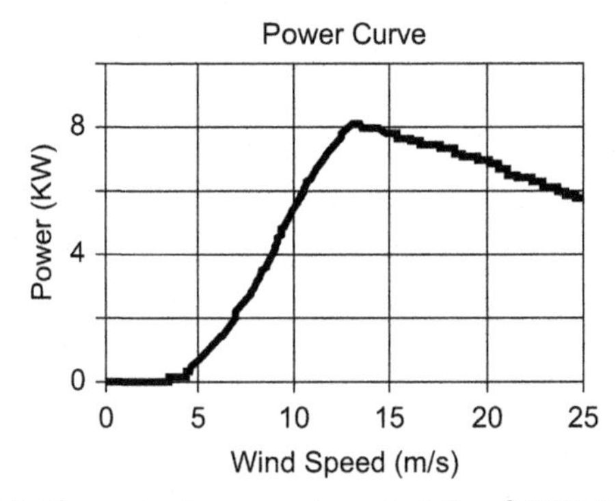

Figure 7.5 The output power characteristic of BWC Excel-R/48.

wind speed measured at hub height. It must be mentioned that exponent law can be used to convert measured data to WG unit installation height, as follows [7,10]:

$$v_W = v_W^{measure} \times \left(\frac{h_{hub}}{hmeasure} \right)^{\alpha} \tag{7.22}$$

where α refers to this coefficient. A low vegetation area, or installation near water or ice covered area, will have a value less than 0.10, while an area with abundant vegetation will have a value over 0.25. An example of a relatively flat surface with a one-seventh power law is grasslands and open terrain, far from big trees/constructions [9].

3.1.2 Electrolyzer, hydrogen storage tank, and fuel cell

Electrolysis functions by passing a direct current through two electrodes (anode and cathode) immersed in water; this produces hydrogen (H_2) and oxygen (O_2). Near the anode, H_2 will be collected, which typically has a pressure of around 30 bars and can be compressed further for storage in a tank [38,44]. Proton exchange membrane FCs (PEMFCs) contain reactants, which their pressure is about 1.2 bar [41]. Due to this, electrolyzers are typically directly connected to hydrogen tanks [27,33,41,43,45]. Although in some cases, an electrolyzer output may be pressurized to 200 bar by a compressor to increase the density of stored energy, this is not as common [38,44]. A second study uses two hydrogen tanks to decrease the energy consumed by the compressor [19]. The hydrogen is injected straight from the electrolyzer into a low-pressure tank and then pumped by compression into a high-pressure tank when the primary tank is full. Consequently, compressors do not run continuously and consume less energy.

However, the developed software allows the hyperelastic electrolyzer to be directly connected to the hydrogen tank, even though the compressor model is not specified in the chapter. The process of transferring power from electrolyzer to H_2 storage tank is modeled by the following equation:

$$P_{el-tank} = P_{ren-el} \times \eta_{el} \quad (7.23)$$

where η_{el} represents efficiency of the electrolyzer. It is assumed that it does not change across all operating ranges [38,43,46,47]. In the tank, the available energy can be measured, as follows:

$$E_{\tan k}(t) = E_{\tan k}(t-1) + \left(P_{ele-\tan k}(t) - \frac{P_{\tan k-FC}(t)}{\eta_{storage}} \right) \times \Delta t \quad (7.24)$$

where P_{tank_FC} is the power injected to the H_2 storage tank. It is assumed that efficiency of storage, i.e., $\eta_{storage}$ is equal to 95% in all working states, regardless of losses due to leakage or pumping [38,48]. For each step t, the mass of hydrogen stored is determined by the following equation:

$$m_{\tan k}(t) = \frac{E_{\tan k}(t)}{HHV_{H_2}} \quad (7.25)$$

where HHV_{H2} is hydrogen higher heating value (39.7 kWh/kg) [49]. The amount of hydrogen stored has lower and upper limits. Hydrogen cannot be stored in a container that exceeds its rated capacity. In addition, there may be some problems, such as a decrease in hydrogen pressure, which limits the amount of hydrogen that can be extracted (here, 5%). In a stored energy system, this is the minimum fraction. So, we have

$$E_{tank,\min} \leq E_{tank}(t) \leq E_{tank,\max} \quad (7.26)$$

Energy from chemical reactions is converted directly into electrical energy using FCs. In the case of intermittent supply, the PEMFC performs reliably and is available at industrial scale in commercial quantities. It can produce large amount of power at a small scale and respond quickly with a power release time of 1−3 s [44]. Input and efficiency (η_{FC}) are considered constant (here, 50%). Hence, the output power is written considering these two attributes, as follows [38,46,47]:

$$P_{FC-inv} = P_{tank-FC} \times \eta_{FC} \quad (7.27)$$

3.1.3 Converter

Considering inverter efficiency (η_{inv}), it is possible to calculate the losses. Inverters are generally intended to work at a constant efficiency (here, 90%) over the entire working range [38,43,50].

$$P_{inv-load} = \left(P_{FC-inv} + P_{ren-inv}\right) \times \eta_{inv} \tag{7.28}$$

3.1.3.1 Operation philosophy

3.1.3.1.1 Power generated by renewable units The WG and PV units generation, injected to the DC bus, can be determined, as follows:

$$P_{ren}\left(n_{WG}^{fail}, n_{PV}^{fail}\right) = \left(N_{WG} - n_{WG}^{fail}\right) \times P_{WG} + \left(N_{PV} - n_{PV}^{fail}\right) \times P_{PV} \tag{7.29}$$

where N_{WG}, N_{PV}, n_{WG}^{fail}, and n_{WG}^{fail} are the number of WG and PV units installed and the number of the same units failed, respectively. Reliability of a system is greatly affected by the availability and inaccessibility of system components. A scheduled (e.g., scheduled maintenance) or forced outage may cause components to be unavailable [51]. The power generated by renewable energy resources is split into two streams: one feeds the inverter to supply the load ($P_{ren-inv}$), and the second one feeds the electrolyzer to produce hydrogen (P_{ren-el}).

3.1.3.1.2 Operation strategy There are three basic conditions for system operation:

- If $P_{ren}(t) = \dfrac{P_{load}(t)}{\eta_{inv}}$, the power generated by the renewable units must directly be fed to the load via the converter.
- If $P_{ren}(t) > \dfrac{P_{load}(t)}{\eta_{inv}}$, a surplus of electricity is sent to the electrolyzer. An electrolyzer power rating can exceed the injected power, which would cause the surplus energy to circulate in a dummy resistive load.
- If $P_{ren}(t) < \dfrac{P_{load}(t)}{\eta_{inv}}$, the FC should supply the shortage of power. FCs cannot deliver the rated power to the grid if the shortage exceeds their rated capacity or the storage of hydrogen cannot afford it. Therefore, a portion of the load must be shed. Due to this, the load is reduced.

In all of these conditions, component limits are considered, so Eqs. (7.20)–(7.29) are the dominant parameters.

3.2 Reliability/cost assessment

Simulations are carried out using 1-h time steps, and then, reliability and cost analyses are performed. The results are extended to 20 years of system operation based on economic factors. There is no consideration for the load growth and the uncertainty associated with solar radiation and wind speed.

3.2.1 Reliability indices

The literature presents a number of reliability indices [26,34,35,44,51,52,53]. There are a variety of indices applied to evaluate generator system reliability, such as loss of load expected (*LOLE*), loss of energy expected (*LOEE*), and expected energy not supplied (*EENS*). The following sections provide definitions of these indices.

3.2.1.1 Loss of load expected

This index is defined as follows:

$$LOLE = \sum_{h=1}^{H} E[LOL(h)] \tag{7.30}$$

Here, $E[LOL(h)]$ stands for the expected (mathematical) amount of loss of load in hth time step that can be written, as follows:

$$E[LOL] = \sum_{s \in S} T(s) \times f(s) \tag{7.31}$$

where, for state s, $T(s)$ is the duration of load loss (in hour), S refers to all possible states set, and $f(s)$ represents the state s encountering probability.

3.2.1.2 Loss of energy expected/Expected energy not supplied

These indexes are presented by the following equation:

$$LOEE = EENS = \sum_{h=1}^{H} E[LOE(h)] \tag{7.32}$$

Here, in h-th time step, $E[LOE(h)]$ represents the probability of loss of energy.

$$E[LOE] = \sum_{s \in S} Q(s) \times f(s) \tag{7.33}$$

The $Q(s)$ here represents the energy lost at state s.

3.2.1.3 Loss of power supply probability

This index is written as follows:

$$LPSP = \frac{LOEE}{\sum_{h=1}^{H} D(h)} = \sum_{h=1}^{H} E[LOE(h)] \tag{7.34}$$

where, in h-th a time step, the load demand (kWh) is defined by $D(h)$.

3.2.1.4 Equivalent loss factor

Equivalent loss factor (ELF) is given by the following equation:

$$ELF = \frac{1}{H} \sum_{h=1}^{H} \frac{E[Q(h)]}{D(h)} \tag{7.35}$$

In Eqs. 7.30–7.35, $H = 8760$. The information on the number of outages and their magnitudes has been provided in Ref. [44]. ELF determines the optimal solution as the primary reliability constraint. All of the aforementioned indices are calculated by the developed software and applied as either an OF or a constraint. Power companies aim at an efficiency level of 0.0001 in developed countries. However, ELF of 0.01, which is acceptable for this study and in rural areas, is acceptable in stand-alone applications as well [44].

3.2.2 System reliability

In this chapter, power failures of PV units, WG systems, and converters are considered. PVs and WGs are assumed to have a forced outage rate of 4% [38,52]. This means 96% of the components will be available. By using the binomial distribution function, one can calculate the probability of occurrence of a state [54]. In this condition, given n_{WG}^{fail} out of total N_{WG} failed WG systems and n_{PV}^{fail} out of all N_{PV} failed PV units, we have

$$f_{ren}\left(n_{WG}^{fail}, n_{PV}^{fail}\right) = \left[\binom{N_{WG}}{n_{WG}^{fail}} \times A_{WG}^{N_{WG}-n_{WG}^{fail}} \times (1 - A_{WG})^{n_{WG}^{fail}}\right] \times \left[\binom{N_{PV}}{n_{PV}^{fail}} \times A_{PV}^{N_{PV}-n_{PV}^{fail}} \times (1 - A_{PV})^{n_{PV}^{fail}}\right] \tag{7.36}$$

where A_{WG} and A_{PV} are availabilities of WG system and PV unit. The outage probabilities of other components, compared with WG systems and PV units, are negligible, as they are stationary indoor devices. In addition, it is crucial to recognize that the inverter is very important for feeding the load, and its failure will cause the entire load outage. There is only one cut-set in the reliability diagram that addresses DC/AC converters [39]. Therefore, despite the small FOR of such a hybrid system, outage probability of the inverter should be considered when reliability evaluations are conducted. The simulation results will demonstrate that this component plays an important role from reliability point of view. However, there are not many comprehensive studies on the reliability of power electronics converters. Therefore, we can roughly calculate the compatibility of a DC/AC converter from the data given in Refs. [7,55,56].

Given failure rate of 2.5×10^{-5} for an inverter [55], since mean time to failure (MTTF) of each equipment is reciprocal of its failure rate [39], the inverter MTTF is equal to 37,037 h (i.e., 4.23 years), Also, Koutroulis et al. [7] have suggested 5 years for MTBF of converters. On the other hand, mean time to repair (MTTR) of each inverter, as shown in Ref. [56], is approximately 40 h. For an inverter, steady-state reliability is determined by the following formula [39]:

$$A_{inv} = \frac{MTTF_{inv}}{MTTF_{inv} + MTTR_{inv}} \tag{7.37}$$

On the other hand, Eq. (7.37) indicates that the converter availability is 99.89%. This means that its reliability is higher than those of renewable energy resources. Ultimately, failure probability of n_{WG}^{fail} WG systems, n_{PV}^{fail} PV units, and n_{inv}^{fail} converters out of N_{WG}, N_{PV}, and N_{inv} (here, $N_{inv} = 1$) installed devices, respectively, is determined as follows:

$$\begin{aligned} f_{system}\left(n_{WG}^{fail}, n_{PV}^{fail}, n_{inv}^{fail}\right) = f_{ren}\left(n_{WG}^{fail}, n_{PV}^{fail}\right) \times \Bigg[\binom{N_{inv}}{n_{inv}^{fail}} \\ \times A_{inv}^{N_{inv} - n_{inv}^{fail}} \times (1 - A_{inv})^{n_{inv}^{fail}} \Bigg] \end{aligned} \tag{7.38}$$

where $f_{ren}(n_{WG}^{fail}, n_{PV}^{fail})$ is the failure probability of n_{WG}^{fail} WG systems and n_{PV}^{fail} PV units determined in Eq. (7.36). It is tedious and time-consuming to conduct calculations when outage of other elements, along with wind speed, solar radiation, and load profile uncertainties, are considered during system reliability assessment. Monte Carlo simulations can be used to accurately assess the reliability [52,53,57,58]. In Ref. [58], it has been used to simulate a WG, PV, and battery system. The results have shown that it can converge to solutions after more than 50 years. In this manner, Monte Carlo simulation is too complex and time-consuming. In addition, it requires detailed information (PDF, probability distribution function), which is mostly unavailable [58].

A 10-year chart of steady-state reliability for a PEMFC is shown in Fig. 7.6 [22]. The FC has a greater than 99% probability of being available for the first 5 years of its useful life. There is no information available about electrolyzers and hydrogen tanks, and reliability evaluations of these equipment do not seem to be of interest. The failure probabilities of FCs, electrolyzers, and hydrogen tanks are not taken into consideration in this study. In contrast to DC/AC converters, these parts do not result in loss of load because these parts are not involved in any cut-set. Additionally, they have a higher reliability than wind turbines and PV arrays. By simplifying reliability calculations and incorporating these facts, it may be possible to justify neglecting component failures.

3.2.2.1 Cost of loss of load

Different methods have been used to estimate the cost of electricity interruptions. Considering, for instance, the willingness of the customer to

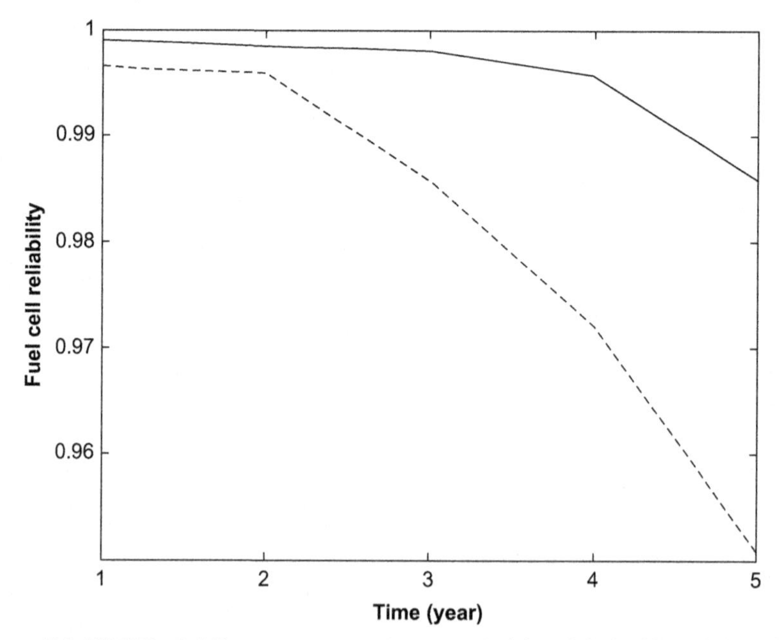

Figure 7.6 PEMFC reliability versus operating years (*solid and dashed lines* for the first 5 years and the second 5 years with battery replacement, respectively [38]. *PEMFC*, proton exchange membrane fuel cells.

pay for expansion, the productivity, or the level of compensation from the government, decreases at industries have been affected, which makes shortages acceptable. For industrial and domestic users, the values are similar: between 5 and 40 US$/kWh in either case [44]. As indicated in Ref. [44], the cost of dissatisfaction is 5.6 US$/kWh due to the loss of load caused by the customer.

3.2.2.2 Approximate method

To evaluate reliability, all possible state transitions of a system should be identified. This enables the determination of reliability indices in agreement with the likelihood of occurrence of each state and resulting loss of load. In this study, a hybrid network with optimal reliability and minimal cost is achieved using the multiple-objective particle swarm optimization (MOPSO) algorithm. In general, this type of algorithms requires a number of simulations. The algorithms are evolutionary or population-based. The study, for example, explores the solution spaces using a swarm of 200 particles over 2000 iterations. Therefore, each iteration supports a simulation of the system for 70 years, i.e., 200 × 2000 × 8760 = 3,504,000,000 h. As there are 199 PV units and 14 WG systems, each simulation time step has to enumerate 2 × 15 × 200 = 6000 states. Then, it is clear that designing a reliable network can be so complicated with a high computational burden. Therefore, attempts to simplify and reduce computation time are well worth it in these types of studies. Failure PDF and equal-probability diagrams of 14 WG systems and 199 PV units (with FOR of 4% for both) are presented in (Fig. 7.4), which can have a nonzero value in a small region. In the approximate method, all the possible states of outage are modeled with an equivalent state for wind turbines and solar arrays. This state is absolutely likely to occur, and renewable energy generated in this state has the same expected value as in all the possible states, as follows:

$$P_{ren}^{eq} = E[P_{ren}] = \sum_{s \in S} P_{ren}(s) \times f_{ren}(s) \tag{7.39}$$

Substituting Eqs. (3.29) and (3.36) in Eq. (3.39), we have

$$P_{ren}^{eq} = \sum_{n_{WG}^{fail}=0}^{N_{WG}} \sum_{n_{PV}^{fail}=0}^{N_{PV}} \left\{ P_{ren}\left(n_{WG}^{fail}, n_{PV}^{fail}\right) \times f_{ren}\left(n_{WG}^{fail}, n_{PV}^{fail}\right) \right\} \tag{7.40}$$

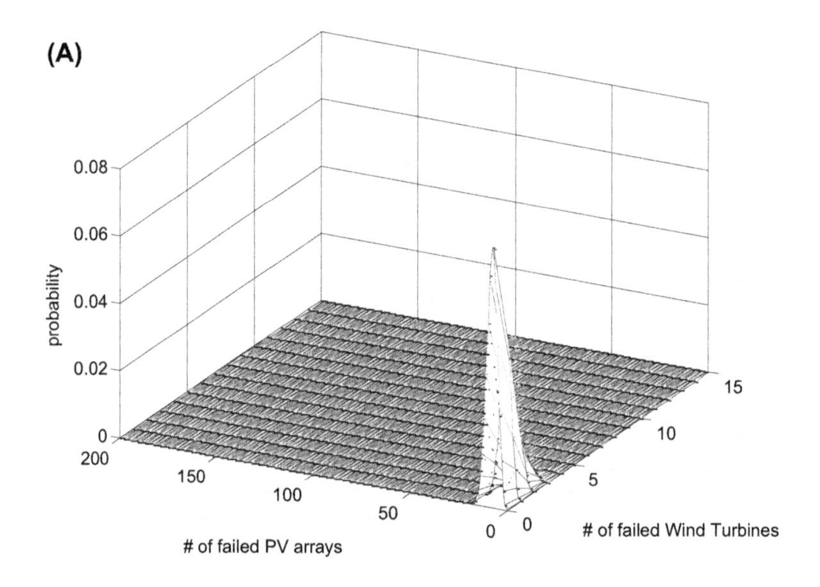

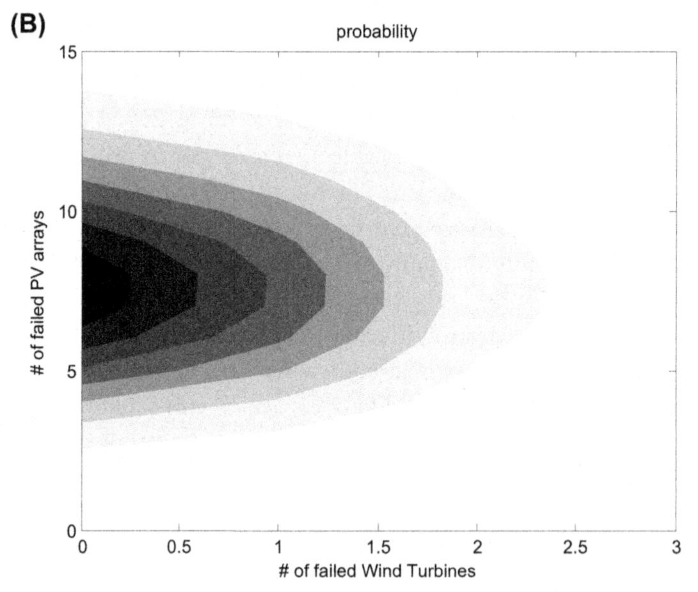

Figure 7.7 Analysis of (A) PDF and (B) curves with equal failures. *PDF*, probability distribution function.

At the end of the day, and with a little new effort, one can prove that (see Appendix):

$$P_{ren}^{eq} = N_{WG} \times P_{WG} \times A_{WG} + N_{PV} \times P_{PV} \times A_{PV} \tag{7.41}$$

It was logically predictable that Eq. (7.41) would exist. We further demonstrate the acceptability of the approximate approach in the coming section. Although the approximate method is less computationally intensive, and therefore, takes less time to compute, the accuracy of reliability evaluations is satisfactory. For example, the number of all possible states during a time step when there are 14 WG systems and 199 PV units can be reduced using this method from 6000 to just two states (Fig. 7.7).

3.2.3 Problem statement

An optimal hydrogen-powered hybrid wind power/PV system is considered in this chapter. There are following optimization variables: wind generators number, PV units number, and installation angle, as well as electrolyzer capacity, H_2 storage tank, FC, and inverter. A reliability index such as LOEE or LOLE is used as an OF. A maximum ELF reliability index is used in this problem. Additionally, simulations are subject to constraints, such as components power and energy limits. The *i*th component annualized cost (AC) can be estimated as [9,27]:

$$AC_i = N_i \times \{[CC_i + RC_i \times K_i(ir, L_i, y_i)] \times CRF(ir, R) + O\&MC_i\} \tag{7.42}$$

where N can be unit number or rating, CC in US$/unit represents capital cost, RC in US$/unit implies each replacement cost, $O\&MC$ is component annual O&M cost in US$/unit-yr, and R refers to project useful lifetime (in this study, 20 years). Also, ir stands for real interest rate, which is 6% in this study and a function of nominal interest rate ($ir_{nominal}$) and annual inflation rate (fr), as follows [25]:

$$ir = \frac{ir_{\text{nominal}} - fr}{1 + fr} \tag{7.43}$$

Also, CRF is capital recovery factor [9]. K is known as present worth of a single payment [43] calculated by the following equations.

$$CRF(ir, R) = \frac{ir \times (1 + ir)^R}{(1 + ir)^R - 1} \tag{7.44}$$

$$K_i(ir, L_i, y_i) = \sum_{n=1}^{y_i} \frac{1}{(1 + ir)^{n \times L_i}} \tag{7.45}$$

In these equations, the useful life and replacement count of a component are determined by L and y, respectively. The replacement of individual components is a function of their useful life and the duration of the project [7].

Using Eq. (7.32), one can calculate the annual loss of load as follows:

$$AC_{loss} = LOEE \times C_{loss} \tag{7.46}$$

where C_{loss} in US$/kWh represents the customer dissatisfaction cost presented in Section 3.2. Last but not least, the OFs are outlined, as follows:

$$J_{Cost} = \min_{X}\left\{\sum_{i} AC_i\right\} \tag{7.47}$$

$$J_{LOEE} = \min_{X}\{LOEE\} \tag{7.48}$$

$$J_{LOLE} = \min_{X}\{LOLE\} \tag{7.49}$$

The component indicator i is represented in Eqs. 7.47–7.49, and X represents the optimization variables. There are the following constraints on the optimization problem.

$$E[ELF] \leq ELF_{\max} \tag{7.50}$$

$$N_i \geq 0 \tag{7.51}$$

$$0 \leq \theta_{PV} \leq \frac{\pi}{2} \tag{7.52}$$

$$E_{\tan k}(0) \leq E_{\tan k}(8760) \tag{7.53}$$

In Eq. (7.52), θ_{PV} is installation angle of PV unit. As stated in Eq. (7.53), the hydrogen tank storage capacity after the first year will be higher than its capacity at the beginning. The reliability is thus evaluated for the worst-case scenario.

4. Particle swarm optimization and multiple-objective particle swarm optimization algorithms

4.1 Particle swarm optimization algorithm

The PSO algorithm probes an area of search space that indicates reliability by utilizing the individual members of the population [59–63] (Fig. 7.8). It is based on the following general principles:

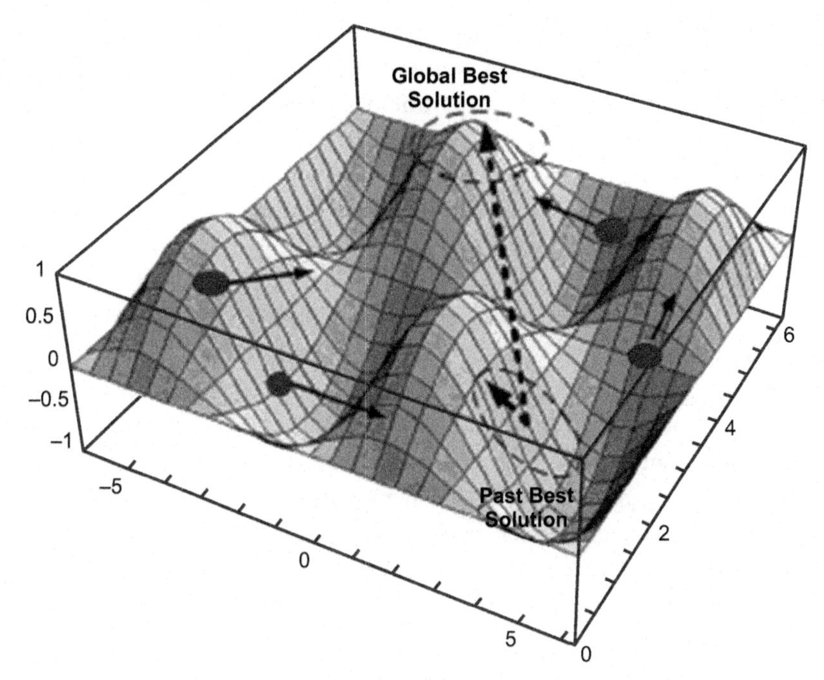

Figure 7.8 PSO searching idea. *PSO*, particle swarm optimization.

In a n-dimensional search space, the ith particle is represented by a n-dimensional vector, $X_i = [x_{i1}, x_{i2}, \ldots, x_{in}]^T$ and velocity $V_i = [v_{i1}, v_{i2}, \ldots, v_{in}]^T$, where $i = 1, 2, \ldots, N$ and N is the population size.

In this algorithm, the particle i remembers the best position it visited so far (P_{best}), i.e., $P_i = [p_{i1}, p_{i2}, \ldots, p_{in}]^T$, and the best position of the best particle in the swarm (G_{best}) is referred as $G_i = [g_1, g_2, \ldots, g_n]^T$ [59–61]. Each particle i adjusts its location in the iteration $t+1$ as follows [59,62,63]:

$$V_i(t+1) = \omega(t) \times V_i(t) + c_1 \times r_1 \times (P_i(t) - X_i(t)) + c_2 \times r_2 \times (G_i(t) - X_i(t)) \tag{7.54}$$

$$X_i(t+1) = X_i(t) + \chi \times V_i(t+1) \tag{7.55}$$

where $\omega(t)$ Omega represents a measure of inertia used to influence future speed by manipulating the history of prior speeds. The constriction factor χ will be applied to restrict the velocity (here $\chi = 0.7$). c_1 and c_2 are the cognitive and social parameters, and r_1 and r_2 denote random real numbers drawn from a uniformly distributed interval $[0, 1]$. $\omega(t)$ omega is the solution to the trade-off between a swarm capacity to explore at a global and local level. According to the experimental results, the inertia coefficient should be initialized at a large value (1 in this case), giving priority to global exploration of the search space, and then slowly decreased to an amount about zero to get an improved result [59,60].

c_1 and c_2 speed up the search algorithm to local and global best orientations, respectively. These parameters should dynamically be adjusted, according to advanced research [59,63]. Experimental results show that c_1 should initially be set at 2.5 and then monotonically decreased to 1.5 during optimization. As an alternative, it makes more sense to track c_2 in an inverse direction. A modulation of search points is indicated in Fig. 7.8.

In GA, the mutation operator developed by Raymond R. Tan prevents premature convergence to suboptimal results and led to a significant increase in successful convergence [59,63]. Thus, mutation computing algorithms are implemented in this chapter as continuous-space PSOs. A fixed probability, in this case 5%, reintroduces every continuous variable into its feasible area.

4.2 Multiobjective optimization

Multiple objectives are usually simultaneously optimized in real-world applications, which are typically noncommensurable and conflicting [59,64–68]. The problem of minimizing multiple objectives can be presented in d-dimension space in the following way: given $x = [x_1, x_2, \ldots, x_d]^T$,

$$\text{Minimize}: f(x) = [f_1(x), f_2(x), \ldots, f_M(x)]^T \tag{7.56}$$

$$\text{Subject to}: \begin{cases} g_j(x) \leq 0; & i = 1, 2, \ldots J \\ h_k(x) = 0; & i = 1, 2, \ldots K \end{cases} \tag{7.57}$$

where $f_i(x)$ represents ith OF, $g_j(x)$ denotes jth inequality constraint, and $h_k(x)$ refers to kth equality constraint [64,69,70]. A feasible region Ω and feasible solution are defined by constraints given by Eq. (7.57) and any point x inside Ω, respectively. The set Ω is mapped by vector function $f(x)$ (representing all possible OFs values) into the set Λ.

In the MOO, the concept of Pareto dominance is used, which is formulated first by an Italian scientist Vilfredo Pareto. The Pareto-optimal set can consist of many solutions that can have the potential to be optimal so that the most proper solution vector will be determined based on some criteria based on them [64,70]:

A vector $u = [u_1, u_2, \ldots, u_M]^T$ is said to dominate a vector $v = [v_1, v_2, \ldots, v_M]^T$ (denoted by $u \preceq v$), for a MOO, if and only if:

$$\forall i \in \{1, .., M\}, \quad u_i \leq v_i \wedge \exists i \in \{1, .., M\} : u_i < v_i \tag{7.58}$$

where M denotes objective space dimension.

A solution $u \in U$, where U is the universe, is said to be *Pareto optimal* if and only if, there is no other solution $v \in U$, such that u is dominated by v. Such solutions (u) are called *nondominated solutions*. The set of all such nondominated solutions forms *Pareto* or *nondominated set* [64,70].

This partition can easily be demonstrated by showing that Pareto dominance induces it. Assume that there is a point f' in F. In its coordinate system, this point indicates the areas in which it dominates f' or is dominated by f'. These areas are shown as $F_{f'}+$ and $F_{f'}-$, respectively. In Fig. 7.9, $F_{f'}+$ and $F_{f'}-$ are illustrated in a two- and three-dimensional spaces, respectively.

With more OFs, the distribution of nondominated vectors becomes extensive, and Fig. 7.9 illustrates this intuitively. As a result, the algorithm that relies on Pareto optimality will perform better. With the Pareto scheme applied to high-dimensional problems, there are many solutions that are not dominated in any generation. It is significant to identify these nondominated solutions to increase the selective pressure [59,71,72].

4.3 Multiobjective particle swarm optimization

When multiple OFs are involved, a global optimum cannot usually be found using the PSO algorithm. So, this algorithm must be modified to deal with MOO problems.

In Ref. [64], whenever and only when the new solution dominates the former P_{best}, it automatically replaces each particle personal best performance (P_{best}). In the update of the global best performance (G_{best}), two major issues should be considered. The first step to achieving Pareto optimality is addressing fitness assignment and selection. Second, the swarm should be managed to provide a Pareto-optimal distribution and prevent premature convergence [59–64].

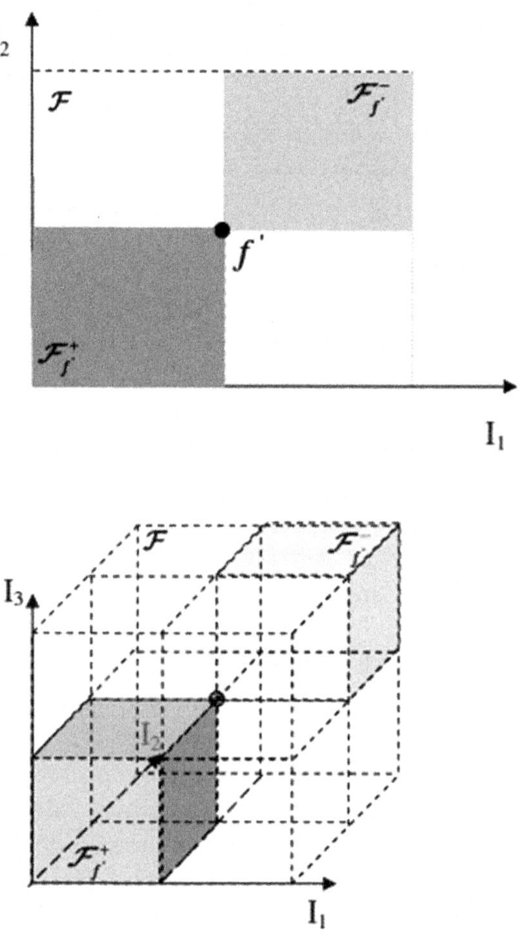

Figure 7.9 Pareto dominance partitions represented in two- and three-dimensional spaces.

In this study, a repository is used that will be limited in number of solutions using an archiving mechanism proposed in Ref. [46]. The approach achieves more diversity and optimality, when compared with conventional archiving, while requiring less complexity. A roulette wheel selection of Refs. [59,73–76] will be used to select the G_{best} from the repository using density values of archive members.

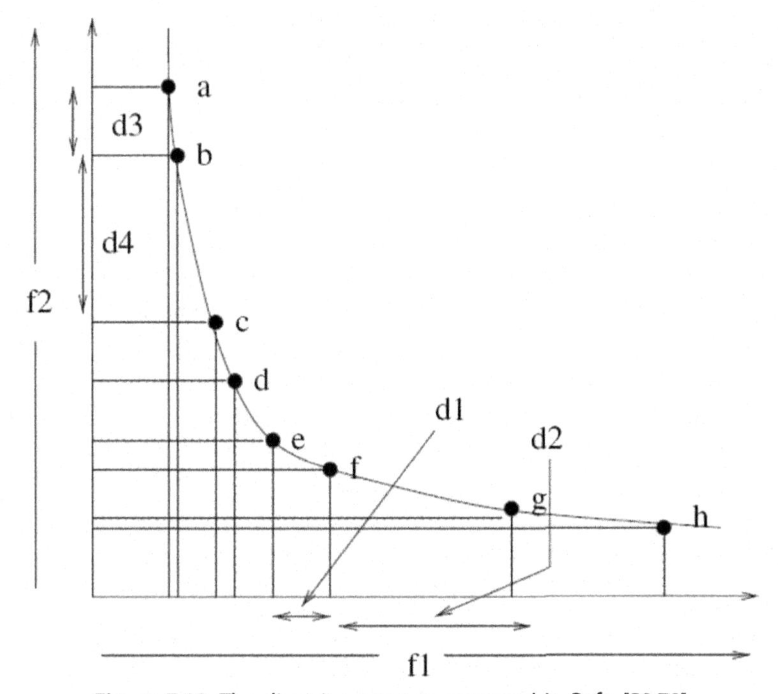

Figure 7.10 The diversity measure proposed in Refs. [59,70].

As a measure of archive diversity, the density parameter (den_i) is introduced for the solution *i* as its distance from its nearest neighbor. Nevertheless, two alternatives may be most similar to one another. For example, in Fig. 7.10, the solution *e* and *f* generate the same distance to the nearest neighbor for minimization a 2-OF problem (i.e., f_1 and f_2); therefore, solutions *e* and *f* will have the same distance to the nearest neighbor since they are the closest to each other. To avoid this problem, the distances from the nearest neighbors of each solution are calculated so that the distance to the nearest neighbor that has not been investigated must be considered. For example, the distance of the nearest neighbor of solution *f* should be calculated in order *a*, *b*, *c*, *d*, *e*, *f*, *g*, and *h*, since this distance, which is between it and solution *e*, has already been considered [59–64].

When computing den_i, the OF amounts for obtained results are ranked in ascending order for each OF. Afterward, we calculate the difference between every solution and its closest neighbor along each objective. den_i represents the density parameter for each solution *i* based on the sum of its differences. So, for the example presented, if the sorting is done based on *f*1, *f* is the solution *e* nearest neighbor, while *g* is the solution *f* nearest neighbor. Consequently, *d*1 and *d*2 correspond to the respective distances along *f*1. For instance, the solution *c* has the distance *d4* from its neighbor *b*, according to objective function *f2*, but the solution *b* has the distance *d3* as it is the nearest neighbor via objective function *f2*. The density can be determined for all the solutions in the archives using the aforementioned method.

Here, the number of swarms is set at 200. The topmost permissible velocity on all the dimensions of the search space is adjusted at 0.5 of its realistic extent to prevent the explosion of the swarm. In a three-dimensional Pareto set containing 100 solutions, the developed software finds the optimal in 2000 iterations. This algorithm is used to obtain the variety of the Pareto set. Algorithm 1 illustrates how the MOPSO algorithm is organized.

4.4 Best trade-off solution

As described earlier, the Pareto-optimal set contains multiple optimal responses, which can be considered as a privilege for this multiobjective optimization. There is not any absolute global optimal solution; therefore, the designer can select a solution that best suits the requirements of various applications. It is unlikely to be possible to precisely define what is needed in most cases due to the fallibility of human judgment. The use of fuzzy sets has been proposed in Refs. [59,71] to resolve this dilemma. In those researches, the following linear membership function has been suggested for each OF (J_i).

$$u_i = \frac{J_i - J_i^{\min}}{J_i^{\max} - J_i^{\min}} \tag{7.59}$$

As the membership function gets lower, it is evident that the OF is getting more fully achieved. A nondominant solution (k) can be defined using the following aggregated membership function:

$$U^k = \sum_{i=1}^{M} u_i^k \tag{7.60}$$

U^k is considered to be the best compromise (best trade-off I) with the smallest membership requirement. Another criterion is proposed for choosing the solution with the best compromise. Firstly, the distance (k) is

Algorithm 1 Archive-based multiple-objective particle swarm optimization

```
1 A:=∅;
2 for i = 1 to N do
3    {x_i, v_i, gbest_i, pbest_i}:=Initialization();
4 end
5 for t = 1 iter do
6    for i = 1 to N do
7       for d = 1 to M do
8          v_id:=χ*(wv_id + c1r1(P_id − x_id) +
     c2r2(G_id − x_id));
9          x_id:=x_id + v_id;
10      end
11      x_i:=CheckConstraints (x_i);
12      If x_i ⋠ a ∀a ∈ A then
13         A:={a ∈ A|a ⊀ x_i};
14         A:=A ∪ x_i
15      end
16   end
17   If x_i ≼ pbest_i ∨ (x_i ⊀ pbest_i ∧ pbest_i ⊀ x_i)
18      pbest_i:=x_i;
19   end
20   gbest_i:=GlobalGuide (x_i;A);
21 end
```

calculated from the origin of each nondominated solution using the following equation:

$$V^k = \left[\sum_{i=1}^{M} \left(u_i^k\right)^2\right]^{½} \tag{7.61}$$

Then, the most appropriate solution (V_k) to the fuzzified origin (best trade-off II) will be selected.

5. Simulation results

5.1 Case study 1: hybrid system including photovoltaic/wind generation/battery energy storage system

In this section, the hybrid wind–solar system will be simulated to analyze its behavior. In the simulations, the data for WG systems and PVs units must be given. The simulated area is Ardebil province in Iran and its data are used. Also, the prices of commercial components of Ref. [1] are used for simulations. For load profile, the IEEE Reliability Test System (IEEE RTS) is used. The peak power is assumed to be 1000 W [43]. The wind speed measured at 40 m height is displayed in Fig. 7.11, and hourly solar irradiation at both vertical and horizontal planes is shown in Fig. 7.12.

5.1.1 Optimal installation angle of photovoltaic panels

In this section, the optimum angle, at which PV panels absorb the maximum amount of energy, should be determined. An angle respect to the horizon is used to calculate the absorbed solar energy (W/m^2) by the following formula:

$$G^i(t, \alpha) = G_V^i(t) \times \cos(\alpha) + G_H^i(t) \times \sin(\alpha) \tag{7.62}$$

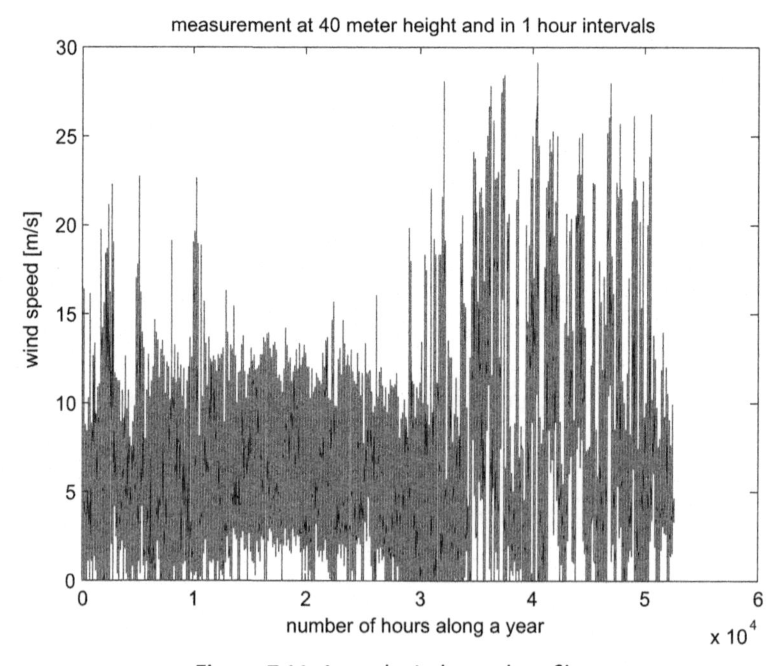

Figure 7.11 Annual wind speed profile.

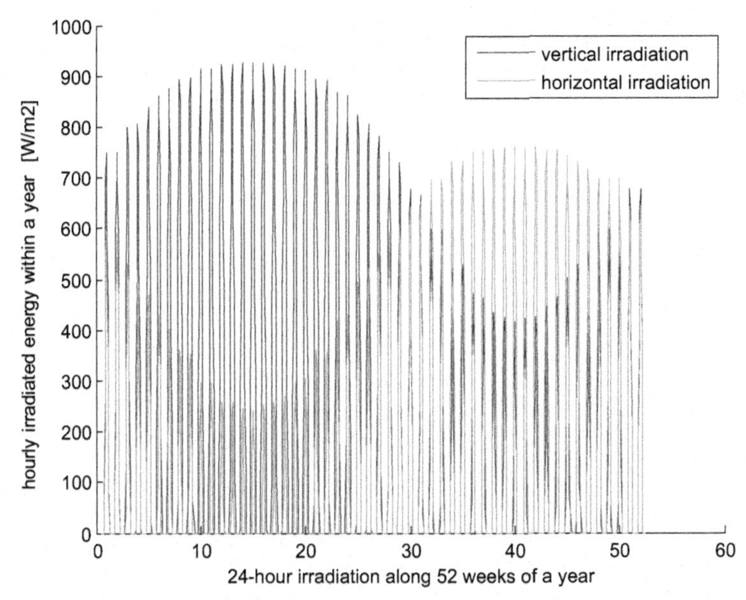

Figure 7.12 Hourly solar irradiation at both vertical and horizontal planes in 1 year.

Assuming $G_H^i(t)$ and $G_V^i(t)$ as horizontal and vertical radiation rates in the tth hour of day i, respectively, the amount of free energy emitted annually will be calculated as follows:

$$E_G(\alpha) = \sum_{i=1}^{365} \sum_{t=1}^{24} G(t, \alpha) \tag{7.63}$$

At the optimal angle, $E_G(\alpha)$ reaches its maximum value. Fig. 7.13 depicts $E_G(\alpha)$ versus α. The Ardebil region latitude and longitude are $48°{:}17'$ and $38°{:}15'$, respectively. A solar array should be installed at a very specific angle, such as $\alpha_{opt} = 34°{:}29'$ as shown in Fig. 7.13.

5.1.2 Hybrid system optimal design

Table 7.1 presents the specifications of elements of the system under study. The simulations have been carried out using MATLAB 7.2 programming environment. There are 30 individual swarms in each generation of hybrid power generation systems, with each individual constituting a potent combination. As a result, each individual is mathematically modeled as a vector of three elements: wind power turbines, solar panels, and batteries. Swarms search the solution space for optimal combinations of components in a system.

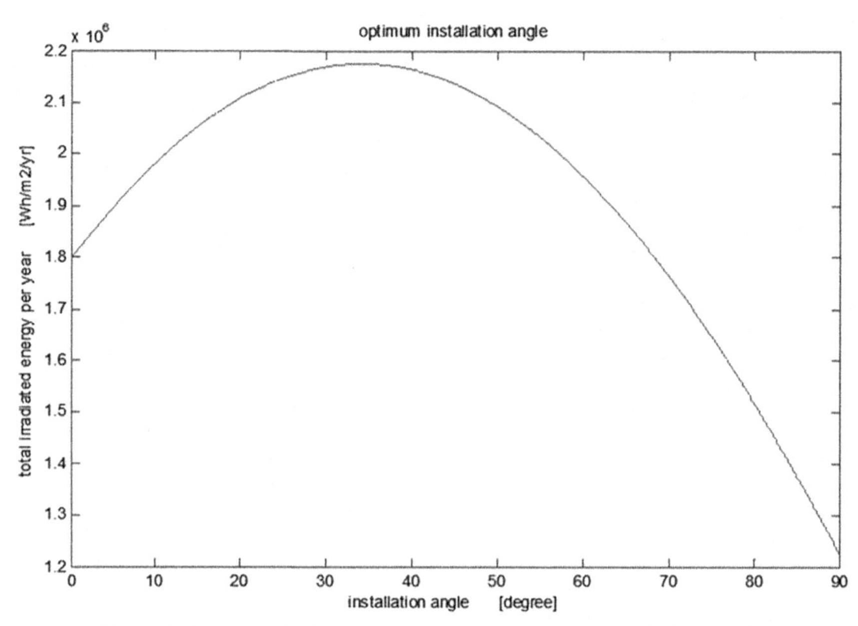

Figure 7.13 Annual absorbed solar energy versus installation angle.

Table 7.1 Elements specifications.

Element	Data
Wind generation system	Power rating = 1000 W, height = 15 m Cut-in, rated and cut-out wind speeds = 2.5, 11 and 24 m/s Value = 2506 €
PV panel	Power rating = 110 W, V_{OC} = 21 V, I_{SC} = 7.22 A, V_{max} = 17 V, I_{max} = 6.47 A, Value = 519.14 €.
BESS	Capacity = 230 Ah, V_{Rated} = 12 V, DOD_{max} = 80%, round-trip efficiency = 85% Lifetime = 3 yr Value = 264 €
Charger	Power rating = 300 W, MPPT used, charging algorithm efficiency = 100% Converter efficiency = 95%, lifetime = 4.5 yr Value = 200 €
Inverter	Power rating = 1500 W, efficiency = 80%, lifetime = 4.5 yr Value = 1942 €

There are 150 generations in the optimization algorithm. The problem has to be solved as integer programming because variable optimizations are only possible in nonnegative integers. As a result, only the integer section of individuals can be applied to simulate the system and calculate the OFs.

Each simulation must consider 4500 years with a population of 30 and 150 generations. In the optimization process, there will be a computationally intensive computation, since each year consists of 8760 simulations of the system. The study will use the average of hourly values around each week to approximate each week as an equivalent day and decrease the problem difficulty and computational burden, which is then modeled for each equivalent day. As a result, the computation requirements are reduced by 1/7.

Implementing the proposed methodology allows the hybrid system to be optimized. As the literature suggests, optimized software should be run many times before choosing the best one as the global solution. In nonconvex and nonlinear optimization problems, there is no guarantee to determine the optimum. The software is run 10 times due to the discontinuity and nonlinearity of the problem, and the cheapest combination is chosen as the global optimal solution.

The best combination consists of 3 wind turbines, 31 solar panels, and 9 batteries (i.e., $X_{opt} = [3\ 31\ 9]$). This combination results in the cost function of $J(Xopt) = 67{,}634.92$ €. A 1.84 GHz CPU 512 MB Pentium IV, Intel PC, needs about 72.5s to carry out the developed software for one time. As shown in Fig. 7.14, the algorithm reaches convergence after 10 independent runs. As this figure shows, PSO usually achieves the best solution after

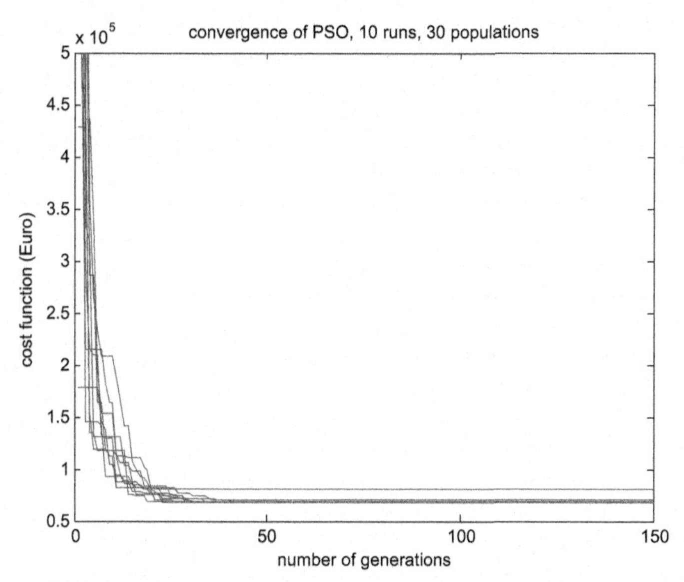

Figure 7.14 PSO algorithm convergence in 10 runs. *PSO*, particle swarm optimization.

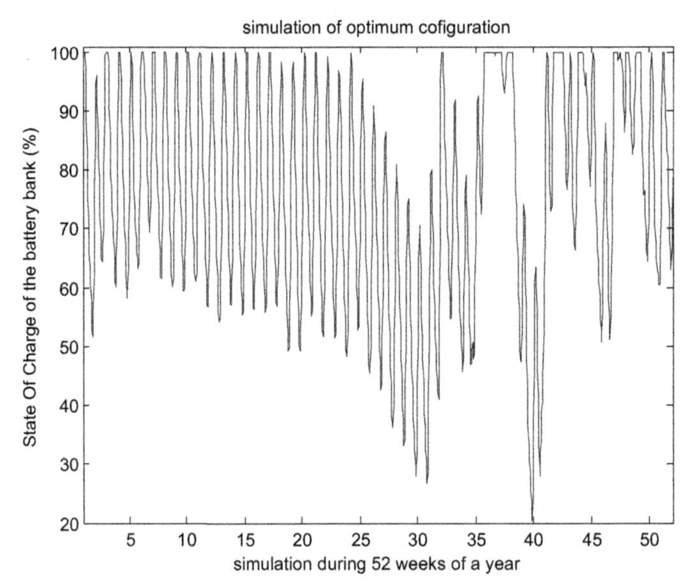

Figure 7.15 Battery bank's state of charge.

50 generations (9 out of 10 runs). A reasonable and acceptable convergence rate of 7 out of 10 runs was observed. Fig. 7.15 illustrates the battery bank status. It can be said that since the SOC does not fall below 20.25, the constraint (Eq. 7.19) is always satisfied.

Furthermore, the developed program determines the optimal combination of WG-type and PV-type systems to illustrate the economic benefits of hybrid system. Table 7.2 presents the results. This table demonstrates how solar and wind energy complement each other to increase solution reliability and decrease the costs. Additionally, it shows that PV units are

Table 7.2 Hybrid and Wg- and PV-type systems optimum combination.

	Elements no.			
Type of network	**WGs**	**PVs**	**Batteries**	**Total costs (€)**
Hybrid	3	31	9	67634.9
PV-type unit	—	57	7	80315.1
WG-type system	24	—	22	123585.0

PV, photovoltaic; *WG*, wind generation.

more economic than WG systems in the region under study, contrary to the results of Ref. [1] because the energy content of irradiation is less than that of wind in region of Ref. [1], which is located in Greece.

5.1.3 Comparison with genetic algorithm

The results of the PSO algorithm and those of the GA toolbox of MATLAB (GA tool) are compared here to determine the effectiveness of the algorithm. The GA is a popular intelligent algorithm applied to many complex applications. Here, it is assumed that the GA population and generations represent those of the PSO population and generations, namely 30 and 150, respectively, and to avoid reaching local minima, 10 attempts are made at running the algorithm. Fig. 7.16 shows that GA converges to optimal solutions within about 100 generations. PSO converges in approximately half of the number of generations compared with GA as indicated in Fig. 7.14. Additionally, GA optimal configuration is related to the cost function of $J(X_{opt}) = 70140.18$ €, which is circa 2500 € more expensive than PSO optimum solution. Table 7.3 shows that PSO outperforms GA in all aspects, including convergence and calculations times.

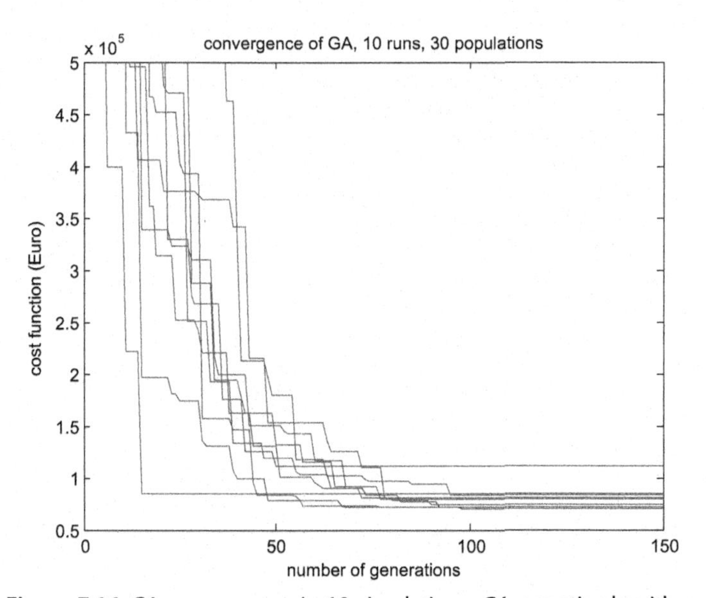

Figure 7.16 GA convergence in 10 simulations. *GA*, genetic algorithm.

Table 7.3 PSO and genetic algorithm evaluation in 10 simulations.

	OF (€)				Calculation time
Optimization by:	**Best**	**Worst**	**Average**	**Std**	**(s)**
PSO	67,634	80,315	69,310	3,980	72.5 s
GA	70,140	110,560	81,228	11,860	146.2 s

GA, genetic algorithm; *OF*, objective function; *PSO*, particle swarm optimization.

5.2 Case study 2: hybrid system including photovoltaic/wind generation/fuel cell

A year-long simulation with 1-h time steps is required for reliability/cost assessments of the region under study. The same data of pervious subsection for wind speed and solar irradiation are used. The IEEE RTS data [51], with a peak power of 50 kW, are used for load profile. The wind is calculated at WG system height of 15 m. Figs. 7.17–7.19 illustrate these data. Table 7.4 summarizes the data of the elements of the network [7,43,52,55,56]. In this study, at the start of the simulation, the stored H_2,

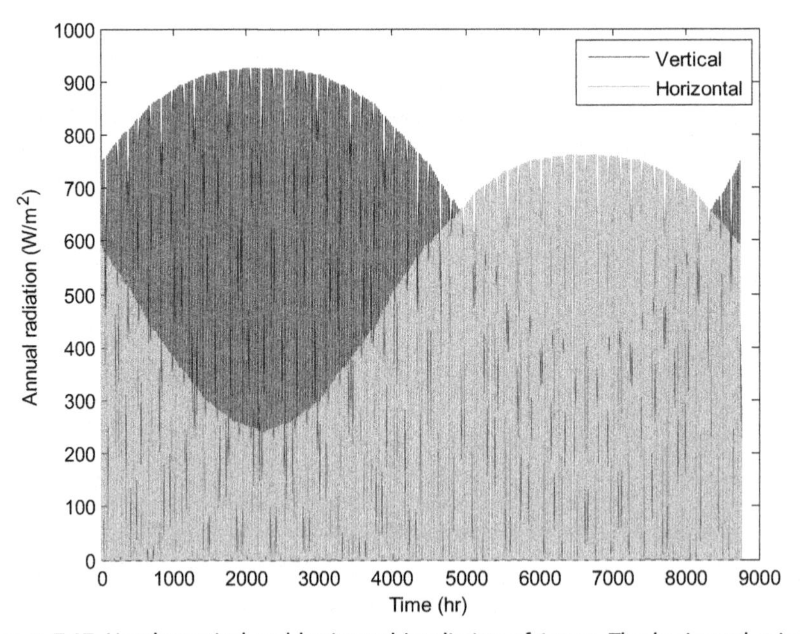

Figure 7.17 Hourly vertical and horizontal irradiation of 1 year. The horizontal axis is derived from the event of March 21, the first hour of the solar year.

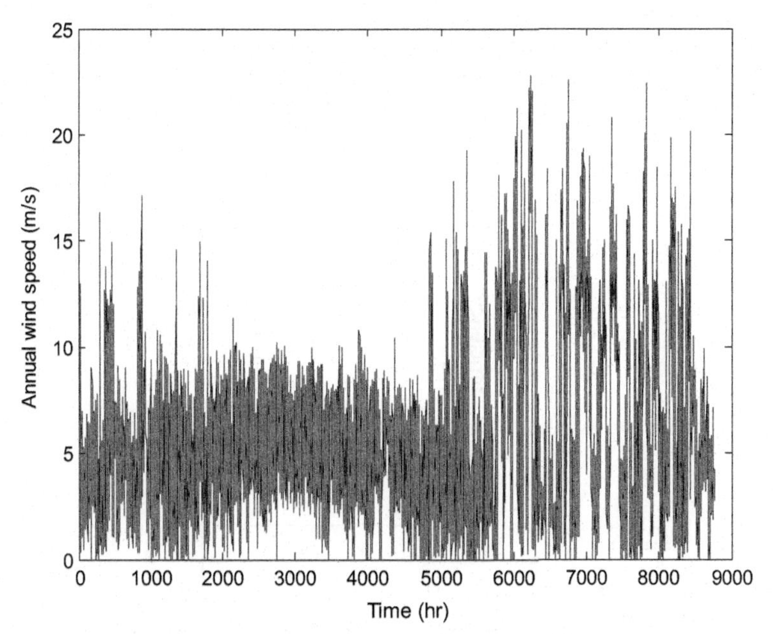

Figure 7.18 Hourly wind speed of 1 year (height = 15 m).

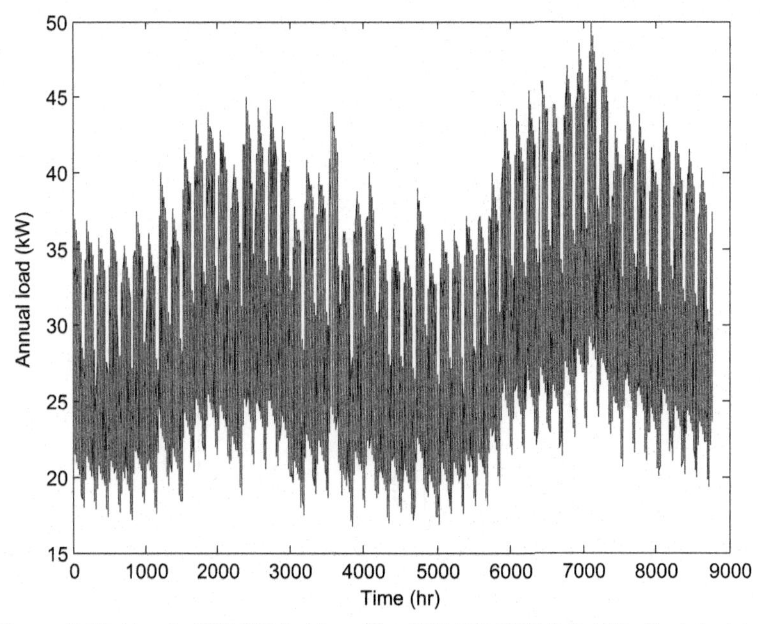

Figure 7.19 Hourly IEEE RTS load profile. *IEEE RTS*, IEEE Reliability Test System.

Table 7.4 Elements data [7,43,52,55,56].

Element	Capital cost (US$ per unit)	Replacement Cost (US$ per unit)	Operation &maintenance cost (US$ per Unit-year)	Lifetime (year)	Efficiency (%)	Availability	Rating
Wind generation system	19,400	1500	75	20	—	96	7.5 kW
Photovoltaic	7000	6000	20	20	—	96	1 kW
Electrolyzer	2000	1500	25	20	0.75	100	1 kW
H_2 tank	1300	1200	15	20	0.95	100	1 kg
Fuel cell	3000	2500	175	5	0.50	100	1 kW
Converter	800	750	8	15	0.90	99.89	1 kW

i.e., E_{tank} $(t-1)$ for $t = 1$ in (625), is 1/2 of its nominal capacity. The initial amount of stored hydrogen is equal to the amount of hydrogen stored at the end of the previous time step for another time step ($t > 1$) that is greater than 1. It is important to mention that the capital cost of the hydrogen tank includes US$1.8/kg of initial hydrogen [48].

5.2.1 Base case

The base case is simulated by a Pentium IV PC, which has 3.2 GHz CPU and 2 GB RAM. State probabilities are 10^{-16} and negligible; they are not enumerated, to save the computation time. For example, in a hybrid system including 14 WG systems and 199 PV units with availabilities of 96%, the number of enumerated states will reduce using this method from 6000 (15×200) to 252, and therefore, about 50% of computation time will be saved. Despite the evident benefits of this method, it cannot be useful for a small hybrid system due to a low number of low probability states. For example, a stateless probability of 10^{-16} can be observed when considering 3 WG systems and 5 PV units. The software finds the optimal combination within 36 h.

Unlike other optimization algorithms, the MOPSO algorithm is not limited by any constraint. To make a proper fitness function, OF and constraints have to be integrated. Also, the software never violates constraints (Eqs. 7.51 and 7.52), because it limits variables within their bounds. In this sense, the fitness function is seen as sum of the OF (Eq. 7.47), together with two penalty factors related to the inequality constraints (Eqs. 7.50 and 7.53). According to this discussion, each violated inequality constraint in the model is multiplied by 1010, and the excess amount is then added to the OF (Eq. 7.47). There are three systems, viz. hybrid system including WG systems and PV units, WG-type systems, and PV-type units. It is assumed that WG systems and PV units are 96% available. The results are presented in Tables 7.5 and 7.6. There are no reliability inequality constraints for mentioned systems of Table 7.6. Actually, due to high costs of loss of load, design of a reliable (as a result, expensive) system makes economic sense.

Figs. 7.20–7.22 show the Pareto frontier and trade-offs of sizing of combination of WG and PV units. Other cases (such as 100% available components) can be tested to determine the Pareto-optimal set and the best trade-off solution. In these figures, it is obvious that the Pareto set is equally scattered along cost, LOLE, and LOEE axes for hybrid and PV-type

Table 7.5 Optimal combination in hybrid, WG-type, and PV-type systems (base case).

		Optimum design						
Network type	**Trade-off solution**	N_{WG}	N_{PV}	P_{el} (kW)	M_{tank} (kg)	P_{FC} (kW)	P_{inv} (kW)	θ_{PV}
Hybrid	Best trade-off I	9	251	126.760	192.2350	48.1520	49.832	29.994
	Best trade-off II			127.019	193.0630	49.0300	50.504	29.970
	Minimum cost			126.155	191.2220	45.4890	48.939	30.075
	Minimum LOEE			127.533	193.8170	50.9084	51.486	29.909
	Minimum LOLE			127.533	193.8170	50.9084	51.486	29.909
WG-type	Best trade-off I	204	—	354.584	1009.295	55.993	53.902	-
	Best trade-off II			354.584	1009.295	55.993	53.902	
	Minimum cost			354.610	1009.212	55.943	53.872	
	Minimum LOEE			354.584	1009.295	55.993	53.902	
	Minimum LOLE			354.584	1009.295	55.993	53.902	
PV-type	Best trade-off I	—	340.016	167.665	200.9810	49.019	49.816	57.230
	Best trade-off II		340.040	167.642	201.8710	49.683	50.146	57.193
	Minimum cost		340.298	167.687	193.4050	45.540	47.930	57.418
	Minimum LOEE		339.936	167.003	212.2130	55.778	53.037	56.862
	Minimum LOLE		339.936	167.003	212.2130	55.778	53.037	56.862

LOEE, loss of energy expected; *LOLE*, loss of load expected; *PV*, photovoltaic; *WG*, wind generation.

Table 7.6 Optimum (reliability and cost) indexes in hybrid, WG-type, and PV-type networks (base case).

		Reliability				Cost			E_{tank}	
Network type	Trade-off solution	LOLE (hour/year)	LOEE (MWh/year)	ELF	LPSP	Investment (MUS$)	J_{COST} (US$/year)	AC_{loss} (US$/year)	$E_{tank}(0)$	$E_{tank}(8760)$
Hybrid	Best trade-off I	34	25	0.00006196	0.00008813	3,183,155	2592947.76	228.73880	3761.05	6923.076
	Best trade-off II	9	5.84835	0.00001473	0.00002173	3,195,813	2597714.82	32.730180	3777.48	6960.692
	Minimum cost	154.93775	229.43648	0.00060050	0.00085257	3,149,071	2581715.00	1285.336	3740.933	6660.267
	Minimum LOEE	0.518042	0.13885	3.40871172e-13	5.15959e-13	3,220,598	2606141.00	0	3792.438	6777.453
	Minimum LOLE	0.518042	0.13885	3.40871172e-13	5.15959e-13	3,220,598	2606141.00	0	3792.438	6777.453
WG-type	Best trade-off I	0	0	0	0	7,222,327	6170553.45	0	19,979.7	31,875.620
	Best trade-off II					7,222,327	6170553.45		19,979.7	31,875.620
	Minimum cost					7,221,651	6170322.00		19,978.03	31,872.990
	Minimum LOEE					7,222,327	6170553.00		19,979.69	31,876.280
	Minimum LOLE					7,222,327	6170553.00		19,979.69	31,876.280
PV-type	Best trade-off I	30	21.360	0.00005043	0.000075	3,792,828	3.16E+06	210.4686	3934.65	7454.6700
	Best trade-off II	15	10.200	0.00002523	0.00003214	3,802,057	3166882.15	82.49374	3952.31	7490.0080
	Minimum cost	227.08183	361.30623	0.00094232	0.00134260	3,739,635	3141763.00	2026.4700	3784.262	7150.9890
	Minimum LOEE	0	0	0	0	3,881,517	3192644.00	0	4157.607	7897.7040
	Minimum LOLE	0	0	0	0	3,881,517	3192644.00	0	4157.607	7897.7040

ELF, equivalent loss factor; *LOEE*, loss of energy expected; *LOLE*, loss of load expected; *LPSP*, loss of power supply probability; *PV*, photovoltaic; *WG*, wind generation.

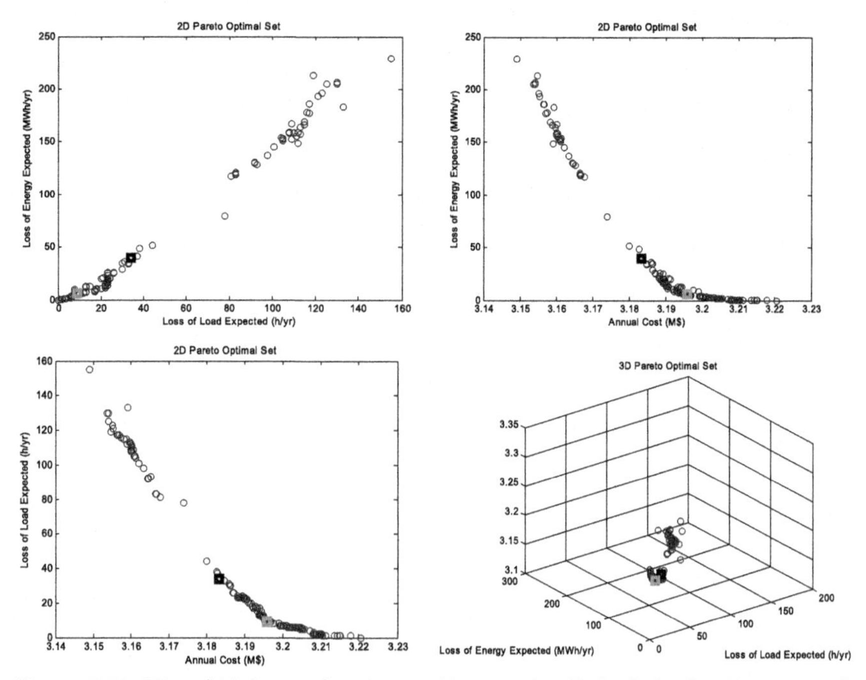

Figure 7.20 2D and 3D Pareto frontiers and best trade-offs for hybrid WG/PV network (base case). *WG*, wind generation; *PV*, photovoltaic.

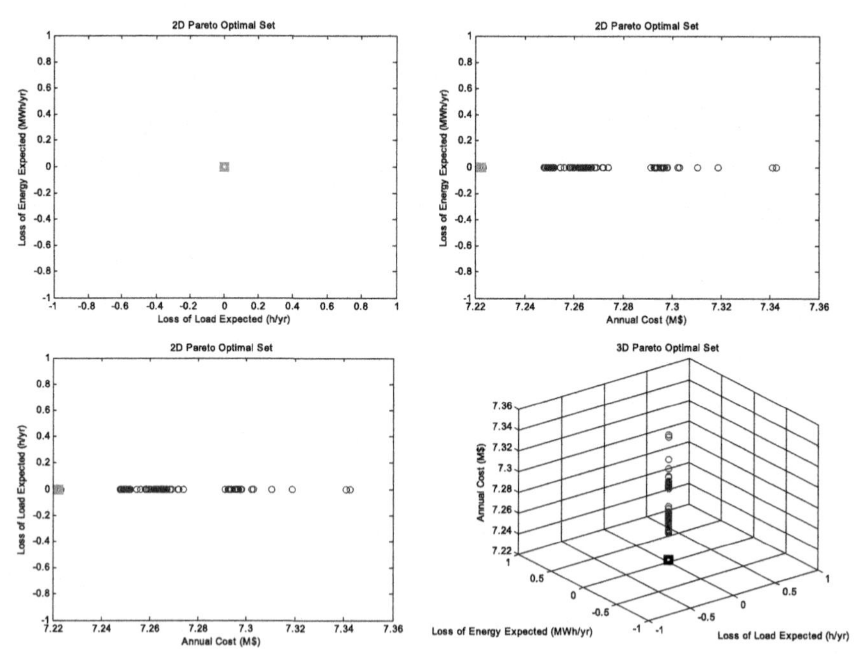

Figure 7.21 2D and 3D Pareto frontiers and best trade-offs for hybrid WG/PV network (base case). *WG*, wind generation; *PV*, photovoltaic.

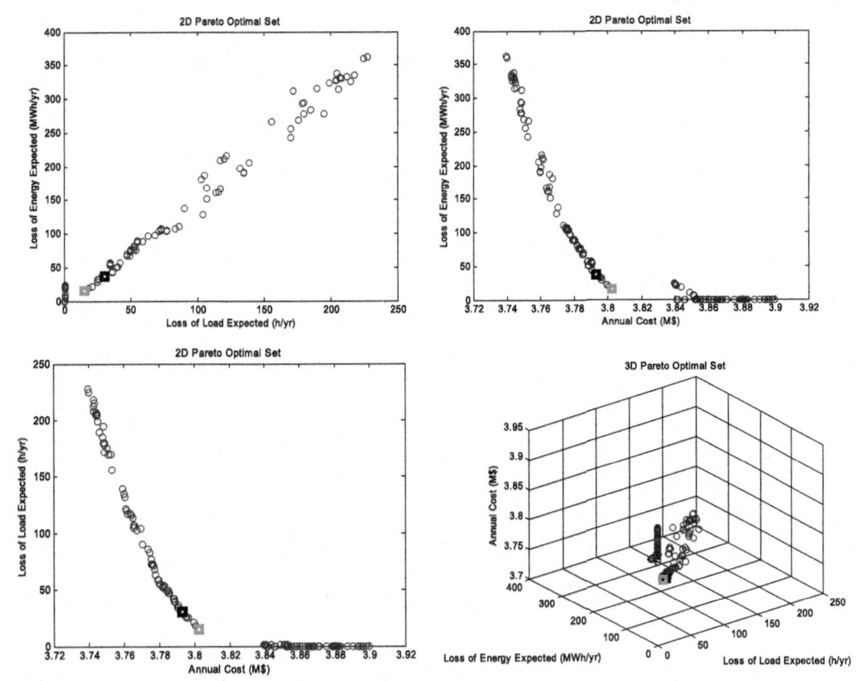

Figure 7.22 2D and 3D Pareto frontiers and best trade-offs for PV-type network (base case). *PV*, photovoltaic.

systems; this means that the trade-off criteria will result in different solutions. Fig. 7.21 illustrates the Pareto frontier for the WG-type system, and the trade-off criteria result in the same solutions along the axes of the cost, LOLE, and LOEE.

Fig. 7.23 shows the hourly expected amounts of the stored energy, during a year, in the hydrogen tank, for hybrid, WG-type, and PV-type systems, since reliable supply at each time step depends greatly on the stored energy. During the hours, when hydrogen is saved to its minimum allowable level, the likelihood of load loss is the greatest. Other factors that impact the reliability include the amount of energy consumed hourly and the conditions of the stored energy.

5.2.2 100% available components

In previous section, it has been mentioned that many researches do not account for outage probabilities for determination of optimum size of

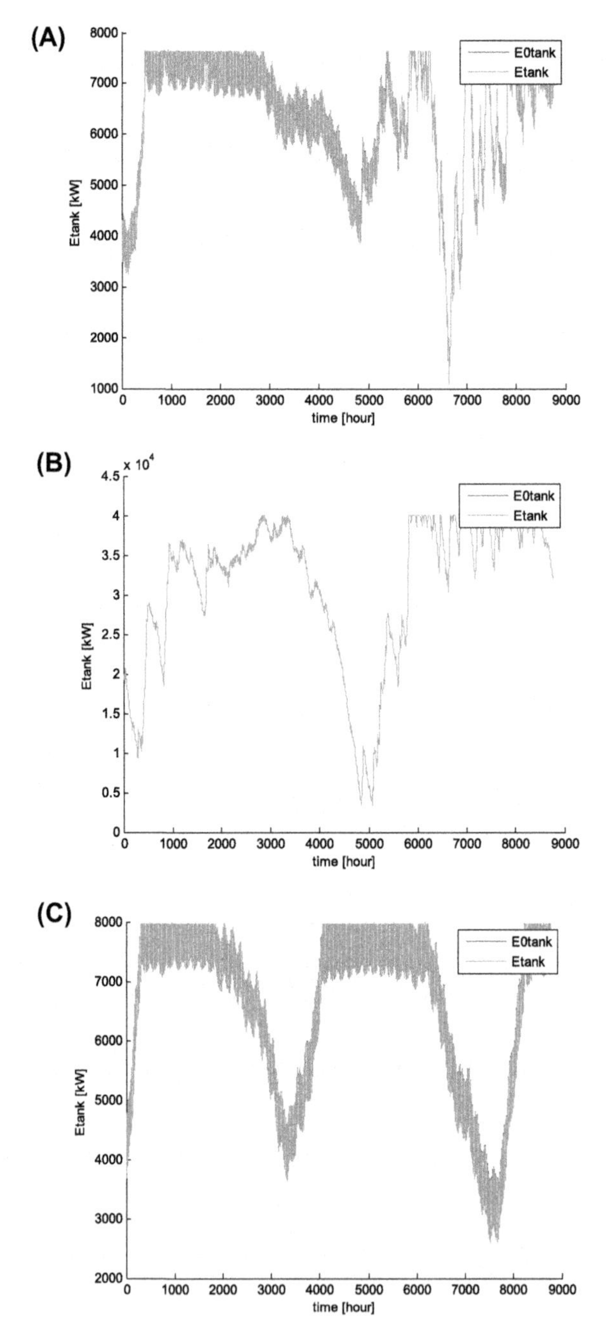

Figure 7.23 Yearly H2 tank expected stored energy in (A) hybrid WG/PV, (B) WG-type, and (C) PV-type networks. *PV*, photovoltaic; *WG*, wind generation.

hybrid WG/PV networks, and the system elements are typically considered to be 100% available. This problem has been solved under the assumption of 100% availability of all components to understand how component outage probabilities affect the cost and reliability. In comparison with the base case, the optimal combination, cost, and reliability are calculated and listed in Tables 7.7 and 7.8. It can be seen in these tables that it is possible to meet the load with higher reliability and less cost using a smaller network. It turns out that due to potential failures in WG systems, PV units, and converters, the system loses 3% in annual cost, but 1% in DC/AC converters and 5% in WG systems. The probability of component failure can increase investment costs by US\$ 64,809, US\$ 182,075, and US\$ 89,930 for hybrid WG/PV, WG-type, and PV-type networks, respectively. As mentioned earlier, failure probabilities deteriorate reliability, a change in the ELF index from −0.00005850 to −0.00001486 for PV-type network.

As mentioned, all the system elements were 100% available, so all methods would yield the same results. There can only be one feasible state for the system at each time step.

5.2.3 Impact of DC/AC convertor on reliability

In this section, the optimal solution is subject to the reliability constraint of ELF = 0.0001, to determine the impact of the inverter on reliability [44]. A run of the program reveals that this considerable big amount of fitness is due to the application of the heavy penalty term assigned to the reliability inequality constraint. Further analysis shows that ELF can drop to zero for the system. This is directly attributed to the fact that the inverter is the unique cut-set of the hybrid network and, therefore, plays a critical role in its reliability diagram. Given A_{inv}, and A_{others} as availability of the converter and all other components including energy resources, respectively, and assuming P_{supply} as probability of supplying the load, its top value can be calculated, as follows:

$$P_{\text{supply}} = A_{others} \times A_{inv} \Rightarrow \lim_{A_{other} \to 1} \left(P_{\text{supply}}\right) = A_{inv} \tag{7.64}$$

Table 7.7 Optimum design of hybrid, WG-type, and PV-type networks (with 100% available elements).

		Optimum design						
Type	**Solution**	N_{WG}	N_{PV}	P_{el} (kW)	M_{tank} (kg)	P_{FC} (kW)	P_{inv} (kW)	θ_{PV}
Hybrid WG//PV	Best trade-off I	7	246	127.259	187.027	50.692	50.012	33.0616
	Best trade-off II		246	127.259	187.027	50.692	50.012	33.0616
	Minimum cost		246	127.362	186.989	50.931	50.180	33.0250
	Minimum LOEE		245	127.199	189.839	53.896	51.141	32.2510
	Minimum LOLE		245	127.199	189.839	53.896	51.141	32.2510
WG-type	Best trade-off I	193	—	355.709	1002.344	62.006	54.744	—
	Best trade-off II			355.709	1002.344	62.006	54.744	
	Minimum cost			353.766	999.374	56.303	52.548	
	Minimum LOEE			355.709	1002.344	62.006	54.744	
	Minimum LOLE			355.709	1002.344	62.006	54.744	
PV-type	Best trade-off I	—	331	169.397	174.897	50.0050	50.510	57.1860
	Best trade-off II			169.579	175.519	50.139	50.577	57.2000
	Minimum cost			169.015	176.185	49.516	50.172	57.1910
	Minimum LOEE			169.624	175.467	50.169	50.598	57.2010
	Minimum LOLE			169.967	174.540	50.524	50.844	57.2000

LOEE, loss of energy expected; *LOLE*, loss of load expected; *PV*, photovoltaic; *WG*, wind generation.

Table 7.8 Optimum cost and reliability indexes for Hybrid, WG-type, and PV-type networks (with 100% available elements).

		Reliability indixes				Cost			E_{tank}	
Type	**Solution**	**LOLE (hour/year)**	**LOEE (MWh/year)**	**ELF**	**LPSP**	**Investment (MUS$)**	**J_{COST} (US$/year)**	**AC_{loss} (US$/year)**	**$E_{tank}(0)$**	**$E_{tank}(8760)$**
Hybrid	Best trade-off I	0	0	0	0	3,134,902	2528139.39	0	3657.67	6853.478
	Best trade-off II					3,134,902	2528139.39	0	3657.67	6853.478
	Minimum cost					3,130,699	2522146.00	304.56680	3656.91	6630.147
	Minimum LOEE					3,182,522	2547590.00	0	3713.481	6731.608
	Minimum LOLE					3,182,522	2547590.00	0	3713.481	6731.608
WG-type	Best trade-off I	0	0	0	0	7,077,828	5988477.93	0	19,841.70	31,542.21
	Best trade-off II					7,077,828	5988477.93		19,841.70	31,542.21
	Minimum cost					7,001,743	5961866.00		19,782.75	31,422.36
	Minimum LOEE					7,077,828	5988478.00		19,841.70	31,542.21
	Minimum LOLE					7,077,828	5988478.00		19,841.70	31,542.21

Continued

Table 7.8 Optimum cost and reliability indexes for Hybrid, WG-type, and PV-type networks (with 100% available elements).—cont'd

		Reliability indixes				Cost			E_{tank}	
Type	**Solution**	**LOLE (hour/year)**	**LOEE (MWh/year)**	**ELF**	**LPSP**	**Investment (MUS$)**	**J_{COST} (US$/year)**	**AC_{loss} (US$/year)**	**$E_{tank}(0)$**	**$E_{tank}(8760)$**
PV-type	Best trade-off I	9.1	6.20000	0.00001484	0.00002300	3,703,402	3073583.20	228.61105	3416.880	6420.145
	Best trade-off II	7	4.90000	0.00001221	0.00002008	3,706,363	3075210.28	28.15963	3429.230	6444.847
	Minimum cost	6	5.02850497	0.00001243	0.00001886	3,691,224	3065754.00	639.08870	3442.451	6468.211
	Minimum LOEE	6	4.86651195	0.00001203	0.00001808	3,706,752	3075339.00	135.39720	3428.188	6439.889
	Minimum LOLE	5	3.02194714	0.00000075	0.00001123	3,710,460	3,076,082	149.1336	3409.787	6403.086

ELF, equivalent loss factor; *LOEE*, loss of energy expected; *LOLE*, loss of load expected; *LPSP*, loss of power supply probability; *PV*, photovoltaic; *WG*, wind generation.

To increase the reliability of such a system, two remedies are usually available. Two options exist: one is using an inverter with a higher reliability (higher availability), and the another one is using multiple inverters in parallel. Tables 7.9 and 7.10 are produced by the developed program in the case of having 100% available inverters.

5.2.4 Accurate and approximate methods comparison

An approximate approach was used to compare the accuracy with that of the accurate one. The optimization process was completed within 7 h in this case, which is much shorter than 20 h of the accurate method. The results are given in Table 7.11. Due to only the differences in reliability calculations between approximate and accurate methods, they both lead to the same cost for investment, replacement, and O&M. According to Table 7.8, the reliability indexes of the approximate method are similar to the reliability indexes of the accurate one. A percentage error magnitude is observed to be almost below 10%. The fact that there is a negative percentage error in the approximate method is evidence that the system is estimated overly optimistically. Other experiments show that the approximate approach consistently produces lower cost and higher reliability in comparison with the accurate one.

5.2.5 Proposed multiple-objective particle swarm optimization and single-objective optimization algorithms comparison

In this subsection, the proposed MOPSO and SOO algorithms [38] are compared to study their efficiency. The results are provided in Tables 7.12 and 7.13. We can see that the proposed one provides a superior solution, but its reliability index is low. This table indicates that with less reliability indices, the MOO algorithm has a better solution. As an example, LPSP minimum is 0.000156810, 0.030 and 0.0050150 in Refs. [38,77] and [34], respectively, but for the base case and the case with 100% availability, it is 0.000088130 and 0.0, respectively.

Also, it can be observed that minimum LOEE (EENS) is 5.8483450 (MWh/year) and 0.0 for the base case and the case with 100% availability, respectively. These values are somewhat less than those of Ref. [78,79]. Also, the MOO performs better in finding a globally optimal solution compared with the conventional search algorithms [34,38,78,79] using the trade-off approaches (best trade-off I and II) [34,38,78,79].

Table 7.9 Optimum design of hybrid, WG-type, and PV-type networks ($A_{inv} = 1$ and $ELF_{max} = 0.0001$).

Type	Solution	Optimum design						
		N_{WG}	N_{PV}	P_{el} (kW)	M_{tank} (kg)	P_{FC} (kW)	P_{inv} (kW)	θ_{PV}
Hybrid	Best trade-off I	7	252	128.598	190.624	47.931	49.635	31.305
	Best trade-off II			128.600	190.635	47.969	49.658	31.305
	Minimum cost			128.586	190.620	47.884	49.612	31.302
	Minimum LOEE			128.599	190.637	47.971	49.659	31.304
	Minimum LOLE			128.598	190.637	47.970	49.659	31.304
WG-type	Best trade-off I	200	—	357.205	1000.669	49.013	49.800	—
	Best trade-off II			357.439	1000.507	49.307	50.270	
	Minimum cost			357.050	1000.460	47.745	49.578	
	Minimum LOEE			357.598	1000.699	52.039	50.253	
	Minimum LOLE			357.594	1000.667	52.038	50.327	
PV-type	Best trade-off I	—	346	172.158	167.281	50.744	50.147	57.224
	Best trade-off II		346	172.601	166.431	51.670	50.487	57.162
	Minimum cost		345	170.380	173.483	48.925	49.063	57.314
	Minimum LOEE		347	173.190	164.526	52.331	50.872	57.125
	Minimum LOLE		347	173.190	164.526	52.331	50.872	57.125

LOEE, loss of energy expected; *LOLE*, loss of load expected; *PV*, photovoltaic; *WG*, wind generation.

Table 7.10 Optimum cost and reliability indexes of hybrid, WG-type, and PV-type networks ($A_{inv} = 1$ and $ELF_{max} = 0.0001$).

		Reliability indexes				Cost			E_{tank}	
Type	**Solution**	**LOLE (hour/year)**	**LOEE (MWh/year)**	**ELF**	**LPSP**	**Investment (MUS$)**	**J_{COST} (US$/year)**	**AC_{loss} (US$/year)**	**$E_{tank}(0)$**	**$E_{tank}(8760)$**
Hybrid	Best trade-off I	44.30959	32.15181	0.000081820	0.0001190	3,176,238	2,588,307	182.5857	3729.073	6712.936
	Best trade-off II	42.30947	30.68983	0.000078068	0.0001140	3,176,722	2,588,459	174.055	3729.278	6713.378
	Minimum cost	48.55628	34.04078	0.000086681	0.0001260	3,175,652	2,588,120	193.934	3728.985	6712.557
	Minimum LOEE	42.30947	30.60183	0.000077843	0.0001140	3,176,750	2,588,467	173.693	3729.318	6713.428
	Minimum LOLE	42.30947	30.62980	0.000077915	0.0001140	3,176,741	2,588,464	173.895	3729.323	6713.429
WG-type	Best trade-off I	9	13.806811	4.0083175e-05	5.1305238e-5	7,069,553	6,082,159	58.9730	19,808.460	31,442.89
	Best trade-off II	6	8.8288060	2.190914e-05	3.280728e-5	7,073,804	6,083,675	48.1930	19,805.230	31,436.44
	Minimum cost	46	37.401437	0.000103290	0.0001390	7,054,331	6,077,595	204.3060	19,804.300	31,434.57
	Minimum LOEE	1	0.9480587	2.254001e-06	3.5229256e-6	7,105,097	6,092,423	5.30900	19,809.060	31,444.09
	Minimum LOLE	1	0.9487957	2.255754e-06	3.5256647e-6	7,105,132	6,092,430	5.31300	19,808.420	31,442.81

Continued

Table 7.10 Optimum cost and reliability indexes of hybrid, WG-type, and PV-type networks ($A_{inv} = 1$ and $ELF_{max} = 0.0001$).—cont'd

Type	Solution	Reliability indexes				Cost			E_{tank}	
		LOLE (hour/year)	LOEE (MWh/year)	ELF	LPSP	Investment (MUS$)	J_{COST} (US$/year)	AC_{loss} (US$/year)	$E_{tank}(0)$	$E_{tank}(8760)$
PV-type	Best trade-off I	4	2.1297600	5.246075e-06	7.91405e-06	3,815,387	3,176,132	11.9690	3265.707	6115.158
	Best trade-off II	0	0	0	0	3,825,983	3,178,964	0	3248.838	6081.447
	Minimum cost	33.	26.94925	6.68782e-05	0.000100	3,791,526	3,167,315	153.9	3388.825	6361.156
	Minimum LOEE	0	0	0	0	3,839,682	3,186,956	0	3211.028	6006.041
	Minimum LOLE	0	0	0	0	3,839,682	3,186,956	0	3211.028	6006.041

ELF, equivalent loss factor; *LOEE*, loss of energy expected; *LOLE*, loss of load expected; *LPSP*, loss of power supply probability; *PV*, photovoltaic; *WG*, wind generation.

Table 7.11 Accurate (Acc.) and approximate (App.) methods comparison in hybrid, WG-type, and PV-type networks.

Type	Solution	Method and error (E%)	Indexes				Cost		
			LOLE (hour/year)	LOEE (MWh/year)	ELF	LPSP	Investment (MUS$)	J_{COST} (US$/year)	AC_{loss} (US$/year)
Base case									
Hybrid	Best trade-off I	Acc.	29	21.50	5.38e-05	7.99e-05	3,792,827.535	3,163,513.154	203.66100
		App.	30	36.40	0.00011	0.000135	3,792,827.535	3,160,000	120.46860
		E%	−3.48	−69.06	−105.080	−69.0600	0	0	40.8484720
	Best trade-off II	Acc.	9.12	5.840	1.47e-05	2.17e-05	3,195,813.245	2,597,714.822	32.75073280
		App.	9.00	5.850	1.47e-05	2.17e-05	3,195,813.245	2,600,000	32.73017597
		E%	1.29	−0.063	−0.06190	−0.06281	0	0	0.062767540
WG-type	Best trade-off I	Acc.	0	0	0	0	7,222,326.555	6,170,553.454	0
		App.					7,222,326.555	6,170,000	
		E%					0	0	
	Best trade-off II	Acc.					7,222,327	6,170,553	
		App.					7,222,327	6,170,553	
		E%					0	0	
PV-type	Best trade-off I	Acc.	29	21.50	5.38e-05	7.99e-05	3,792,827.535	3,163,513.154	203.6610
		App.	30	36.40	0.000110	0.00013	3,792,827.535	3,160,000	120.4686
		E%	−3.45	−69.06	−105.080	−69.060	0	0	40.84850
	Best trade-off II	Acc.	14	+9.37	2.33e-05	3.48e-05	3,802,057.042	3,166,882.150	90.63400
		App.	15	16.20	4.92e-05	6.01e-05	3,802,057.042	3,170,000	52.49400
		E%	−7.14	−72.66	−111.230	−72.660	0	0	42.08100

Continued

Table 7.11 Accurate (Acc.) and approximate (App.) methods comparison in hybrid, WG-type, and PV-type networks.—cont'd

			Indexes				Cost		
Type	**Solution**	**Method and error (E%)**	**LOLE (hour/year)**	**LOEE (MWh/year)**	**ELF**	**LPSP**	**Investment (MUS$)**	**J_{COST} (US$/year)**	**AC_{loss} (US$/year)**
Full availability (100%)									
Hybrid	Best	Acc.	0	0	0	0	3,134,902	2,528,139	0
	trade-	App.					3,134,902	2,528,139	
	off I	E%					0	0	
	Best	Acc.					3,134,902	2,528,139	
	trade-	App.					3,134,902	2,528,139	
	off II	E%					0	0	
WG-type	Best	Acc.	0	0	0	0	7,077,828	5,988,478	0
	trade-	App.					7,077,828	5,988,478	
	off I	E%					0	0	
	Best	Acc.					7,077,828	5,988,478	
	trade-	App.					7,077,828	5,988,478	
	off II	E%					0	0	
PV-type	Best	Acc.	9.00	6.00	1.49e-05	2.23e-05	3,703,402	3,073,583	35.417800
	trade-	App.	10.10	5.60	1.53e-05	0.0000218	3,703,402	3.07E+06	33.611050
	off I	E%	−12.22	6.666667	−2.68456	2.242150	0	0	5.7024600
	Best	Acc.	6.00	5.03	1.24e-05	1.87e-05	3,706,363	3,075,210	139.67260
	trade-	App.	6.60	4.90	1.32e-05	1.72e-05	3,706,363	3.08E+06	128.15963
	off II	E%	−10.0	−7.54	−6.45161	8.021390	0	0	7.9136690

ELF, equivalent loss factor; *LOEE*, loss of energy expected; *LOLE*, loss of load expected; *LPSP*, loss of power supply probability; *PV*, photovoltaic; *WG*, wind generation.

Table 7.12 Optimum design comparison between suggested MOPSO and SOO algorithms [38] for hybrid case.

		Optimum design						
		N_{WG}	N_{PV}	P_{el} (kW)	M_{tank} (kg)	P_{FC} (kW)	P_{inv} (kW)	θ_{PV}
Approach	**Solution**	**Base case**						
Suggested MOPSO	Best trade-off 1 Best trade-off 2	7	246	127.2590	187.0270	50.6920	50.01210	33.06160
SOO	—	8	224	119.440	144.190	43.4310	43.72500	34.12900
		Full availability						
Suggested MOPSO	Best trade-off 1 Best trade-off 2	7	246	127.2590	187.0270	50.6920	50.01210	33.06160
SOO	—	9	214	117.2700	127.100	43.7360	46.46400	33.9930

MOPSO, multiple-objective particle swarm optimization; *SOO*, single-objective optimization.

Table 7.13 Reliability and cost comparison between proposed MOPSO and SOO algorithms [38] for hybrid case.

Approach	Solution	Indexes				Cost		
		LOLE (hour/year)	LOEE (MWh/year)	ELF	LPSP	J_{COST} (US$/year)	Investment (MUS$)	AC_{loss} (US$/year)
		Base case						
Suggested MOPSO	Best trade-off 1	34	25	0.000061960	0.000088130	3,183,155	2592947.760	228.73880
	Best trade-off 2	9	5.8483450	0.000014730	0.000021730	3,195,813	2597714.820	32.730,180
SOO	–	335.960	2.3726670	0.008360360	0.008815270	250,410	2321000	13,287
		Full availability						
Suggested MOPSO	Best trade-off 1 Best trade-off 2	0	0	0	0	3,134,902	2528139.39	0
SOO	–	303	2,202,768.70	0.00788,500	0.008184040	242,648	2,247,700	12,335

ELF, equivalent loss factor; *LOEE*, loss of energy expected; *LOLE*, loss of load expected; *LPSP*, loss of power supply probability; *MOPSO*, multiple-objective particle swarm optimization; *SOO*, single-objective optimization.

6. Conclusion

The primary purpose of a hybrid wind–solar-generating system is to supply power reliably in various atmospheric conditions as low cost as possible. In this chapter, a hybrid system was designed for 20 years' period of operation. A novel application of PSO was used to optimally design the combination of components. A comparison was made between obtained results and those provided by the GA. In terms of accuracy, convergence speed, and complexity, the comparison illustrated that PSO has several advantages over GA. PSO, in summary, provided satisfactory results in terms of hybrid system sizing and is well suited for the task.

Another case study was investigated, which includes a hybrid WG/PV/FC system and a hydrogen tank. This system was designed to operate for 20 years. An optimization algorithm based on a MOPSO was applied to achieve the optimum combination. OFs used for the optimization were the LOLE, LOEE, and LPSP, and an extra reliability constraint was the ELF index. This chapter addressed the reliability of three key elements of a hybrid system, i.e., WG systems, PV units, and inverters, and showed that the hybrid system costs are directly associated with the reliability of those components.

Including failures of another component and uncertainties of wind speed, solar irradiation, and load in reliability and cost assessment will require time-consuming and computationally intensive methods. These aspects can be subject to future studies of researchers. In this chapter, comprehensive software developed in MATLAB environment was presented, which can carry out all the mentioned vast calculations, hourly simulations, assessment, and optimization of reliability indices. Besides wind speeds, solar radiation, and load requirements, the software also requires specifications of system components.

Based on empirical data from a case study, component failures had a significant effect on system reliability and economy. In the design process of any renewable energy system, it is important to consider the possibility of failures, which may result in longer generation outages. Also, the hybrid system reliability diagram revealed that the inverter was very important as it was the unique cut-set. There is evidence that the reliability of the inverter was an upper limit for reliability of this system, therefore impacting how much power it should generate. Also, an approximate reliability assessment method for this case was proposed, which reduced the computations time by a considerable amount.

Using the MOO approach to analyze Pareto-optimal global solutions, it was shown that this algorithm was more effective in determining globally optimized Pareto-optimal solutions as compared with conventional SOO algorithms. A MOO algorithm could obtain a reliable and cost-effective solution for different combinations of the hybrid system.

Appendix

Proof of Eq. (7.4):

Replacing Eq. (7.29) in Eq. (7.40), it can be written as

$$E[P_{ren}] = \overbrace{\sum_{n_{WG}^{fail}=1}^{N_{WG}} \sum_{n_{PV}^{fail}=1}^{N_{PV}} \left\{ \left(N_{WG} - n_{WG}^{fail}\right) \times P_{WG} \times f_p\left(n_{WG}^{fail}, n_{PV}^{fail}\right) \right\}}^{A}$$

$$+ \overbrace{\sum_{n_{WG}^{fail}=1}^{N_{WG}} \sum_{n_{PV}^{fail}=1}^{N_{PV}} \left\{ \left(N_{PV} - n_{PV}^{fail}\right) \times P_{PV} \times f_p\left(n_{WG}^{fail}, n_{PV}^{fail}\right) \right\}}^{B} \tag{A.1}$$

Using Eq. (A.1), we have

$$A = P_{WG} \times \sum_{n_{WG}^{fail}=1}^{N_{WG}} \left\{ \left(N_{WG} - n_{WG}^{fail}\right) \times \sum_{n_{PV}^{fail}=1}^{N_{PV}} f_p\left(n_{WG}^{fail}, n_{PV}^{fail}\right) \right\} \tag{A.2}$$

$$A = P_{WG} \times \sum_{n_{WG}^{fail}=1}^{N_{WG}} \left\{ N_{WG} \times \sum_{n_{PV}^{fail}=1}^{N_{PV}} f_p\left(n_{WG}^{fail}, n_{PV}^{fail}\right) - n_{WG}^{fail} \times \sum_{n_{PV}^{fail}=1}^{N_{PV}} f_p\left(n_{WG}^{fail}, n_{PV}^{fail}\right) \right\} \tag{A.3}$$

$$A = P_{WG} \times N_{WG} \times \overbrace{\sum_{n_{WG}^{fail}=1}^{N_{WG}} \sum_{n_{PV}^{fail}=1}^{N_{PV}} f_p\left(n_{WG}^{fail}, n_{PV}^{fail}\right)} - P_{WG}$$

$$\times \sum_{n_{WG}^{fail}=1}^{N_{WG}} \left\{ n_{WG}^{fail} \times \overbrace{\sum_{n_{PV}^{fail}=1}^{N_{PV}} f_p\left(n_{WG}^{fail}, n_{PV}^{fail}\right)}^{C} \right\}$$

Now substituting Eq. (7.17) in Eq. (A.4), it can be written as

$$C = \binom{N_{WG}}{n_{WG}^{fail}} \times A_{WG}^{N_{WG}-n_{WG}^{fail}} \times (1 - A_{WG})^{n_{WG}^{fail}}$$

$$\times \overbrace{\sum_{n_{PV}^{fail}}^{N_{PV}} \left\{ \binom{N_{PV}}{n_{PV}^{fail}} \times A_{PV}^{N_{PV}-n_{PV}^{fail}} \times (1 - A_{PV})^{n_{PV}^{fail}} \right\}}^{D} \tag{A.5}$$

Based on binomial distribution, $D = 1$ and

$A = P_{WG} \times N_{WG} - P_{WG}$

$$\times \overbrace{\sum_{n_{WG}^{fail}=1}^{N_{WG}} \left\{ n_{WG}^{fail} \times \binom{N_{WG}}{n_{WG}^{fail}} \times A_{WG}^{N_{WG}-n_{WG}^{fail}} \times (1 - A_{WG})^{n_{WG}^{fail}} \right\}}^{E} \tag{A.6}$$

Based on the same distribution for E, and using Eq. (A.7) [39], we have

$$\sum_{x=0}^{n} x \times \binom{n}{x} \times p^{x} \times q^{n-x} = n \times p \tag{A.7}$$

Then the terms A and B are obtained as follows:

$$A = P_{WG} \times N_{WG} \times A_{WG} \tag{A.8}$$

$$B = P_{PV} \times N_{PV} \times A_{PV} \tag{A.9}$$

Thus, Eq. (7.22) has been proved.

References

[1] H. Kamankesh, V.G. Agelidis, A. Kavousi-Fard, Optimal scheduling of renewable micro-grids considering plug-in hybrid electric vehicle charging demand, Energy 100 (1) (2016) 285–297.

[2] X. Fang, S. Ma, Q. Yang, J. Zhang, Cooperative energy dispatch for multiple autonomous microgrids with distributed renewable sources and storages, Energy 99 (15) (2016) 48–57.

[3] H.R. Baghaee, M. Mirsalim, G.B. Gharehpetian, Real-time verification of new controller to improve small/large-signal stability and fault ride-through capability of multi-DER microgrids, IET Gen. Trans. Dist. 10 (12) (2016) 3068–3084, https://doi.org/10.1049/iet-gtd.2016.0315.

[4] H.R. Baghaee, M. Mirsalim, G.B. Gharehpetian, Performance improvement of multi-DER microgrid for small and large-signal disturbances and nonlinear loads: novel complementary control loop and fuzzy controller in a hierarchical droop-based control scheme, IEEE Syst. J. 99 (1) (2016) 1–8, https://doi.org/10.1109/JSYST.2016.2580617.
[5] L. Barelli, G. Bidini, F. Bonucci, A micro-grid operation analysis for cost-effective battery energy storage and RES plants integration, Energy 113 (15) (2016) 831–844.
[6] H.R. Baghaee, M. Mirsalim, G.B. Gharehpetian, Power calculation using RBF neural networks to improve power sharing of hierarchical control scheme in multi-DER microgrids, IEEE J. Emerg. Select. Topics Power Elect. 99 (1) (2016) 1–8, https://doi.org/10.1109/JESTPE.2016.2581762.
[7] E. Koutroulis, D. Kolokotsa, A. Potirakis, K. Kalaitzakis, Methodology for optimal sizing of stand-alone photovoltaic/wind-generator systems using genetic algorithms, Solar Energy 80 (1) (2006) 1072–1088.
[8] R. Cozzolino, L. Tribioli, G. Bella, Power management of a hybrid renewable system for artificial islands: a case study, Energy 106 (1) (2016) 774–789.
[9] Z. Shi, R. Wang, T. Zhang, Multi-objective optimal design of hybrid renewable energy systems using preference-inspired co-evolutionary approach, Solar Energy 118 (1) (2015) 96–106.
[10] A. Kashefi Kaviani, G.H. Riahy, S.H.M. Kouhsari, Optimal design of a reliable hydrogen-based stand-alone wind/PV generation system, in: Proceeding of 11th International Conference on Optimization of Electrical and Electronic Equipment (OPTIM'08), Brasov, Romania; May 22–24, 2008.
[11] W. Wu, V.I. Christiana, S.A. Chen, J.J. Hwang, Design and techno-economic optimization of a stand-alone PV(photovoltaic)/FC (fuel cell)/battery hybrid power system connected to a wastewater-to-hydrogen processor, Energy 84 (1) (2016) 462–472.
[12] A. Maleki, F. Pourfayaz, M.A. Rosen, A novel framework for optimal design of hybrid renewable energy-based autonomous energy systems: a case study for Namin, Iran, Energy 98 (1) (2016) 168–180.
[13] M.G. Rodrigues, R. Godina, S.F. Santos, A.W. Bizuayehu, J. Contreras, J.P.S. Catalão, Energy storage systems supporting increased penetration of renewables in islanded systems, Energy 75 (1) (2016) 265–280.
[14] S. Ozlu, I. Dincer, Performance assessment of a new solar energy-based multi-generation system, Energy 112 (1) (2016) 164–178.
[15] A. Kashefi Kaviani, H.R. Baghaee, G.H. Riahy, Design and optimal sizing of a photovoltaic/wind-generator system using particle swarm optimization, in: Proceedings of the 22nd Power System Conference (PSC), Tehran, Iran; December 19–21, 2007.
[16] H. Saboori, R. Hemmati, M. Ahmadi-Jirdehi, Reliability improvement in radial electrical distribution network by optimal planning of energy storage systems, Energy 93 (2) (2015) 2299–2312.
[17] D. Pavković, M. Hoić, J. Deur, J. Petrić, Energy storage systems sizing study for a high-altitude wind energy application, Energy 76 (1) (2014) 91–103.
[18] R. Dufo-López, J.L. Bernal-Agustín, Multi-objective design of PV–wind–diesel–hydrogen–battery systems, Renew. Energy 33 (12) (2008) 2559–2572.
[19] A. Mills, S. Al-Hallaj, Simulation of hydrogen-based hybrid systems using hybrid2, Int. J. Hydrogen Energy 29 (2004) 991–999.
[20] A.S. Raj, P.C. Ghosh, Standalone PV-diesel system vs. PV-H2 system: an economic analysis, Energy 42 (1) (2012) 270–280.
[21] R. Belfkira, L. Zhang, G. Barakat, Optimal sizing study of hybrid wind/PV/diesel power generation unit, Solar Energy 85 (1) (2011) 100–110.

[22] A. Rajabi-Ghahnavieh, S.A. Nowdeh, Optimal PV–FC hybrid system operation considering reliability, Int. J. Electr. Power Energy Syst. 60 (2014) 325–333.
[23] W.D. Kellogg, M.H. Nehrir, G. Venkataraman, V. Gerez, Generation unit sizing and cost analysis for stand alone wind, photovoltaic and hybrid wind/PV systems, IEEE Trans. Energy Convers. 13 (1) (March 1998) 70–75.
[24] S. Roy, Optimal planning of wind energy conversion systems over an energy scenario, IEEE Trans. Energy Convers. 12 (3) (September 1997) 248–254.
[25] G. Mere, C. Berger, D.U. Sauer, Optimization of an off-grid hybrid PV–Wind–Diesel system with different battery technologies using genetic algorithm, Solar Energy 97 (1) (2013) 460–473.
[26] D. Xu, L. Kang, L. Chang, B. Cao, Optimal sizing of standalone hybrid wind/PV power systems using genetic algorithms, in: Canadian Conference on Electrical and Computer Engineering, vols. 1–4, 2005, pp. 1722–1725.
[27] S.M. Hakimi, S.M. Tafreshi, A. Kashefi Kaviani, Unit sizing of a stand-alone hybrid power system using particle swarm optimization (PSO), in: Proceeding of the International Conference on Automation and Logistics, Jinan, China, August 2007.
[28] Kaabeche, M. Belhamel, R. Ibtiouen, Sizing optimization of grid-independent hybrid photovoltaic/wind power generation system, Energy 36 (2) (2011) 1214–1222.
[29] S. Ghaem-Sigarchian, R. Paleta, A. Malmquist, A. Pina, Feasibility study of using a biogas engine as backup in a decentralized hybrid (PV/wind/battery) power generation system – case study Kenya, Energy 90 (2) (2015) 1830–1841.
[30] C. Li, X. Ge, Y. Zheng, C. Xu, Y. Ren, C. Song, C. Yang, Techno-economic feasibility study of autonomous hybrid wind/PV/battery power system for a household in Urumqi, China, Energy 55 (15) (2013) 263–272.
[31] C. Shang, D. Srinivasan, T. Reindl, Generation-scheduling-coupled battery sizing of stand-alone hybrid power systems, Energy 114 (1) (2016) 671–682.
[32] H. Ren, Q. Wu, W. Gao, W. Zhou, Optimal operation of a grid-connected hybrid PV/fuel cell/battery energy system for residential applications, Energy 113 (1) (2016) 702–712.
[33] I. Janghorban-Esfahani, P. Ifaei, J. Kim, C.K. Yoo, Design of hybrid renewable energy systems with battery/hydrogen storage considering practical power losses: a MEPoPA (modified extended-power pinch analysis), Energy 100 (1) (2016) 40–50.
[34] A. Maleki, F. Pourfayaz, Optimal sizing of autonomous hybrid photovoltaic/wind/battery power system with LPSP technology by using evolutionary algorithms, Solar Energy 115 (1) (2015) 471–483.
[35] A. Maleki, F. Pourfayaz, Artificial bee swarm optimization for optimum sizing of a stand-alone PV/WT/FC hybrid system considering LPSP concept, Solar Energy 107 (1) (2014) 227–235.
[36] N. Destro, A. Benato, A. Stoppato, A. Mirandola, Components design and daily operation optimization of a hybrid systemwith energy storages, Energy (2016) (in press), Corrected Proof, Available online 16 June 2016.
[37] D.P. Clarke, Y.M. Al-Abdeli, G. Kothapalli, Multi-objective optimization of renewable hybrid energy systems with desalination, Energy 88 (1) (2015) 457–468.
[38] A. Kashefi Kaviani, G.H. Riahy, S.H.M. Kouhsari, Optimal design of a reliable hydrogen-based stand-alone wind/PV generating system, considering component outages, Renew. Energy 34 (Issue 11) (November 2009) 2380–2390.
[39] R. Billinton, R.N. Allan, Reliability Evaluation of Engineering Systems: Concepts and Techniques, second ed., Plenum Press, New York, 1992.
[40] A. Mawardi, R. Ptchumani, Effects of parameter uncertainty on the performance variability of proton exchange membrane (PEM) fuel cells, J. Power Sources 160 (2006) 232–245.

[41] A.T.D. Perera, R.A. Attalage, K.K.C.K. Perera, V.P.C. Dassanayake, A hybrid tool to combine multi-objective optimization and multi-criterion decision making in designing standalone hybrid energy systems, Appl. Energy 107 (July 2013) 412–425.

[42] H. Ren, W. Zhou, K. Nakagami, W. Gao, Q. Wu, Multi-objective optimization for the operation of distributed energy systems considering economic and environmental aspects, Appl. Energy 87 (12) (2010) 3642–3651.

[43] M.J. Khan, M.T. Iqbal, Dynamic modeling and simulation of a small wind-fuel cell hybrid energy system, Renew. Energy 30 (2005) 421–439.

[44] R.S. Garcia, D. Weisser, A wind–diesel system with hydrogen storage: joint optimization of design and dispatch, Renew. Energy 31 (2006) 2296–2320.

[45] D.B. Nelson, M.H. Nehrir, C. Wang, Unit sizing and cost analysis of stand-alone hybrid wind/PV/fuel cell power generation system, Renew. Energy 31 (2006) 1641–1656.

[46] J.P. Stempien, M. Ni, Q. Sun, S.H. Chan, Production of sustainable methane from renewable energy and captured carbon dioxide with the use of Solid Oxide Electrolyzer: a thermodynamic assessment, Energy 82 (1) (2015) 714–721.

[47] B. Zakeri, S. Syri, Electrical energy storage systems: a comparative life cycle cost analysis, Renew. Sustain. Energy Rev. 533 (42) (2015) 569–596.

[48] M.Y. El-Sharkh, M. Tanrioven, A. Rahman, M.S. Alam, Cost related sensitivity analysis for optimal operation of a grid-parallel PEM fuel cell power plant, J. Power Sources 161 (2006) 1198–1207.

[49] K. Strunz, E.K. Brock, Stochastic energy source access management: infrastructure-integrative modular plant for sustainable hydrogen-electric cogeneration, Int. J. Hydrogen Energy 31 (2006) 1129–1141.

[50] R. Scozzari, M. Santarelli, Techno-economic analysis of a small size short range EES (electric energy storage) system for a PV (photovoltaic) plant serving a SME (small and medium enterprise) in a given regulatory context, Energy 71 (1) (2014) 180–193.

[51] R. Billinton, R.N. Allan, Reliability Evaluation of Power Systems, Plenum Press, New York, 1984.

[52] R. Karki, R. Billinton, Reliability/cost implications of PV and wind energy utilization in small isolated power systems, IEEE Trans. Energy Convers. 16 (4) (December 2001) 368–373.

[53] B.R. Bagen, Evaluation of different operating strategies in small standalone power systems, IEEE Trans. Energy Convers. 20 (3) (September 2005) 654–660.

[54] S. Nomura, Y. Ohata, T. Hagita, H. Tsutsui, S. Tsuji-Iio, R. Shimada, Wind farms linked by SMES systems, IEEE Trans. Appl. Supercond. 15 (2005) 1951–1954.

[55] M.K. Khairil, S. Javanovic, Reliability modeling of uninterruptible power supply systems using fault tree analysis method, Eur. Trans. Electr. Power 19 (2) (2009) 258–273.

[56] M. Marchesoni, S. Savio, Reliability analysis of a fuel cell electric city car, in: IEEE 2005 European Conference on Power Electronics and Applications; 11–14, September 2005, p. 10.

[57] R. Karki, R. Billinton, Cost-effective wind energy utilization for reliable power supply, IEEE Trans. Energy Convers. 19 (2004) 435–440.

[58] G. Tina, S. Gagliano, S. Raiti, Hybrid solar/wind power system probabilistic modeling for long-term performance assessment, Solar Energy 80 (2006) 578–588.

[59] H.R. Baghaee, M. Mirsalim, G.B. Gharehpetian GB, A. Kashefi Kaviani, Security/cost-based optimal allocation of multi-type FACTS devices using multi-objective particle swarm optimization, Simulation Int. Trans. Soc. Model. Simul. 88 (8) (2012) 999–1010, https://doi.org/10.1177/0037549712438715.

[60] K.E. Parasopoulos, M.N. Vrahatis, On the computation of all global minimizers through particle swarm optimization, IEEE Trans. Evol. Comput. 8 (3) (June 2004).
[61] A. Shukla, S.N. Singh, Advanced three-stage pseudo-inspired weight-improved crazy particle swarm optimization for unit commitment problem, Energy 96 (1) (2016) 23–36.
[62] J. Zhang, Q. Tang, Y. Chen, S. Lin, A hybrid particle swarm optimization with small population size to solve the optimal short-term hydro-thermal unit commitment problem, Energy 109 (1) (2016) 765–780.
[63] M. Nafar, G.B. Gharehpetian, T. Niknam, Improvement of estimation of surge arrester parameters by using modified particle swarm optimization, Energy 36 (8) (2011) 4848–4854.
[64] L. Wang L, C. Singh, Environmental/economic power dispatch using a fuzzified multi-objective particle swarm optimization algorithm, Elec. Power Syst. Res. 77 (2007) 1654–1664.
[65] X. Li, A. Malkawi, Multi-objective optimization for thermal mass model predictive control in small and medium size commercial buildings under summer weather conditions, Energy 112 (1) (2016) 1194–1206.
[66] E. Khorasaninejad, H. Hajabdollahi, Thermo-economic and environmental optimization of solar assisted heat pump by using multi-objective particle swam algorithm, Energy 72 (1) (2014) 680–690.
[67] H. Doagou-Mojarrad, G.B. Gharehpetian, H. Rastegar, J. Olamaei, Optimal placement and sizing of DG (distributed generation) units in distribution networks by novel hybrid evolutionary algorithm, Energy 54 (1) (2013) 129–138.
[68] A. Taghipour-Rezvan, N. Shams-Gharneh, G.B. Gharehpetian, Robust optimization of distributed generation investment in buildings, Energy 48 (1) (2012) 455–463.
[69] P. Kumar Tripathi, S. Bandyopadhyay, S.K. Pal, Multi-objective particle swarm optimization with time variant inertia and acceleration coefficients, Inf. Sci. 177 (2007) 5033–5049.
[70] P. Zhang, H. Chen, X. Liu, Z. Zhang, An iterative multi-objective particle swarm optimization-based control vector parameterization for state constrained chemical and biochemical engineering problems, Biochem. Eng. J. 103 (2015) 138–151.
[71] Y. Wang, Y. Yang, Particle swarm with equilibrium strategy of selection for multi-objective optimization, Eur. J. Oper. Res. 200 (1) (January 1 , 2010) 187–197.
[72] X. Wei, A. Kusiak, M. Li, F. Tang, Y. Zeng, Multi-objective optimization of the HVAC (heating, ventilation, and air conditioning) system performance, Energy 83 (1) (2015) 294–306.
[73] J. Zhong, X. Hu, J. Zhang, M. Gu, Comparison of performance between different selection strategies on simple genetic algorithms, in: International Conference on Intelligent Agents, Computational Intelligence for Modeling, Control and Automation, and International Conference on Web Technologies and Internet Commerce, vol. 1, 2005, pp. 1115–1121.
[74] B. Wang, S. Wang, X. Zhou, J. Watada, Multi-objective unit commitment with wind penetration and emission concerns under stochastic and fuzzy uncertainties, Energy 111 (1) (2016) 18–31.
[75] A. Deihimi, B. Keshavarz-Zahed, R. Iravani, An interactive operation management of a micro-grid with multiple distributed generations using multi-objective uniform water cycle algorithm, Energy 106 (1) (2016) 482–509.

[76] B. Bahmani-Firouzi, E. Farjah, R. Azizipanah-Abarghooee, An efficient scenario-based and fuzzy self-adaptive learning particle swarm optimization approach for dynamic economic emission dispatch considering load and wind power uncertainties, Energy 50 (1) (2013) 232–244.
[77] D. Feroldi, D. Zumoffen, Sizing methodology for hybrid systems based on multiple renewable power sources integrated to the energy management strategy, Int. J. Hydrogen Energy 39 (16) (2014) 8609–8620.
[78] R. Hosseinalizadeh, G.H. Shakouri, M.S. Amalnick, P. Peyman Taghipour, Economic sizing of a hybrid (PV–WT–FC) renewable energy system (HRES) for stand-alone usages by an optimization-simulation model: case study of Iran, Renew. Sustain. Energy Rev. 54 (1) (2016) 139–150.
[79] S. Ahmadi, S. Abdi, Application of the hybrid Big Bang–Big Crunch algorithm for optimal sizing of a stand-alone hybrid PV/wind/battery system, Solar Energy 134 (1) (2016) 366–374.

CHAPTER 8

Power management in hybrid microgrids

1. Introduction

Stochastic charging and discharging behavior of plug-in hybrid electric vehicles (PHEVs) and intermittent power generation pattern of renewable energy resources (RERs) add demand side and supply side uncertainties in microgrids (MGs), respectively, and therefore, result in new challenges [1,2]. To overcome these problems, the MG operator can use a few solutions such as the application of dispatchable DGs, charge or discharge of ESSs [3–5], and also, in the case of grid-connected operation mode, energy exchange with the main grid. The energy exchange with the main grid depends on the operating conditions of MGs and the retail energy market. The MG operator should consider technical issues, economical operation, and reliability indexes of the MG to determine the best operating condition. Therefore, the role of the power management system of an MG is vital. There are differences between the control system of conventional power systems and MGs. Furthermore, the penetration level of RERs in MGs is higher than in bulk power systems. Thus, MGs will have more complicated power management systems [6].

To solve the mentioned issues in the MG power management system, considering RERs and PHEVs uncertainties, various solutions have been suggested in the literature. In Ref. [7], the combination of ESSs and wind farms has been proposed to alleviate the power flow variations. A centralized control system for a DC smart home has been suggested in Ref. [4], to reduce power fluctuations and costs of charging and discharging each battery used in the smart home. A smart PHEV charging and discharging strategy has been used in Ref. [8] to provide grid support and mitigate PV unit generation variations. A single-phase residential MG, including electric vehicles (EVs), has been investigated in Ref. [9], and a power management system has been proposed for the integration of PV

Microgrids and Methods of Analysis
ISBN 978-0-12-816172-2
https://doi.org/10.1016/B978-0-12-816172-2.00008-0

units and battery energy storage systems (BESSs). The energy losses of MGs considering different penetration levels of PHEVs have been reduced in Refs. [10] and [11] using two optimization algorithms. The effect of optimal PHEV charging and demand-side power management on different applications has been discussed in Refs. [12–15] for peak load reduction. To charge PHEVs using PV units, a smart charging strategy has been suggested in Ref. [16]. In this study, the size of the BESS has been determined to reduce the MG need to the main grid. Therefore, the exchanged energy with the main grid has been reduced, which cannot guarantee an economical solution for the MG. For example, considering the energy price, it is possible to have an economical solution for minimum energy drawn from the main grid and maximum energy fed to it.

In this chapter, the well-known strategies used in MG power management will be addressed. It is easy to divide them into centralized and decentralized methods, but in some cases, it is difficult to distinguish the differences. For instance, the MG with local controllers and a supervisory controller has been known by some authors as decentralized control because it needs a low-bandwidth communication system. However, the disadvantages of the power management methods reported in the literature will be discussed. In this type of studies, the following features are important:

- In the MG under study, there must be a coordinated operation among the PV system, a wind generation unit, diesel generators, BESS, and other components of the system considering operational constraints.
- The power management system must consider the effect of two operation modes of PHEVs, i.e., the modes known as vehicle to grid (V2G) and grid to the vehicle (G2V).
- The performance of the power management system must be assessed considering the effect of the penetration level of PHEVs and RERs.

In the power management method, the exchanged energy with the main grid should usually be reduced. Also, the main charging energy of PHEVs should be supplied by RERs, and the rest of the consumption must be fed by MG other distributed energy resources (DERs), to reduce the energy exchange with the grid. To make the power management strategy compared with the other ones, different cases must be studied for different penetration levels of PHEVs in V2G and G2V operation modes. These two operation modes can be managed to reduce the challenges of the integration of RERs into MGs.

2. Basic control and management structure of microgrid

The structure of the MG control system is shown in Fig. 8.1 [18]. In this figure, the physical layer includes DERs and their converters loads and distribution system components such as switchgear, lines, transformers, circuit breakers, etc.

The local generation and consumption control and ESS management are realized in the local control layer. As shown in this figure, the secondary control level sends the signals to the local controllers using the communication system. At the highest level, we have the supervisory control and data acquisition system, known as MG central controller, MG supervisory controller (MGSC), or energy management system (EMS). The EMS should control power quality, manage ancillary services, obtain energy market participation, and optimize system operation [17]. The interaction with the main grid is obtained through the market operator (MO) and the distribution network operator (DNO). The information exchange between

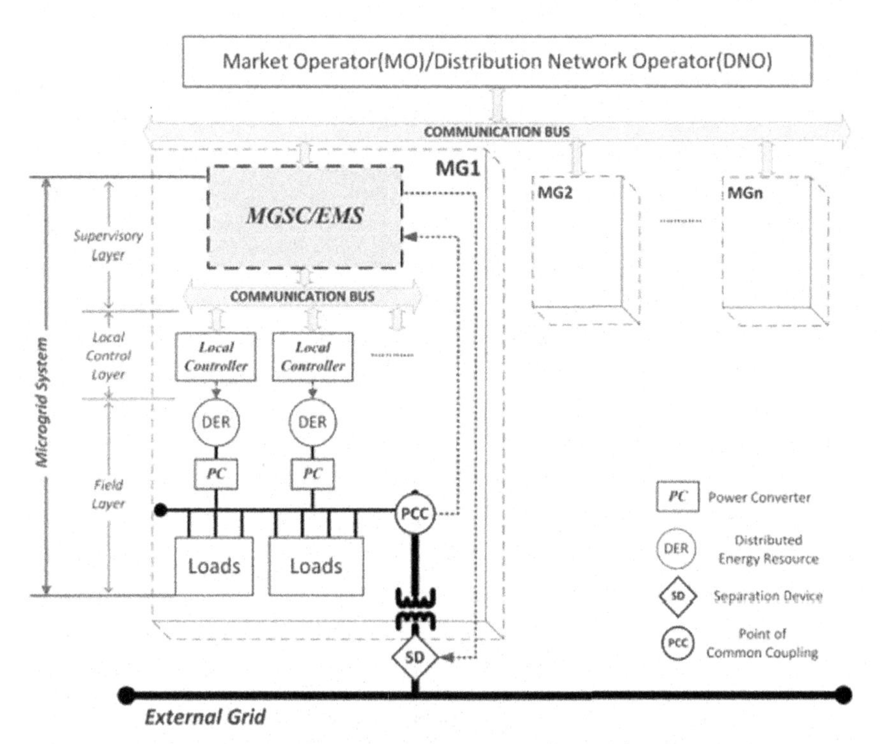

Figure 8.1 General structure of a microgrid control system [20].

the EMS and the main grid MO and DNO is possible using a communication network. In MGs, the realization of new features and characteristics has resulted in challenges for EMSs [17]. For example, EMS must obtain a smooth operation mode change between on-grid and off-grid modes due to intentional or unintentional islanding, incorporate demand response programs, and overcome the challenges of integrating RERs considering their uncertainty and low inertia.

3. Hierarchical power management

As mentioned in previous chapters and shown in Fig. 8.2, a three-level hierarchical structure can be used in the MG power management system [18]. In addition to the voltage and frequency control, the first level must control the power sharing among DERs according to their features. At the secondary level, the voltage and frequency must be recovered, and the power quality and reliability should be enhanced. The highest level of this hierarchy, economic, and environmental optimized must be optimized, discussed in the next sections.

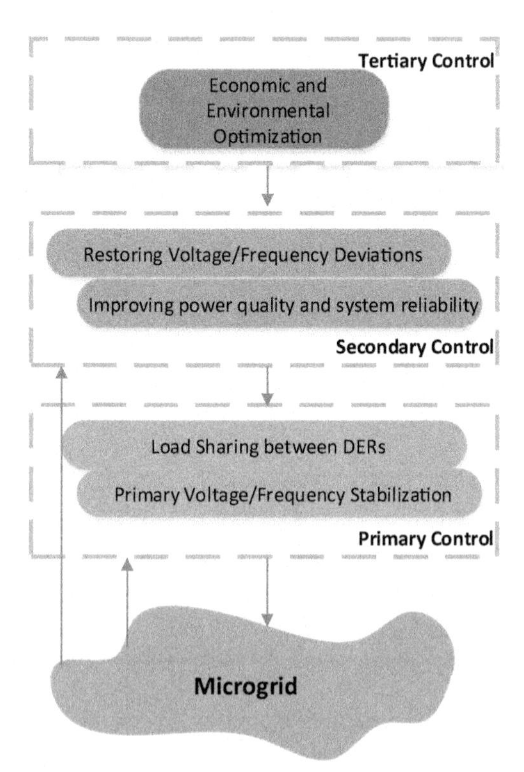

Figure 8.2 MG power management hierarchical structure [6]. *MG*, microgrid.

3.1 Primary control

To achieve the main aim of the primary control level, two approaches have been presented. The first one is the active load sharing method, and the second one is the droop-based control method.

- *Active load sharing method*: As discussed in Ref. [21]; in this method, the individual module output current is compared with the average or maximum current, and the error voltage of the module is generated and used to control the voltage of the reference or output feedback. To share power error, maximum power, or average active and reactive powers, a high-band communication system must be used. The set-points, i.e., current, active power, and reactive power references, can be specified by centralized, master–slave, and average load sharing methods [19,22,23]. The advantage of this method is its accuracy, and its disadvantage is the need for a communication system, which can result in low reliability.
- *Droop-based control method*: In this method, an increase in the output current results in a decrease in the output voltage, which is also called the autonomous, independent, or wireless control method, because it does not need any communication system. It must be mentioned that the secondary controller in the hierarchical structure uses a communication system to send control orders to the primary control level. The main characteristics of this method are modeled by two equations. The first one shows a frequency reduction in an active power increase; in the second equation, we have a voltage reduction in a reactive power increase. Considering the importance of the second method, more details will be given in the following.

The connection of inverter-interfaced distributed energy resources (IIDERs) to the MG is through a filter. Therefore, the output impedance can be considered to be inductive, and their droop-controlled equations for their active and reactive powers are as follows:

$$\omega = \omega_0 - m_p(P_G - P_0) \tag{8.1}$$

$$|V| = |V_o| - n_q(Q_G - Q) \tag{8.2}$$

In the case of a resistive connection to the MG (line with high R/X ratio), these equations change their form, as follows [24]:

$$\omega = \omega_o - m_p(Q_G - Q_o) \tag{8.3}$$

$$|V| = |V_o| - n_q(P_G - P_o) \tag{8.4}$$

In this chapter, Eqs. (8.1) and (8.2) are applied to systems under study considering the IEEE standard 1547.7. However, it must be noted that Eqs. (8.1) and (8.2), and also Eqs. (8.3) and (8.4), assume and represent a decoupling between active and reactive powers. Still, in practice, the output impedance of an IIDER is a complex number. As a result, the active and reactive powers are not decoupled, and they depend on frequency and voltage. In this case, the $P-V-\omega$ and $Q-V-\omega$ equations must be considered as follows [25]:

$$\omega = \omega_o - m_p(P_G - Q_G) \tag{8.5}$$

$$|V| = |V_o| - n_q(P_G + Q_G) \tag{8.6}$$

3.2 Secondary control

As mentioned before, the deviations of voltage and frequency must be restored at this level. The control system compares the measured values, i.e., voltage and frequency, with their reference values, and then, the error signals are applied to proportional integral (PI) controllers to compensate them. Afterward, the compensated signals will be added to the reference values [18]. Here, the main problem is the determination of PI controller coefficients, especially in MGs with a complex and nonlinear model. The compensated signals can be applied to other controllers such as field logic controllers, or potential function-based controllers. Also, it must be mentioned that the intermittent output power of RERs can be compensated by the ESSs such as BESS. But the increase in the number of charging and discharging cycles of BESS results in lifetime reduction and increases in costs (BESS replacement).

3.3 Power management in tertiary control

The optimal power management in the tertiary control level should manage the MG to have a minimum cost, the maximum contribution of RERs, minimum loss of a lifetime of BESSs, and minimum greenhouse gas emission [26]. In the literature, this type of management is known as optimal scheduling or planning of MGs.

Two modules from the optimal power management. The first module is the "forecasting module," which must forecast the RERs generation and load consumption. To develop a weather forecast model, this module uses current weather conditions and historical weather data, as shown in Fig. 8.3. In Refs. [27] and [28], the RERs generation has been forecasted by

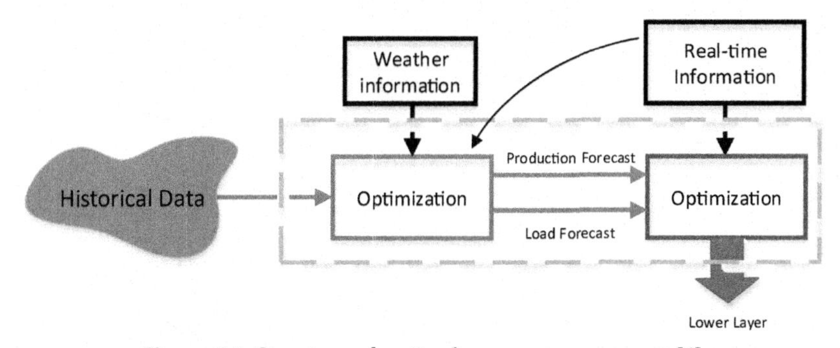

Figure 8.3 Structure of optimal power management [6].

an ANN-based on real-time data. The ANN has been trained using generation data and weather data. The second module is the "optimization module," which must optimize single-objective or multiobjective problem (performance index) considering MG constraints such as DER limitations and make optimal decisions considering the forecasted demand, generation, and price of energy in each interval. To have accurate results, the MG uncertainties must be modeled by a well-known method such as the point estimate method [29] merged with chance-constrained programming [30]. Neglecting the uncertainties results in an optimization problem in the form of a set of linear or nonlinear equations, which can be solved by deterministic methods.

4. Generalization of power management concepts to DC and hybrid AC/DC microgrids

The connection methods of DERs and loads to the MG and the configurations of AC and DC buses are different. As shown in Fig. 8.4, they can be categorized into three architectures:

- AC-coupled MG (ACMG)
- DC-coupled MG (DCMG)
- AC–DC-coupled MG (AC/DC MG)

In ACMGs, a coupling AC bus connects DC loads, DGs and ESSs via their interfacing converters. Also, AC loads can be directly connected. In DCMGs, a coupling DC bus connects DC loads, DGs, and ESSs. AC loads can be connected to this bus using an interfacing converter (IFC), and an IFC connects the DC bus to the AC bus of the main grid. In AC/DC MGs, some parts of loads, DGs, and ESSs are connected to the AC bus, and another part of them has a connection with the DC bus. An interlinking

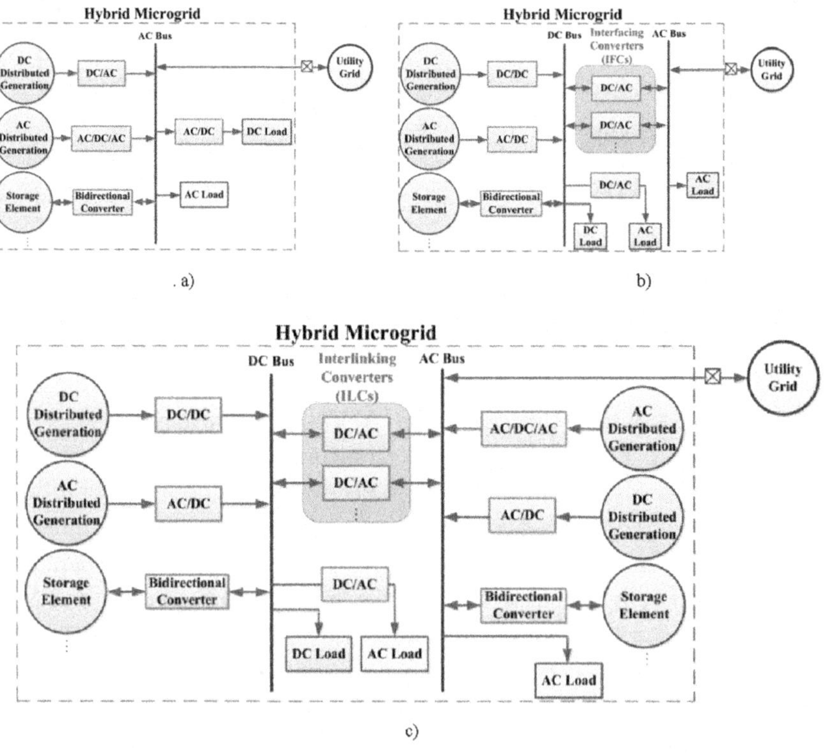

Figure 8.4 (A) AC-coupled MG (ACMG), (B) DC-coupled MG (DCMG), and (C) AC–DC-coupled MG (AC/DC MG) [31]. *MG*, microgrid.

converter (ILC) connects these two buses [31]. As shown in Fig. 8.4, a circuit breaker is between the MG AC bus and the main grid. As a result, it is possible to have an islanded operation mode.

The power management and control strategies of these MGs determine DGs and ESSs output active and reactive powers and variations of bus voltages and frequency, respectively. In the following, these strategies will be discussed for ACMGs, DCMGs, and AC/DC MGs.

4.1 AC microgrids

The main task of the AC MG, power management system is obtaining a balance between generation and consumption. Also, the AC bus voltage and frequency should be controlled, particularly in the islanded operation mode. To study these MGs, the DGs and ESSs can be modeled as voltage or current sources connected to the AC bus.

Fig. 8.5 indicates the classification of AC-coupled MGs power management strategies divided into two main groups: strategies for grid-connected mode and strategies for islanded mode. The power management strategies in grid-connected mode are divided into strategies for dispatched output power and strategies for undispatched/nondispatched output power.

In the grid-connected dispatched mode, a higher control level in the main grid, such as the DNO shown in Fig. 8.1, dispatches the power exchange between the ACMG and the main grid. From the main grid point of view, the MG is considered a controllable energy sink or source, which can play an important role in energy management systems. Therefore, the output power of DERs must be controlled by adjusting their current or voltage. Using the DERs output current, their power can be controlled by controlling their output current considering the given power reference. The output voltage and system frequency are specified by the main grid. In the case of controlling the DERs output voltage, to adjust their power, they can be considered as a synchronous generator [32]. It must be mentioned that this is valid for islanded mode as well. The dispatched power can be shared among ACMG resources using power balancing schemes [20]. For example, RERs can operate at their

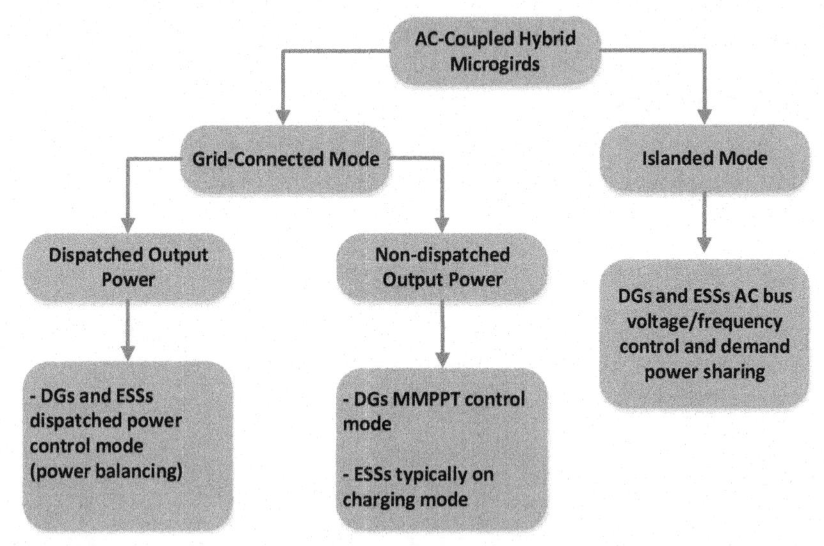

Figure 8.5 Classification of ACMGs power management strategies [31]. *ACMG*, AC-coupled microgrid.

maximum power point (MPP), and the other resources must cover the shortage of power considering operational limitations of resources, response time, etc. [33,34].

When the ACMG operates in on-grid nondispatched mode, the generation of DERs will supply the system, RERs will work at their MPP, and ESSs are usually charged [35,36] or used to smooth the output power of RERs. In grid-connected nondispatched mode, the ACMG can control the active and reactive powers exchanged with the main grid, and therefore, support the main grid the same as the dispatched mode [37,38]. As a result, the grid is supported by the controllers of DER converters.

As shown in Fig. 8.5, the parallel branch of the strategies for the grid-connected mode is the strategies for the islanded mode of operation. In this case, the voltage of the AC bus and the frequency of the system are important system variables, which should be controlled. Also, the balance between the generation and the consumption in ACMG must be considered, and the system must have an acceptable power sharing among DERs. To meet these circumstances, different solutions have been presented in the literature. The well-known droop-based control method is in this group. In this control method, the output active and reactive power of each DER must be determined so that the voltage and the frequency must stay in an acceptable range [18,32]. As another solution, it is possible to have RERs at their MPP, but the ESSs must control AC bus voltage and frequency [39,40].

4.2 DC microgrids

The classification of the power management strategies in DC-coupled MGs (DCMGs) is presented in Fig. 8.6. The same as the ACMGs, it can be seen

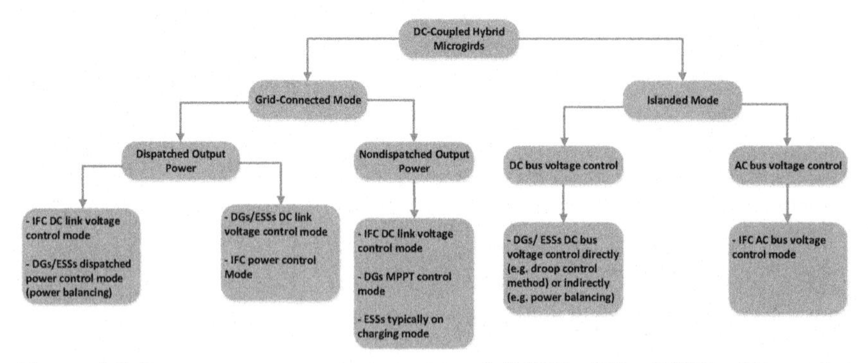

Figure 8.6 Power management strategies of DCMGs [31]. *DCMGs*, DC-cpupled microgrid.

two main groups; strategies for on-grid and islanded modes. Here, we have strategies for dispatched and nondispatched output power in case of having a grid connection. The IFC, shown in Fig. 8.4, can control the bidirectional power exchange between DC bus and AC bus, which the latter has a direct connection to the main grid. It is obvious that the IFC output power can be controlled by adjusting the converter output current or voltage. Also, the IFC may operate in other operation modes such as DC link voltage control mode or AC bus voltage control mode [41−43].

In the grid-connected mode with the dispatched output power, we have two strategies. In the first strategy, the DC link voltage is set at the reference voltage by the IFC, i.e., the IFC is in DC bus voltage control mode, and DERs control the exchanged/dispatched power of the DCMG with the main grid using a power balancing control [50−52]. In the second strategy, the dispatched power of the DCMG is controlled by the IFC, i.e., the IFC is in power control mode, the DC link voltage is adjusted by the ESSs on the DC bus based on the droop-based control method [45−47], and the DGs are controlled by the droop-based control method or can operate at MPP [44,48−51].

In the grid-connected mode with the nondispatched output power [52,53], the DC link voltage is adjusted by the IFC, i.e., it works in DC bus voltage control mode, the operation point of RERs must be set at MPP, and ESSs are charged or should smooth the RERs output power.

In the DCMG islanded mode of operation, there is no support from the main grid. Therefore, the voltage of the DC bus and the voltage and frequency of the AC bus must simultaneously be regulated. In this case, the IFC controls the voltage and frequency of the AC bus, i.e., the IFC is in AC bus control mode [17,54]. To control the DC bus voltage, it can be realized directly or indirectly. In the directly controlled DC bus method, DERs sets the DC bus voltage at its reference voltage. In the indirectly controlled DC bus control method, we have parallel IFCs; some of them are in the AC bus control mode, and the others adjust the voltage of the DC bus and should balance the generation and consumption of the DCMG.

In DCMGs, it is possible to have a direct connection of ESSs to the DC bus. As a result, the DC bus voltage is fixed, and in this case, the power management at the DC bus can be performed by a current control method such as access-oriented storage [55−58] and capacity-oriented storage [59]. Also, it can be realized by the DGs that are connected to the DC bus.

4.3 AC/DC microgrids

An AC/DC MG consists of a DC subsystem, an AC subsystem, and, in the case of a grid-connected operation mode, a connection to the main grid. Therefore, the power management strategies must simultaneously control the DC bus voltage, AC bus voltage, frequency, and system power balance. As can be seen in Fig. 8.4, the ILC connects the DC bus to the AC bus, and at the same time, has a connection to the main grid. Considering the conditions of the AC/DC MG, the ILC can operate in the following operation modes:

- Bidirectional power control mode
- DC bus control mode
- AC bus voltage and frequency control mode

Fig. 8.7 shows the strategies, which can be used for power management in AC/DC MGs. The same as the previous cases, the main groups are grid-connected and islanded modes strategies. The strategies in the grid-connected operation mode are divided into two groups: strategies with dispatched and nondispatched output power.

Two strategies can be applied to the AC/DC MG in the grid-connected mode with the dispatched output power. In the first strategy, the DC bus voltage is controlled by the ILC. The dispatched output power is provided by generation coordination between the DERs of the DC subsystem and the DERs of the AC subsystem. In the second strategy, the DC bus voltage is controlled by DERs of the DC subsystem, and the dispatched output power is jointly provided by the ILC and DERs of the AC subsystem [45,46] Therefore, the ILC is in power control operation mode.

In the grid-connected mode with the nondispatched output power, the RERs of two subsystems are operating at MPP [60,61], and the ESSs in AC and DC subsystems can be charged or used to smooth the power, which must be delivered to the main grid. Also, the DC bus voltage is controlled by the ILC, which transfers the power from the DC side to the main grid.

In the islanded mode, the voltage of the DC bus and the voltage and frequency of the AC bus must be adjusted by coordinated operation of the ILC, DERs of AC subsystem, and DERs of DC subsystem. Also, the power balance between generation and consumption in the AC/DC MG must be preserved. The voltage and frequency of the AC subsystem and its power sharing can be obtained by power management strategies such as droop-based [17,32,62–65] and master–slave [61] control methods. The DERs of the DC subsystem adjust the voltage of the DC bus. This task can be

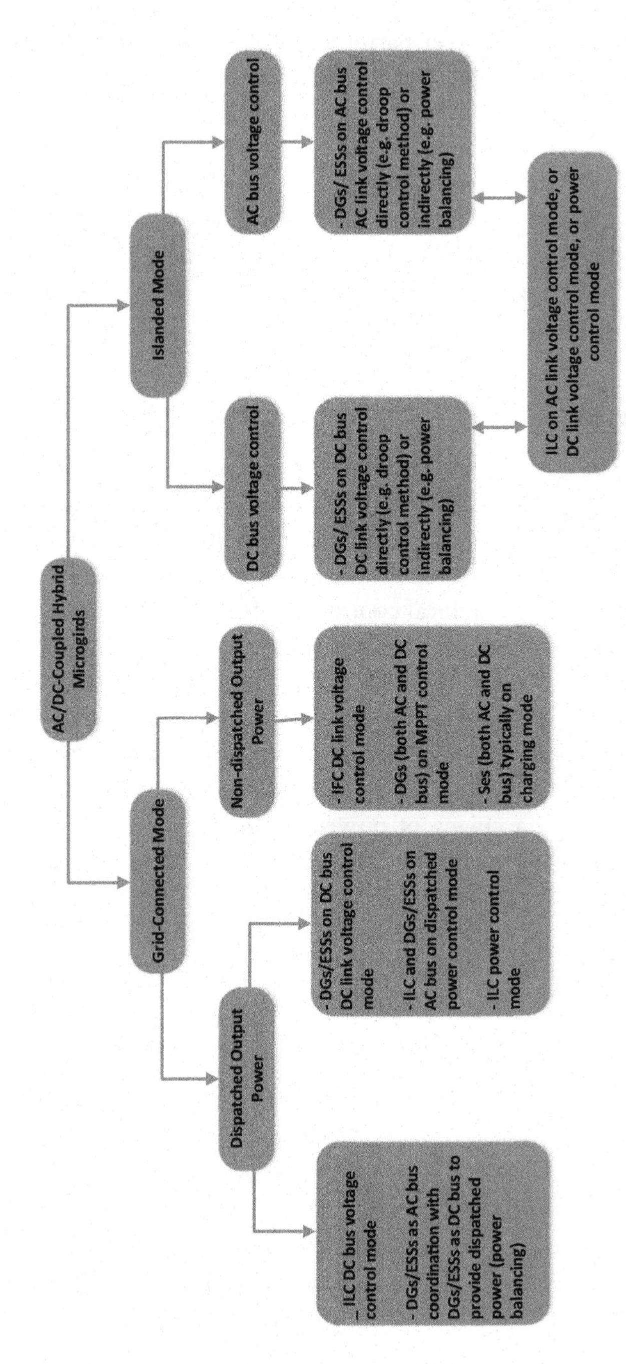

Figure 8.7 Power management strategies of AC/DC MGs [31].

realized directly [45,46] (e.g., using the droop-based control strategy), or indirectly (e.g., using the power balancing control method) [4,60]. However, in this mode, the role of the ILC is vital for power management [45,46]. It can be used in the DC bus, AC bus, or power control modes. For example, in the case of control of the DC bus by DC subsystem DERs and control of the AC subsystem by its DERs, the ILC should manage the power exchange between these two subsystems to have a power balance in the AC/DC MG. Also, if the AC/DC MG has two parallel ILCs, one can operate in the DC bus control mode, and the other one can operate in the AC bus or power control mode.

5. Summary

The development of the smart grid can be expedited by the application of well-designed MGs. The MGs can be considered as a managed and controlled component of a smart grid. Therefore, their power management and control is very important for power system engineers.

In this chapter, the hierarchical control architecture was discussed for the MGs, which have various elements and different topologies. To control the physical layer of the MG without using a communication system, the droop-based control was used for the local controllers as the primary control level. Also, the secondary and tertiary control levels were discussed in the hierarchical control structure of MGs. In this structure, the researchers are trying to improve the accuracy of power sharing of this control method. Also, the importance of forecasting and optimization algorithms was investigated for an optimal power management system. In the last part of this chapter, the power management strategies were classified for different architectures of MGs, including AC MGs, DC MGs, and AC/DC MGs. For each of these topologies, the proposed solutions were presented for grid-connected and islanded operation modes.

References

[1] M. Tabari, A. Yazdani, Stability of a dc distribution system for power system integration of plug-in hybrid electric vehicles, IEEE Trans. Smart Grid 5 (5) (September 2014) 2564–2573.

[2] H.R. Baghaee, M. Mirsalim, G.B. Gharehpetian, H.A. Talebi, A decentralized power management and sliding mode control strategy for hybrid AC/DC microgrids including renewable energy resources, IEEE Trans. Ind. Inform. (2017), pp. 1–1.

[3] K.S. Kook, K. McKenzie, Y. Liu, S. Atcitty, A study on applications of energy storage for the wind power operation in power systems, in: 2006 IEEE Power Engineering Society General Meeting, IEEE, 2006.
[4] K. Tanaka, K. Uchida, K. Ogimi, T. Goya, A. Yona, T. Senjyu, T. Funabashi, C. Kim, Optimal operation by controllable loads based on smart grid topology considering insolation forecasted error, IEEE Trans. Smart Grid 2 (3) (September 2011) 438–444.
[5] M. ElNozahy, T.K. Abdel-Galil, M. Salama, Probabilistic ESS sizing and scheduling for improved integration of PHEVs and PV systems in residential distribution systems, Elec. Power Syst. Res. 125 (August 2015) 55–66.
[6] N. Yang, D. Paire, F. Gao, A. Miraoui, Power management strategies for microgrid - a short review, in: Proc. IEEE Industry Applications Society Annual Meeting, Lake Buena Vista, FL, USA, October 2013, pp. 1–6.
[7] K.J. McKenzie, S. Atcitty, A study on applications of energy storage for the wind power operation in power systems, in: 2006 IEEE Power Engineering Society General Meeting, June 2006, pp. 5 pp.–.
[8] M.J.E. Alam, K.M. Muttaqi, D. Sutanto, Effective utilization of available PEV battery capacity for mitigation of solar PV impact and grid support with integrated v2g functionality, IEEE Trans. Smart Grid 7 (3) (May 2016) 1562–1571.
[9] N. Saxena, I. Hussain, B. Singh, A.L. Vyas, Implementation of a grid-integrated PV-battery system for residential and electrical vehicle applications, IEEE Trans. Ind. Electron. 65 (8) (August 2018) 6592–6601.
[10] H. Nafisi, S.M.M. Agah, H.A. Abyaneh, M. Abedi, Two-stage optimization method for energy loss minimization in microgrid based on smart power management scheme of PHEVs, IEEE Trans. Smart Grid 7 (3) (May 2016) 1268–1276.
[11] E. Sortomme, M.M. Hindi, S.D.J. MacPherson, S.S. Venkata, Coordinated charging of plug-in hybrid electric vehicles to minimize distribution system losses, IEEE Trans. Smart Grid 2 (1) (March 2011) 198–205.
[12] Y. Mou, H. Xing, Z. Lin, M. Fu, Decentralized optimal demand-side management for PHEV charging in a smart grid, IEEE Trans. Smart Grid 6 (2) (March 2015) 726–736.
[13] P. Goli, W. Shireen, PV integrated smart charging of PHEVs based on DC link voltage sensing, IEEE Trans. Smart Grid 5 (3) (May 2014) 1421–1428.
[14] Y. Ota, H. Taniguchi, T. Nakajima, K.M. Liyanage, J. Baba, A. Yokoyama, Autonomous distributed v2g (vehicle- to-grid) satisfying scheduled charging, IEEE Trans. Smart Grid 3 (1) (March 2012) 559–564.
[15] Y. Ma, T. Houghton, A. Cruden, D. Infield, Modeling the benefits of vehicle-to-grid technology to a power system, IEEE Trans. Power Syst. 27 (2) (May 2012) 1012–1020.
[16] G.C. Mouli, P. Bauer, M. Zeman, System design for a solar powered electric vehicle charging station for workplaces, Appl. Energy 168 (2016) 434–443.
[17] F. Katiraei, R. Iravani, N. Hatziargyriou, A. Dimeas, Microgrids management, IEEE Power Energy Mag. 6 (3) (May 2008) 54–65.
[18] J.M. Guerrero, J.C. Vasquez, J. Matas, L.G. de Vicuna, M. Castilla, Hierarchical control of droop-controlled AC and DC microgrids—a general approach toward standardization, IEEE Trans. Ind. Electron. 58 (1) (January 2011) 158–172.
[19] A. Bidram, A. Davoudi, Hierarchical structure of microgrids control system, IEEE Trans. Smart Grid 3 (4) (December 2012) 1963–1976.
[20] L. Meng, E.R. Sanseverino, A. Luna, T. Dragicevic, J.C. Vasquez, J.M. Guerrero, Microgrid supervisory controllers and energy management systems: a literature review, Renew. Sustain. Energy Rev. 60 (July 2016) 1263–1273.
[21] W. Qiu, Z. Liang, Practical design considerations of current sharing control for parallel PWM applications, in: Proc. IEEE Applied Power Electronics Conf & Expo., vol. 1, 2005, pp. 281–286.

[22] A. Ghazanfari, M. Hamzeh, H. Mokhtari, H. Karimi, Active power management of multi-hybrid fuel cell/supercapacitor power conversion system in a medium voltage microgrid, IEEE Trans. Smart Grid 3 (4) (December 2012) 1903–1910.

[23] S. Luo, Z. Ye, R.-L. Lin, F. Lee, A classification and evaluation of paralleling methods for power supply modules, in: Proc. IEEE Annual Power Electronics Specialists Conf (PESC), vol. 2, 1999, pp. 901–908.

[24] J. Guerrero, J. Matas, L.G. de-Vicuna, M. Castilla, J. Miret, Decentralized control for parallel operation of distributed generation inverters using resistive output impedance, IEEE Trans. Ind. Electron. 54 (2) (2007) 994–1004.

[25] W. Yao, M. Chen, J. Matas, J. Guerrero, Z.M. Qian, Design and analysis of the droop control method for parallel inverters considering the impact of the complex impedance on the power sharing, IEEE Trans. Ind. Electron. 58 (2) (2011) 576–588.

[26] S. Pourmousavi, M. Nehrir, C. Colson, C. Wang, Real-time energy management of a stand-alone hybrid wind-microturbine energy system using particle swarm optimization, IEEE Trans. Sustain. Energy 1 (3) (October 2010) 193–201.

[27] S. Chakraborty, M.D. Weiss, M.G. Simoes, Distributed intelligent energy management system for a single-phase high-frequency AC microgrid, IEEE Trans. Ind. Electron. 54 (1) (Febuary 2007) 97–109.

[28] A. Chaouachi, R. Kamel, R. Andoulsi, K. Nagasaka, Multiobjective intelligent energy management for a microgrid, IEEE Trans. Ind. Electron. 60 (4) (April 2013) 1688–1699.

[29] S. Mohammadi, B. Mozafari, S. Solimani, T. Niknam, An adaptive modified frefly optimization algorithm based on hong's point estimate method to optimal operation management in a microgrid with consideration of uncertainties, Energy 51 (2013) 339–348.

[30] Z. Wu, W. Gu, R. Wang, X. Yuan, W. Liu, Economic optimal schedule of CHP microgrid system using chance-constrained programming and particle swarm optimization, in: Proc. IEEE Power & Energy Society General Meeting, July 2011, pp. 1–11.

[31] F. Nejabatkhah, Y.W. Li, Overview of power management strategies of hybrid AC/DC microgrid, IEEE Trans. Power Electron. 30 (12) (December 2015) 7072–7089.

[32] Y.W. Li, D.M. Vilathgamuwa, P.C. Loh, Design, analysis and real-time testing of controllers for multi-bus microgrid system, IEEE Trans. Power Electron. 19 (September 2004) 1195–1204.

[33] K.T. Tan, P.L So, Y.C. Chu, M.Z.Q Chen, Coordinated control and energy management of distributed generation inverters in a microgrid, IEEE Trans. Power Deliv. 28 (2) (April 2013) 704–713.

[34] K.T. Tan, X.Y. Peng, P.L. So, Y.C. Chu, M.Z.Q. Chen, Centralized control for parallel operation of distributed generation inverters in microgrids, IEEE Trans. Smart Grid 3 (4) (December 2012) 1977–1987.

[35] C.T. Rodríguez, D.V. de la Fuente, G. Garcerá, E. Figueres, J.A.G. Moreno, Reconfigurable control scheme for a PV microinverter working in both grid-connected and island modes, IEEE Trans. Ind. Electron. 60 (4) (April 2013) 1582–1595.

[36] S. Jiang, W. Wang, H. Jin, D. Xu, Power management strategy for microgrid with energy storage system, in: Proc. IEEE IECON- 37th Annual Industrial Electronics Society Conf, 2011, pp. 1524–1529.

[37] C.K. Sao, P.W. Lehn, Control and power management of converter fed microgrids, IEEE Trans. Power Syst. 23 (3) (August 2008) 1088–1098.

[38] J. C. Vasquez, R. A. Mastromauro, J. M. Guerrero, M. Liserre, Voltage support provided by a droop-controlled multifunctional inverter, IEEE Trans. Ind. Electron. 56 (11) (November 2009) 4510–4519.

[39] B. Belvedere, M. Bianchi, A. Borghetti, C.A. Nucci, M. Paolone, A. Peretto, A microcontroller-based power management system for standalone microgrids with hybrid power supply, IEEE Trans. Sustain. Energy 3 (3) (July 2012) 422–431.

[40] C. Wang, M.H. Nehrir, Power management of a standalone wind/photovoltaic/fuel-cell energy system, IEEE Trans. Energy Convers. 23 (3) (September 2008) 957–967.

[41] F. Blaabjerg, R. Teodorescu, M. Liserre, A.V. Timbus, Overview of Controland Grid Synchronization for Distributed power Generation Systems, IEEE Trans. Ind. Electron. 53 (October 2006) 1398–1409.

[42] A. Yazdani, R. Iravani, Voltage-Sourced Converter in Power Systems: Modelling, Control, And Application, WIELY, IEEE PRESS, 2010.

[43] J.M. Carrasco, L.G. Franquelo, J.T. Bialasiewicz, E. Galvan, R.C.P. Guisado, M.A.M. Prats, J.I. Leon, N. MorenoAlfonso, Power-electronic systems for the grid integration of renewable energy sources: a survey, IEEE Trans. Power Electron. 53 (August 2006) 1002–1016.

[44] T. Zhou, B. François, Energy management and power control of a hybrid active wind generator for distributed power generation and grid integration, IEEE Trans. Ind. Electron. 58 (1) (January 2011) 95–104.

[45] P.C. Loh, D Li, Y.K Chai, F Blaabjerg, Autonomous operation of hybrid microgrid with AC and DC subgrids, IEEE Trans. Power Electron. 28 (5) (May 2013) 2214–2223.

[46] P.C. Loh, D. Li, Y.K. Chai, F. Blaabjerg, Autonomous control of interlinking converter with energy storage in hybrid AC–DC microgrid, IEEE Trans. Ind. App. 49 (3) (May 2013) 1374–1383.

[47] D. Chen, L. Xu, Autonomous DC Voltage Control of a DC microgrid With multiple Slack Terminals, IEEE Trans. Power Syst. 27 (4) (November 2012) 1897–1905.

[48] H. Fakham, D Lu, B. Francois, Power control design of a battery charger in a hybrid active PV generator for load following applications, IEEE Trans. Ind. Electron. 58 (1) (January 2011) 85–94.

[49] A. Hajizadeh, M.A. Golkar, Fuzzy neural control of a hybrid fuel cell/battery distributed power generation system, IET Renew. Power Gener. 3 (4) (2009) 402–414.

[50] N. Gyawali, Y. Ohsawa, O. Yamamoto, Power management of double-fed induction generator-based wind power system with integrated smart energy storage having superconducting magnetic energy storage/fuel-cell/electrolyzer, IET Renew. Power Gener. 5 (6) (2011) 407–421.

[51] H. Kakigano, Y. Miura, T. Ise, Distribution voltage control for DC microgrids using fuzzy control and gain scheduling technique, IEEE Trans. Power Electron. 28 (5) (May 2013) 2246–2258.

[52] D.V. de la Fuente, C.L.T. Rodríguez, G. Garcerá, E. Figueres, R.O. González, Photovoltaic power system with battery backup with grid-connection and islanded operation capabilities, IEEE Trans. Ind. Electron. 60 (4) (April 2013) 1571–1581.

[53] L. Xu, D. Chen, Control and operation of a DC microgrid with variable generation and energy storage, IEEE Trans. Power Deliv. 26 (4) (October 2011) 2513–2522.

[54] N. Pogaku, M. Prodanovic, T.C. Green, Modeling, analysis and testing of autonomous operation of an inverter based microgrid, IEEE Trans. Power Electron. 22 (March 2007) 613–625.

[55] J. Moreno, M. E Ortúzar, J.W. Dixon, Energy management system for a hybrid electric vehicle, using ultra capacitors and Neural networks, IEEE Trans. Ind. Electron. 53 (2) (April 2006) 614–623.

[56] M.B. Camara, H. Gualous, F Gustin, A Berthon, Design and new control of DC/DC converters to share energy between supercapacitors and batteries in hybrid vehicles, IEEE Trans. Veh. Technol. 57 (5) (September 2008) 2721–2735.

[57] M.B. Camara, H. Gualous, F. Gustin, A. Berthon, B. Dakyo, DC/DC converter design for supercapacitor and battery power management in hybrid vehicle applications—polynomial control strategy, IEEE Trans. Ind. Electron. 57 (2) (Febuary 2010) 587–597.
[58] T. Azib, O. Bethoux, G. Remy, C. Marchand, E. Berthelot, An innovative control strategy of a single converter for hybrid fuel cell/supercapacitor power source, IEEE Trans. Ind. Electron. 57 (12) (December 2010) 4024–4031.
[59] A. Ravey, B. Blunier, A. Miraoui, Control strategies for fuel-cell-based hybrid electric vehicles: from offline to online and experimental results, IEEE Trans. Veh. Technol. 61 (6) (July 2012) 2452–2458.
[60] X. Liu, P Wang, P.C Loh, A hybrid AC/DC microgrid and its coordination control, IEEE Trans. Smart Grid 2 (2) (July 2011) 278–286.
[61] Z. Jiang, X. Yu, Hybrid DC- and AC-linked microgrids: towards Integration of Distributed Energy resources, in: IEEE Energy 2030 Conf, November 2008, pp. 1–8.
[62] J. He, Y.W. Li, An enhanced microgrid load demand sharing strategy, IEEE Trans. Power Electron. 27 (9) (September 2012) 3984–3995.
[63] Y. Li, Y.W. Li, Power management of inverter interfaced autonomous microgrid based on virtual frequency-voltage frame, IEEE Trans. Smart Grid 2 (1) (March 2011) 30–41.
[64] Y. Chung, W. Liu, D.A. Cartes, E.G. Collins, S. Moon, Control methods of Inverter-Interfaced distributed generators in a microgrid System, IEEE Trans. Ind. App. 46 (3) (May 2010) 1078–1089.
[65] E. Barklund, N. Pogaku, M. Prodanovic, C.H. Aramburo, T.C. Green, Energy management in autonomous microgrid using stability-constrained droop control of inverters, IEEE Trans. Power Electron. 23 (5) (September 2008) 2346–2352.

Further reading

[1] M.B. Delghavi, A. Yazdani, Sliding-mode control of AC voltages and currents of dispatchable distributed energy resources in master-slave-organized inverter-based microgrids, IEEE Trans. Smart Grid 10 (1) (January 2019) 980–991.
[2] P. Piagi, R. Lasseter, Autonomous control of microgrids, in: The IEEE Power Engineering Society General Meeting. Montreal, Que., Canada, June 2006, pp. 1–8.
[3] IEEE Standard 1547, IEEE Standard for Interconnecting Distributed Resources with Electric Power Systems, 2003.
[4] J.Y. Kim, J.H. Jeon, S.K. Kim, C. Cho, J.H. Park, H.M. Kim, K.Y. Nam, Cooperative control strategy of energy storage system and microsources for stabilizing the microgrid during islanded operation, IEEE Trans. Power Electron. 25 (12) (December 2010) 3037–3048.
[5] B. Wang, M. Sechilariu, F. Locment, Intelligent DC microgrid with smart grid communications: control strategy consideration and design, IEEE Trans. Smart Grid 3 (4) (December 2012) 2148–2156.
[6] L.N. Khanh, J.J. Seo, Y.S. Kim, D.J. Won, Power management strategies for a grid-connected PV-FC hybrid system, IEEE Trans. Power Deliv. 25 (3) (Jul. 2010) 1874–1882.
[7] H. Kanchev, D. Lu, F. Colas, V. Lazarov, B. Francois, Energy management and operational planning of a microgrid with a PV-based active generator for smart grid applications, IEEE Trans. Ind. Electron. 58 (10) (October 2011) 4583–4593.

Index

Note: 'Page numbers followed by "*f*" indicate figures and "*t*" indicate tables.'

G

H

W

Printed in the United States
by Baker & Taylor Publisher Services